AF813501

de Gruyter Series in Nonlinear Analysis and Applications 11

Editors

A. Bensoussan (Paris)
R. Conti (Florence)
A. Friedman (Minneapolis)
K.-H. Hoffmann (Munich)
L. Nirenberg (New York)
A. Vignoli (Rome)

Managing Editor

J. Appell (Würzburg)

Stanislav V. Emelyanov
Sergey K. Korovin
Nikolai A. Bobylev
Alexander V. Bulatov

Homotopy of Extremal Problems

Theory and Applications

Walter de Gruyter · Berlin · New York

Authors

Stanislav V. Emelyanov
Department of Computational Mathematics
and Cybernetics
Moscow State University
Vorob'evy Gory
119899 Moscow
Russia

Sergey K. Korovin
Department of Computational Mathematics
and Cybernetics
Moscow State University
Vorob'evy Gory
119899 Moscow
Russia

Nikolai A. Bobylev
Institute of Control Problems of RAS
65 Profsoyiznaya St.
117997 Moscow
Russia

Alexander V. Bulatov
Institute of Control Problems of RAS
65 Profsoyiznaya St.
117997 Moscow
Russia

Keywords: homotopy method, variational calculus, control theory, nonlinear programming, bifurcation theory, Conley index, nonlinear functional analysis, stability

Library of Congress Cataloging-in-Publication Data

A CIP catalogue record for this book is available from the Library of Congress.

ISBN 978-3-11-018942-1

ISSN 0941-813X

Bibliographic information published by the Deutsche Nationalbibliothek

The Deutsche Nationalbibliothek lists this publication in the Deutsche Nationalbibliografie; detailed bibliographic data are available in the Internet at http://dnb.d-nb.de.

© Copyright 2007 by Walter de Gruyter GmbH & Co. KG, 10785 Berlin, Germany.

All rights reserved, including those of translation into foreign languages. No part of this book may be reproduced or transmitted in any form or by any means, electronic or mechanical, including photocopy, recording, or any information storage and retrieval system, without permission in writing from the publisher.

Printed in Germany.

Cover design: Thomas Bonnie, Hamburg

Editing: Natalia and John Wilson, Oxford, UK

Printing and binding: Hubert & Co. GmbH & Co. KG, Göttingen

Preface

This monograph is devoted to the applications of the homotopy method to the investigation of variational problems. The authors have attempted not only to describe applications of the homotopy method to the analysis of general variational problems, but also to include applications to specific problems of analysis, the calculus of variations, mathematical physics, nonlinear programming, etc.

The main constructions of this monograph are based on the following observation: if, when a variational problem is deformed, a critical point remains isolated, and, for some value of the parameter describing the deformation, this critical point is a minimizer, then the critical point is a minimizer for the variational problem for all values of the parameter.

The book consists of an introduction and five chapters.

The first chapter is of an introductory character. It contains information from topology, classical functional analysis, convex and nonsmooth analysis, the theory of differential equations, and the theory of extremal problems.

The second and third chapters are devoted to applications of the homotopy method to the investigation of variational problems. Finite-dimensional problems are studied in the second chapter and infinite-dimensional problems in the third chapter.

Chapter 4 is an exposition of the theory of Conley index. The main results of this chapter are theorems on the homotopy invariance of Conley index.

The final chapter contains a wide variety of applications of the homotopy method. There are applications to problems of classical analysis (proofs of various inequalities, and generalizations and improvements, determination of exact constants, proof of a criterion for quadratic forms to be positive definite), to nonlinear programming problems, to multicriteria problems, to problems of variational calculus and optimal control, to stability theory and to bifurcation theory.

The treatment in the monograph is self-contained. All prerequisite results and definitions of a general character are either given in the first chapter or described when needed. The book is intended to be accessible to beginning graduate students.

The authors are grateful to the scientific editor Academician E.F. Mishchenko for very fruitful discussions and recommendations, to the referees Yu.S. Popkov and E.S. Pyatnitskii for their very helpful suggestions, to V.I. Skalyga for his help in preparing the monograph and to L.A. Selivanova for her highly professional work in designing the book.

The authors are grateful to the editors of de Gruyter, in particular to Dr R. Plato, Prof. J.S. Wilson and Mrs. N. Wilson, for the help in preparing the English translation of the book.

Moscow, June 2007

A.V. Bulatov
S.V. Emelyanov
S.K. Korovin

Table of Contents

Preface v

Introduction 1

1 Preliminaries 5
1.1 Topological, Metric, and Normed Spaces 5
1.1.1 Topological Spaces 5
1.1.2 Metric Spaces 7
1.1.3 Vector Spaces 8
1.1.4 Normed and Banach Spaces 9
1.1.5 Hilbert Spaces 10
1.1.6 Concrete Spaces of Functionals 11
1.2 Compactness 13
1.2.1 Compact Sets 13
1.2.2 Compactness Criteria in Function Spaces 14
1.3 Linear Functionals and Dual Spaces 15
1.3.1 Linear Functionals 15
1.3.2 The Hahn–Banach Theorem 16
1.3.3 The Dual Space 16
1.3.4 Reflexive Spaces 17
1.3.5 Weak Convergence 17
1.3.6 Weak Compactness 17
1.4 Linear Operators 18
1.4.1 Definitions 18
1.4.2 Convergence of Linear Operators 19
1.4.3 Inverse Operators 19
1.4.4 Unbounded Operators 20
1.4.5 The Spectrum of a Linear Operator 20
1.4.6 Adjoint Operators 21
1.4.7 The Space of Linear Operators 21
1.4.8 Completely Continuous Operators 21
1.4.9 Embedding Operators 22
1.5 Nonlinear Operators and Functionals 22
1.5.1 Continuity 22
1.5.2 Differentiability 22

1.5.3 Lipschitzian and Convex Functionals 25
1.5.4 Some Special Nonlinear Operators and their Properties 29
1.5.5 Extensions of Mappings and the Partition of Unity Theorem 30
1.6 Contraction Mappings, the Implicit Function Theorem, and Differential Equations in a Banach Space 31
1.6.1 Contraction Mappings 31
1.6.2 The Implicit Function Theorem 32
1.6.3 The Cauchy Problem 33
1.7 Minimizers of Nonlinear Functionals 34
1.7.1 Critical Points 34
1.7.2 Second-Order Necessary and Sufficient Conditions for a Minimizer 35
1.7.3 Conditional Extrema 36
1.7.4 Extrema of Lipschitzian and Convex Functionals 37
1.7.5 Weierstrass Theorems 39
1.8 Monotonicity 41
1.8.1 Definitions 41
1.8.2 Potential Operators 42
1.8.3 Monotonicity and Convexity 42

2 Finite-Dimensional Problems 45
2.1 Nondegenerate Deformations of Smooth Functions 45
2.1.1 Invariance of Local Minimizers 45
2.1.2 Invariance of a Global Minimizer 49
2.1.3 Linear Deformations 54
2.2 Nondegenerate Deformations of Nonsmooth Functions 56
2.2.1 Generalized Derivatives and Generalized Gradients ... 56
2.2.2 The Deformation Theorem 56
2.2.3 Global Deformations of Lipschitzian Functions 64
2.2.4 Linear Deformations of Lipschitzian Functions 68
2.2.5 Regular and Critical Points of Continuous Functions .. 69
2.2.6 Deformations of Continuous Functions 73
2.3 Converses of Deformation Theorems 76
2.3.1 Real Polynomials 76
2.3.2 Real Analytic Functions 78
2.3.3 Smooth Functions 79
2.4 Theorems of Hopf and Parusinski 86
2.4.1 Statement of the Problem 86
2.4.2 The Degree of a Mapping and Rotation of a Vector Field 87
2.4.3 Hopf's Theorem 91
2.4.4 Parusinski's Theorem 97
2.4.5 Preparatory Lemmas 98
2.4.6 Proof of Theorem 2.4.11 99

3 **Infinite-Dimensional Problems** 105
3.1 Deformations of Functionals on Hilbert Spaces 105
3.1.1 H-Regular Functionals 105
3.1.2 The Deformation Principle for Minimizers 105
3.1.3 Preparatory Lemmas 106
3.1.4 Proof of the Main Theorem 107
3.1.5 The Property of H-Regularity 111
3.2 Deformations of Functionals on Banach Spaces 112
3.2.1 E-Regular Functionals 112
3.2.2 Preparatory Lemmas 113
3.2.3 Proof of the Deformation Theorem 115
3.3 Global Deformations of Functionals 116
3.3.1 Growing Functionals 116
3.3.2 Global Deformations 119
3.3.3 Generalizations 121
3.4 Deformations of Lipschitzian Functionals 121
3.4.1 (P, S)-Regular Functionals 121
3.4.2 The Deformation Theorem 122
3.4.3 Preparatory Lemmas 122
3.4.4 Proof of the Deformation Theorem 125
3.5 Deformations of Nonsmooth Problems with Constraints 127
3.5.1 Statement of the Problem 127
3.5.2 The Deformation Theorem 129
3.5.3 Linear Deformations and Conditions for Global Minimizers 132

4 **Conley Index** 135
4.1 Conley Index in Finite-Dimensional Problems 135
4.1.1 Flows in Finite-Dimensional Spaces 135
4.1.2 Index Pairs for Invariant Sets 139
4.1.3 Conley Index 142
4.1.4 Homotopy Invariance of the Conley Index 154
4.1.5 The Conley Index of Nondegenerate Critical Points 158
4.2 Conley Index in Infinite-Dimensional Problems 163
4.2.1 (E, H)-Regular Functionals 163
4.2.2 The Conley Index of Critical Points of (E, H)-Regular Functionals 164
4.2.3 Homotopy Invariance of the Conley Index 170

5 **Applications** 173
5.1 Problems of Classical Analysis 173
5.1.1 Proving Inequalities (General Principles) 173
5.1.2 Sylvester's Criterion 174
5.1.3 Young's Inequality 177
5.1.4 Minkowski's Inequality 178

5.1.5 Jensen's Inequality 179
5.1.6 Cauchy's Inequality (Inequality of the Arithmetic and Geometric Means) 180
5.1.7 Improvements and Extensions 182
5.2 Nonlinear Programming Problems 183
5.2.1 Extremals of Lipschitz Nonlinear Programming Problems 183
5.2.2 Extremals of Classical Nonlinear Programming Problems 184
5.2.3 The Deformation Theorem 185
5.2.4 Linear Deformations of Nonlinear Programming Problems and Invariance of Global Minimizers 191
5.2.5 Sufficient Conditions for Minimizers in Nonlinear Programming Problems 193
5.2.6 Inequalities with Constraints 199
5.2.7 Bernstein's Inequality 204
5.3 Multicriteria Problems 207
5.3.1 Definitions 207
5.3.2 The Deformation Theorem 208
5.3.3 Multicriteria Problems with Constraints 210
5.4 Problems in the Calculus of Variations 216
5.4.1 One-Dimensional Problems 216
5.4.2 Multidimensional Integral Functionals 218
5.4.3 The Deformation Theorem 219
5.4.4 Deformations of Integral Functionals in the Problem of Weak Minimizers 221
5.4.5 Functional Inequalities 229
5.4.6 Solvability of Boundary Value Problems and Criteria for Minimizers of Integral Functionals 231
5.4.7 Investigation of Critical Points for a Minimizer 235
5.5 Stability of Solutions of Ordinary Differential Equations 236
5.5.1 Stability of Gradient Systems 236
5.5.2 Stability of Hamiltonian Systems 238
5.5.3 Stability of Gradient Systems in the Large 239
5.5.4 Conley Index and the Stability of Dynamical Systems . 240
5.6 Optimal Control Problems 244
5.6.1 Statement of the Problem 244
5.6.2 Auxiliary Results 245
5.6.3 Deformation Theorems 252
5.7 Bifurcation of Critical Points in Variational Problems 255
5.7.1 Statement of the Problem and a Necessary Condition for Bifurcation 255
5.7.2 Deformation Principle for Minimizers in the Study of Bifurcation Points 256

5.7.3 Conley Index and Bifurcation Points 259
5.7.4 Bifurcation Points of Critical Points in One-Dimensional Variational Problems 265
5.7.5 Analysis of Bifurcation Values of a Parameter 267
5.7.6 Bifurcations of Solutions of the Ginzburg–Landau Equations 269

Additional Remarks and Bibliographic Comments 273

References 281

Notation 293

Name Index 297

Subject Index 299

Introduction

The homotopy method (or continuation method), which dates back to the nineteenth century, plays an active part today in various branches of mathematics. The general idea is geometrically visual and simple: if we are given an equation (algebraic, differential, integral, integro-differential, operator, etc.) and we want information about its solutions (existence, their local properties, construction of approximate solutions, etc.), then we include this equation in a specially constructed one-parameter family of equations which constitute a homotopy (or deformation) from the equation to some equation which has a known solution, and we "deform this solution with respect to the parameter" to obtain a solution of the original equation. Here is a more formal explanation of this idea.

Suppose that we are given an equation

$$A(x) = 0 . \tag{1}$$

Assume that we can include Eq. (1) in a one-parameter family

$$A(x; \lambda) = 0 \qquad (0 \leqslant \lambda \leqslant 1) \tag{2}$$

in such a way that Eq. (2) has a solution $x(\lambda)$ which depends smoothly on λ. Suppose that the equation

$$A(x; 0) = 0$$

has a solution x_0 and that

$$A(x; 1) = A(x) .$$

Differentiating the identity

$$A(x(\lambda); \lambda) \equiv 0$$

with respect to λ, we obtain

$$A'_x(x(\lambda); \lambda)\frac{dx(\lambda)}{d\lambda} + A'_\lambda(x(\lambda); \lambda) \equiv 0 .$$

Thus $x(\lambda)$ is a solution of the Cauchy problem

$$\begin{cases} A'_x(x; \lambda)\dfrac{dx}{dt} + A'_\lambda(x; \lambda) = 0, \\ x(0) = x_0. \end{cases} \tag{3}$$

If this solution can be extended to the interval [0,1], then $x(1)$ is a solution of Eq. (1).

We shall now describe another way to investigate Eq. (2) which is of a discrete character.

We divide the interval $[0, 1]$ into subintervals by choosing points

$$0 = \lambda_0 < \lambda_1 < \cdots < \lambda_n = 1 .$$

Let

$$\delta = \max_{0 \leqslant i \leqslant n-1} (\lambda_{i+1} - \lambda_i) .$$

If δ is sufficiently small, then it is reasonable to expect that x_0 is close to a solution $x(\lambda_1)$ of the equation

$$A(x; \lambda_1) = 0 .$$

Taking this as an initial approximation for an iterative procedure (say, Newton's method), we can find, with sufficient accuracy, an approximation of x_1 to $x(\lambda_1)$. We may regard the point x_1, in turn, as an initial condition imposed for the approximate construction of a solution $x(\lambda_2)$ of the equation

$$A(x; \lambda_2) = 0 ,$$

and so on. At the last step we obtain, with the required degree of accuracy, a solution $x(1)$ of Eq. (1).

Of course these procedures have to be justified. Thus, for instance, we need existence theorems for solutions of Eq. (2) for all λ. The assumption that the solution $x(\lambda)$ depends smoothly on λ is rather restrictive since in specific problems the set of solutions of Eq. (2) may turn out to be very complicated. Moreover it is not *a priori* clear how to obtain the one-parameter family of Eqs. (2), although in practice the parameter often λ enters into the equation under study in a natural way. It should be pointed out here that the commonest way to construct a one-parameter family of Eqs. (2) is to take equations of the form

$$\lambda A(x) + (1 - \lambda) B(x) = 0 ,$$

where the standard equation

$$B(x) = 0$$

is constructed using *a priori* information about the equation Eq. (1) under study.

One of the most general and effective ways to apply the homotopy method to the qualitative investigation of operator equations of the form

$$x - C(x) = 0 \tag{4}$$

where C is a completely continuous operator, was developed by Leray and Schauder [159]. In this method, the parameter λ enters into the equations linearly, i.e., the family of equations has the form

$$x - \lambda C(x) = 0 \quad (0 \leqslant \lambda \leqslant 1) . \tag{5}$$

If for all $\lambda \in [0, 1]$ the solutions $x(\lambda)$ of Eq. (5) satisfy a general *a priori* inequality

$$\|x(\lambda)\| \leqslant r \quad (0 \leqslant \lambda \leqslant 1) ,$$

then Eqs. (5) (and, in particular, Eq. (4), the equation of interest) are solvable. The proof of this result is based on a topological invariant introduced by Leray and Schauder, namely, the degree of a mapping. The Leray–Schauder method has been generalized and developed in many theoretical works and also in works of an applied character. Noteworthy here are the works [31, 139, 145, 151, 153, 204, 214] and the bibliography therein.

The homotopy method was used by Gavurin [123] to establish the solvability of operator equations in Banach spaces (cf. also Rosenbloom [195], Polyak [191], Li [161], Zhang De-Tong [229], Wacker [228], Allgower and Georg [5], Smale [216], Hirsch and Smale [133], Chow, Mallet-Paret and Yorke [62], and Kellogg, Li and Yorke [141]).

The homotopy method was apparently first used for the numerical solution of equations by Lahaye [154] (see also [184]). The method was developed in the works by Freudenstein and Roth [121], Shidlovskaya [207], Davidenko [79–83], Roberts and Shipman [194], and Bosarge [48]. For a more extensive bibliography see the monographs [184, 206].

This monograph gives an account of the applications of the homotopy method to variational problems.

1 Preliminaries

This chapter contains the material from functional analysis that is needed in the monograph. No proofs are given since most of the results are well known and can be considered to be classical.

1.1 Topological, Metric, and Normed Spaces

In this section, we introduce the notions of topological, metric, Banach and Hilbert spaces and give some specific examples of spaces which play a role in the monograph.

1.1.1 Topological Spaces

A *topological space* is a set X together with a family $\mathcal{T}$ of subsets of X satisfying the following three conditions:

(1) $\varnothing \in \mathcal{T}$, $X \in \mathcal{T}$;
(2) the union of any collection of sets from $\mathcal{T}$ belongs to $\mathcal{T}$;
(3) the intersection of any two sets from $\mathcal{T}$ belongs to $\mathcal{T}$.

A family of subsets satisfying these three conditions is called a *topology* on X. The sets belonging to $\mathcal{T}$ are called *open* sets. A subset F of X is called *closed* if its complement $X \setminus F$ is open. An *open neighborhood* of a subset Y of X is an open set containing Y, and a neighborhood of Y is a set containing an open neighborhood of Y.

Let Y be a subset of X. The family of sets $\{U \cap Y : U \in \mathcal{T}\}$ is a topology on Y, called the *induced* or *subspace* topology on Y.

An *equivalence relation* on a set X is a set of pairs

$$R \subset \{(x, y) : x, y \in X\}$$

satisfying the following conditions:

(1) $(x, x) \in R$ for all $x \in X$;
(2) $(x, y) \in R$ implies $(y, x) \in R$;
(3) $(x, y) \in R$ and $(y, z) \in R$ imply $(x, z) \in R$.

If $(x, y) \in R$, then we say that x and y are *equivalent*. The equivalence relation gives rise to a partition of X into pairwise disjoint subsets, called *equivalence classes*, consisting of equivalent elements. The set of equivalence classes corresponding to R is called the *quotient space* of X with respect to R and is denoted by X/R. We write $[x]$ for the equivalence class containing an element x. The *quotient topology* on X/R is defined as follows: a subset $U \subset X/R$ is *open* if the set

$$\bigcup_{[x]\in U} \{x\}$$

is open in X. Let X_1 and X_2 be topological spaces and consider the Cartesian product $X_1 \times X_2 = \{(x_1, x_2) \mid x_1 \in X_1, x_2 \in X_2\}$. We introduce a topology, called the *product topology* on $X_1 \times X_2$, by taking as open sets all unions of sets of the form $U_1 \times U_2$ with U_1 open in X_1 and U_2 open in X_2.

Let X_1 and X_2 be again two topological spaces. A mapping $f : X_1 \to X_2$ is said to be *continuous* if the complete preimage $f^{-1}(U)$ of any open set $U \subset X_2$ is an open set in X_1. Two mappings f_0, $f_1 : X_1 \to X_2$ are *homotopic* (and we write $f_0 \sim f_1$) if there exists a continuous mapping

$$f : X_1 \times [0,1] \to X_2$$

satisfying the conditions

$$f(x,0) = f_0(x) \quad (x \in X_1)\,,$$
$$f(x,1) = f_1(x) \quad (x \in X_1)\,.$$

The spaces X_1 and X_2 are *homotopy equivalent* if there exist continuous mappings $f : X_1 \to X_2$ and $g : X_2 \to X_1$ such that the composite mappings $f \circ g$ and $g \circ f$ are homotopic to the respective identity mappings. We write $X_1 \sim X_2$ to indicate that the spaces X_1 and X_2 are homotopy equivalent.

A *topological pair* is an ordered pair (X, A), where X is a topological space and A is an arbitrary subset of X. A mapping $f : X_1 \to X_2$ is a *mapping of the topological pair* (X_1, A_1) *into the topological pair* (X_2, A_2) (and we write $f : (X_1, A_1) \to (X_2, A_2)$) if $f(A_1) \subset A_2$. There is an obvious notion of homotopy equivalence for topological pairs similar to that introduced above for ordinary spaces. Homotopy equivalence is an equivalence relation on any set of topological pairs.

It is sometimes convenient to consider a topological space X as the topological pair $(X, \varnothing)$. In the theory of Conley index, use is made of topological pairs (X, A) in which A contains just one point; topological pairs of this kind are called *topological spaces with base point* (or *pointed spaces*).

Here are two properties of continuous mappings of topological pairs.

Proposition 1.1.1. *Consider mappings* f_0, $f_1 : (X_1, A_1) \to (X_2, A_2)$, g_0, $g_1 : (X_2, A_2) \to (X_3, A_3)$. *If* $f_0 \sim f_1$ *and* $g_0 \sim g_1$, *then* $g_0 \circ f_0 \sim g_1 \circ f_1$.

Proposition 1.1.2. *If the composite mappings* $g \circ f$ *and* $h \circ g$ *in the sequence of mappings*

$$(X_1, A_1) \xrightarrow{f} (X_2, A_2) \xrightarrow{g} (X_3, A_3) \xrightarrow{h} (X_4, A_4)$$

are homotopy equivalences, then f, g and h are homotopy equivalences.

To each each pair (X_1, X_2) of topological spaces we associate a topological space X_1/X_2 with base point, called the *quotient space* X_1 with respect to X_2, as follows: if $X_1 \cap X_2 = \varnothing$, then

$$X_1/X_2 = \{\{x\} : x \in X_1\} \cup \{\varnothing\}$$

with base point $\{\varnothing\}$. If $X_1 \cap X_2 \neq \varnothing$, then we define the equivalence relation $R = \{(x, y) : x = y \text{ or } x,\ y \in X_2\}$ on X_1 and set

$$X_1/X_2 = X_1/R$$

with base point $X_1 \cap X_2$. The *product* of two quotient spaces X_1/X_2 and Y_1/Y_2 is the quotient space

$$X_1/X_2 \wedge Y_1/Y_2 = (X_1 \times Y_1)/((X_2 \times Y_1) \cup (X_1 \times Y_2)) .$$

We denote by S^p $(p = 0, 1, 2, \ldots)$ the p-dimensional sphere with base point. The following statements hold (see, e.g., [116]).

Proposition 1.1.3. *$S^p \wedge S^q \sim S^{p+q}$ for $p = 0, 1, \ldots,$ and $q = 0, 1, \ldots$.*

Proposition 1.1.4. *$S^p \not\sim S^q$ for $p \neq q$.*

1.1.2 Metric Spaces

Metric spaces are topological spaces of a special type.

A set M is a *metric space* if for all $x,\ y \in M$ a nonnegative real number $d(x, y)$ is defined (called the *distance* between x and y) satisfying the following conditions:

(1) $d(x, y) = 0$ if and only if $x = y$;
(2) $d(x, y) = d(y, x)$ for all $x, y \in M$;
(3) $d(x, y) \leqslant d(x, z) + d(z, y)$ for all $x, y, z \in M$.

A function $d : M \times M \to \mathbb{R}_+$ satisfying these conditions is called a *metric* on M, and together with d the set M becomes a *metric space*. (Here $\mathbb{R}_+$ denotes the nonnegative semiaxis in the real line $\mathbb{R}$.) Conditions (1), (2), (3) for the metric d are often called the *identity axiom*, the *symmetry axiom* and the *triangle inequality*.

Let M be a metric space with metric d. Each subset M_0 of M becomes a metric space with the same metric d, and M_0 is called a *subspace* of the metric space M. A sequence (x_n) in M is said to be *convergent* if there exists a point $x \in M$ for which

$$\lim_{n \to \infty} d(x, x_n) = 0 .$$

The unique point $x \in M$ for which this holds is called the *limit* of (x_n).

The *open* (resp. *closed*) ball $\mathring{B}(r,y)$ (resp. $B(r,y)$) in M is the collection of points $x \in M$ satisfying the inequality $d(x,y) < r$ (resp. $(d(x,y) \leqslant r)$. The number r is the *radius* of the ball and the point y is its *center*.

A *neighborhood* of a point $y \in M$ is a subset of M that contains $\mathring{B}(r,y)$ for some $r > 0$.

A subset $M_0 \subset M$ is said to be *bounded* if it is contained in some ball.

A point $y \in M$ is a *limit point* of the subset M_0 of M if every neighborhood of y contains at least one point of M_0 which is different from y.

A subset of M is *closed* if it contains all of its limit points. The set obtained by adding to a subset M_0 all of its limit points is a closed set; it is called the *closure* of M_0 and is denoted by $\overline{M_0}$.

The operation of closure has the following properties:

$$\overline{M_0 \cup M_1} = \overline{M_0} \cup \overline{M_1}\,, \quad M \subseteq \overline{M}\,, \quad \overline{(\overline{M})} = \overline{M}\,.$$

A subset M_0 of M is *open* if its complement $M \setminus M_0$ is closed. Evidently the family of open sets is a topology on M.

A metric (or topological) space M is said to be *connected* if it does not have nonempty open subsets $M_0, M_1 \subset M$ such that $M_0 \cup M_1 = M$ and $M_0 \cap M_1 = \varnothing$.

A subset M_0 is said to be *dense* in M if $\overline{M_0} = M$, and M is *separable* if M has a countable dense subset.

A sequence (x_n) in M is a *Cauchy sequence* if

$$\lim_{\min(n,m)\to\infty} d(x_n, x_m) = 0\,.$$

A metric space M is *complete* if every Cauchy sequence in M is convergent. The following result is often used to prove that a metric space is complete.

Theorem 1.1.1. *Let M be a complete metric space and $M_0 \subset M$. Then M_0 is complete if and only if it is closed in M.*

1.1.3 Vector Spaces

Let K be either the set $\mathbb{R}$ of real numbers or the set $\mathbb{C}$ of complex numbers. Let E be a set whose elements satisfy the following two groups of axioms.

1 (addition in E). The sum $x+y$ is defined for all $x, y \in E$, and the sum operation has the following properties:

(a) $x + y = y + x$ for all $x, y \in E$;

(b) $x + (y + z) = (x + y) + z$ for all $x, y, z \in E$;

(c) there exists a unique element 0 such that $x + 0 = x$ for all $x \in E$;

(d) for every $x \in E$ there exists a unique element $-x \in E$ such that $x + (-x) = 0$.

2 (scalar multiplication in E). For each $\lambda \in K$ a multiplication mapping $x \mapsto \lambda x$ from E to E is defined and the following conditions hold:

(a) $\lambda(\mu x) = (\lambda\mu)x$ for all $\lambda, \mu \in K$ and $x \in E$;
(b) $\lambda(x + y) = \lambda x + \lambda y$ for all $\lambda \in K$ and $x, y \in E$;
(c) $(\lambda + \mu)x = \lambda x + \mu y$ for all $\lambda, \mu \in K$ and $x \in E$;
(d) $1 \cdot x = x$ for all $x \in E$.

A set E satisfying these two groups of axioms is called a *vector space* over K and the elements of K are called scalars. If $K = \mathbb{R}$ then we have a *real vector space* and if $K = \mathbb{C}$ we have a *complex vector space.*

A nonempty subset E_0 of a vector space E is called a *vector subspace* if every linear combination $\lambda_1 x_1 + \cdots + \lambda_n x_n$ of elements $x_1, \ldots, x_n \in E_0$ belongs to E_0.

Let $E_1, \ldots, E_n$ be vector subspaces of the vector space E. If each element $x \in E$ can be uniquely represented in the form

$$x = x_1 + \cdots + x_n \quad (x_i \in E_i \text{ for } i = 1, \ldots, n) ,$$

then we say that E is the *direct sum* of the vector subspaces E_i, and we write

$$E = \bigoplus_{i=1}^{n} E_i .$$

Let E_0 be a vector subspace of a vector space E. We consider the equivalence relation on E in which elements x_1, x_2 are equivalent if and only if $x_1 - x_2 \in E_0$. The equivalence classes L are called cosets; thus distinct cosets are disjoint. The collection of all cosets forms a *quotient vector space* E/E_0, with natural operations of addition and scalar multiplication defined as follows. If L_1, L_2 are cosets and $x_1 \in L_1$, $x_2 \in L_2$, then $L_1 + L_2$ is the coset containing $x_1 + x_2$; this is independent of the choices of $x_1 \in L_1$, $x_2 \in L_2$ and so addition of elements of E/E_0 is well defined. Similarly, the product λL of a coset L by a scalar λ is defined to be the coset containing λx, where x is any element of L.

1.1.4 Normed and Banach Spaces

Let E be a vector space over K and suppose that for each element $x \in E$ there is defined a real number $\|x\|$, called the *norm* of x, satisfying the following conditions (axioms for a norm):

(1) $\|x\| \geqslant 0$ for all $x \in E$, with $\|x\| = 0$ only if $x = 0$;
(2) $\|\lambda x\| = |\lambda| \, \|x\|$ for all $\lambda \in K$ and $x \in E$;
(3) $\|x + y\| \leqslant \|x\| + \|y\|$ for all $x, y \in E$.

In this case, we say that E is a *normed vector space.* We can then define a metric d in E by writing

$$d(x, y) = \|x - y\| .$$

It is easy to verify that the axioms for a metric hold. The notion of convergence of sequences defined by this metric d is known as *convergence in norm* or *uniform convergence.*

A normed vector space that is complete with respect to convergence in norm is called a *Banach space.* All vector spaces arising in the text are Banach spaces (unless otherwise specified). Since Banach spaces are in particular metric spaces, all notions introduced above for metric spaces (balls, closed and open sets, connected sets, etc.) are applicable.

1.1.5 Hilbert Spaces

Let H be a real vector space, and suppose that for each pair of elements $x, y \in H$ there is associated a real number (x, y), called the *inner product* of x, y, satisfying the following conditions:

(1) $(x, y) = (y, x)$ for all $x, y \in H$;
(2) $(x_1 + x_2, y) = (x_1, y) + (x_2, y)$ for all $x_1, x_2, y \in H$;
(3) $(\lambda x, y) = \lambda(x, y)$ for all $\lambda \in \mathbb{R}$, $x, y \in H$;
(4) $(x, x) \geqslant 0$ for all $x \in H$, with $(x, x) = 0$ if and only if $x = 0$.

It is easy to verify that we can regard H as a normed space with norm $\|x\| = (x, x)^{1/2}$. If H is complete in this norm, then H is called a *Hilbert space.*

Two elements $x, y \in H$ are called *orthogonal* if $(x, y) = 0$. An element x is *orthogonal to a subspace* H_0 of H if x is orthogonal to every element $y \in H_0$.

Let H_0 be a closed subspace of the Hilbert space H.

Theorem 1.1.2. *Each element $x \in H$ can be written uniquely in the form*

$$x = y + z, \tag{1.1.1}$$

with $y \in H_0$ and z orthogonal to H_0.

The element y in (1.1.1) is the *orthogonal projection* of x onto H_0. The collection of elements orthogonal to H_0 is a closed vector subspace H^0 known as the *orthogonal complement* of H_0.

The space H is the direct sum of H_0 and its orthogonal complement H^0. In this case we say that H is the *orthogonal sum* of the subspaces H_0 and H^0 and write $H = H_0 \dotplus H^0$.

A family $h_1, \dots, h_n, \dots$ of elements of a Hilbert space H is called orthonormal if $(h_i, h_j) = \delta_{ij}$, where δ_{ij} is the *Kronecker delta,* defined by $\delta_{ij} = 0$ if $i \neq j$ and $\delta_{ij} = 1$ if $i = j$.

An orthonormal family $\{h_i\} \subset H$ is called *complete* if there is no nonzero element orthogonal to all elements of $\{h_i\}$. An orthonormal family $\{h_i\}$ is *closed* if the subspace that it generates is H. A closed orthonormal family is called an *orthonormal basis* of H.

1.1.6 Concrete Spaces of Functionals

The space $C(\overline{\Omega})$. Suppose that Ω is a domain in $\mathbb{R}^N$ with closure $\overline{\Omega}$. We denote by $C(\overline{\Omega})$ the set of continuous and bounded real functions $u = u(x)$ on $\overline{\Omega}$, with pointwise linear operations, defined by

$$(u+v)(x) = u(x) + v(x), \quad (\lambda u)(x) = \lambda u(x),$$

and norm

$$\|u\|_{C(\overline{\Omega})} = \sup_{x\in\overline{\Omega}} |u(x)| \ .$$

Convergence in norm in $C(\overline{\Omega})$ is uniform convergence on $\overline{\Omega}$. Since the uniform limit of a sequence of continuous functions is continuous, $C(\overline{\Omega})$ is a complete (i.e., Banach) space.

The space $L_p(\Omega)$ $(1 \leqslant p < \infty)$. This is the set of Lebesgue measurable real functions u on Ω that are summable with degree p, with addition and scalar multiplication defined pointwise and norm defined by the equation

$$\|u\|_{L_p(\Omega)} = \left(\int_\Omega |u(x)|^p \, dx \right)^{1/p} .$$

Like $C(\overline{\Omega})$, $L_p(\Omega)$ is a Banach space.

The *Hölder inequality* expresses an important relation between functions from the spaces $L_p(\Omega)$: if p, $q > 1$ and $p^{-1} + q^{-1} = 1$ and if $u \in L_p(\Omega)$, $v \in L_q(\Omega)$, then

$$\int_\Omega u(x)v(x)\, dx \leqslant \|u\|_{L_p(\Omega)} \|v\|_{L_q(\Omega)} \ .$$

The space $L_\infty(\Omega)$. We need a definition before we can describe the elements of this space.

We write $\mathcal{E}$ for the family of all subsets E of Ω of measure zero. For each Lebesgue measurable function $\alpha : \Omega \to \mathbb{R}$ we set

$$\mu_0(\alpha) = \inf_{E\in\mathcal{E}} \sup_{\Omega\setminus E} \alpha(x) \ .$$

The number $\mu_0(\alpha)$ is called the *essential maximum* of α on Ω and it is denoted by

$$\operatorname*{vrai\,max}_{x\in\Omega} \alpha(x) \ .$$

The space $L_\infty(\Omega)$ consists of all Lebesgue measurable functions $u : \Omega \to \mathbb{R}$ for which the essential maximum of the absolute value of u is finite. The norm on $L_\infty(\Omega)$ is defined by

$$\|u\|_{L_\infty(\Omega)} = \operatorname*{vrai\,max}_{x\in\Omega} |u(x)| \ .$$

Suppose that Ω is a bounded domain in $\mathbb{R}^N$ and $x = (x_1, \dots, x_N)$ is a point in $\mathbb{R}^N$. For each vector $\alpha = (\alpha_1, \dots, \alpha_N)$ with nonnegative integer coordinates write $|\alpha| = \alpha_1 + \dots + \alpha_N$, and set

$$\mathcal{D}^\alpha = \left(\frac{\partial}{\partial x_1}\right)^{\alpha_1} \dots \left(\frac{\partial}{\partial x_N}\right)^{\alpha_N}, \quad D^k u = \{\mathcal{D}^\alpha u : |\alpha| = k\} .$$

The space $C^k(\overline{\Omega})$. This space consists of all k-times continuously differentiable functions defined on $\overline{\Omega}$. We define a norm on $C^k(\overline{\Omega})$ by the equation

$$\|u\|_{C^k(\overline{\Omega})} = \sum_{|\alpha| \leqslant k} \max_{x \in \overline{\Omega}} |\mathcal{D}^\alpha u(x)| .$$

The space $C^k(\overline{\Omega})$ is a separable Banach space with respect to this norm.

The space $\mathring{C}^k(\overline{\Omega})$. This is the vector subspace of $C^k(\overline{\Omega})$ consisting of the functions $u(x)$ vanishing on $\partial\Omega$, with norm induced by the norm on $C^k(\overline{\Omega})$.

The space $C^{k,\delta}(\overline{\Omega})$. The elements of this Banach space are the k-times continuously differentiable functions $u(x)$ on $\overline{\Omega}$ whose kth derivatives satisfy the Hölder condition with exponent $\delta \in (0,1)$. The norm in this space is defined by

$$\|u\|_{C^{k,\delta}(\overline{\Omega})} = \|u\|_{C^k(\overline{\Omega})} + \sum_{|\alpha|=k} \sup_{\substack{x,y \in \overline{\Omega} \\ x \neq y}} \frac{|\mathcal{D}^\alpha u(x) - \mathcal{D}^\alpha u(y)|}{|x-y|^\delta} .$$

The Sobolev space $W_p^m(\Omega)$. Let m be a positive integer and $p \in [1, \infty)$. Consider the family of all infinitely differentiable functions $u : \overline{\Omega} \to \mathbb{R}$. The *Sobolev space* $W_p^m(\Omega)$ is the completion of this space with respect to the norm

$$\|u\|_{W_p^m(\Omega)} = \left(\int_\Omega \sum_{|\alpha| \leqslant m} |\mathcal{D}^\alpha u(x)|^p \, dx \right)^{1/p} ;$$

its elements are the functions $u \in L_p(\Omega)$ having generalized derivatives up to the kth derivative which are summable with degree p.

The Sobolev space $\mathring{W}_p^m(\Omega)$. This is the vector subspace of $W_p^m(\Omega)$ which is the closure of the space of all infinitely differentiable functions with supports in Ω.

Both $W_2^m(\Omega)$ and $\mathring{W}_2^m(\Omega)$ are Hilbert spaces with respect to the inner product

$$(u,v)_{W_2^m(\Omega)} = \left(\int_\Omega \sum_{|\alpha| \leqslant m} \mathcal{D}^\alpha u(x) \mathcal{D}^\alpha v(x) \, dx \right)^{1/2} .$$

Note that an inner product in $\overset{\circ}{W}{}_2^m(\Omega)$ can be also defined by

$$(u,v)_{\overset{\circ}{W}{}_2^m(\Omega)} = \left(\int_\Omega \sum_{|\alpha|=m} \mathcal{D}^\alpha u(x) \mathcal{D}^\alpha v(x)\, dx\right)^{1/2}.$$

The norm on $\overset{\circ}{W}{}_2^m(\Omega)$ corresponding to this inner product is equivalent to the original norm on this space.

If the boundary of Ω is sufficiently smooth, then, for $0 \leqslant k \leqslant m-1$, we have embeddings of Sobolev spaces:

$$W_p^m(\Omega) \subset W_q^k(\Omega) \quad \text{if} \quad \frac{1}{q} \geqslant \frac{1}{p} - \frac{m-k}{N} > 0, \tag{1.1.2}$$

$$W_p^m(\Omega) \subset W_q^k(\Omega) \quad \text{if} \quad q = \infty,\ \frac{1}{p} = \frac{m-k}{N}, \tag{1.1.3}$$

$$W_p^m(\Omega) \subset C^{k,\delta}(\overline{\Omega}) \quad \text{if} \quad \frac{N}{p} < m - (k+\delta). \tag{1.1.4}$$

1.2 Compactness

Compactness is one of the most important concepts in functional analysis. In this section, we discuss the main properties of compact sets and give criteria for compactness of sets in various function spaces.

1.2.1 Compact Sets

A subset K of a metric space M is said to be *compact* if every sequence in K has a subsequence that converges to an element of K. A compact set is sometimes said to be *compact in itself.* It is clear that a compact space is a complete metric space. A subset K of M is called *precompact* if its closure in M is compact.

A subset $\mathcal{M}$ of a metric space M is called an *ε-net* for a subset K (where $\varepsilon > 0$) if for each point $x \in K$ there exists a point $y \in \mathcal{M}$ such that

$$\rho(x,y) < \varepsilon .$$

Theorem 1.2.1 (Hausdorff). *A closed subset K of a complete metric space M is compact if and only if for every $\varepsilon > 0$ there exists a finite ε-net for K in M.*

Here are two more criteria for compactness.

A family $\{G_\alpha\}$ of open sets in a metric space M is an *open cover* of a subset K of M if every point $x \in K$ belongs to at least one set G_α.

Theorem 1.2.2. *A closed subset K of a metric space M is compact if and only if every open cover of K contains a finite subcover.*

A family of sets $\{F_\alpha\}$ is *centered* (or has the *finite intersection property*) if each of its finite subfamilies has nonempty intersection.

Theorem 1.2.3. *A closed subset K of a metric space M is compact if and only if every centered system of closed subsets of K has nonempty intersection.*

1.2.2 Compactness Criteria in Function Spaces

Every compact set in a metric space is closed and bounded. Therefore, in the compactness criteria for subsets of function spaces below, the subsets are assumed to be closed and bounded.

Compactness criterion in $C(\overline{\Omega})$. The functions in a set $K \subset C(\overline{\Omega})$ are *equicontinuous* if, for all $\varepsilon > 0$, there exists a $\delta > 0$, depending only on ε, such that for all $x_1, x_2 \in \overline{\Omega}$ satisfying $|x_1 - x_2| < \delta$ and for all $u \in K$ the inequality

$$|u(x_1) - u(x_2)| < \varepsilon$$

holds.

Theorem 1.2.4 (Arzelà–Ascoli). *Let Ω be a bounded domain in $\mathbb{R}^N$. A closed and bounded subset K of $C(\overline{\Omega})$ is compact if and only if the functions from K are equicontinuous.*

Compactness criterion in the spaces $L_p(\Omega)$ $(1 \leqslant p < \infty)$. In order to formulate the compactness criterion, it is convenient to regard all functions $u \in L_p(\Omega)$ as extended to the whole space $\mathbb{R}^N$ and to be zero outside Ω.

Let $K \subset L_p(\Omega)$. The functions $u \in K$ are *equicontinuous in the mean* if for all $\varepsilon > 0$ there exists a $\delta = \delta(\varepsilon) > 0$ such that for all $|h| < \delta$ we have

$$\left(\int_\Omega |u(x+h) - u(x)|^p \, dx \right)^{1/p} < \varepsilon$$

for all $u \in K$.

Theorem 1.2.5 (M. Riesz). *Let Ω be a bounded domain in $\mathbb{R}^N$. A closed and bounded subset K of $L_p(\Omega)$ is compact if and only if the functions of K are equicontinuous in the mean.*

Let M be a subset of a Banach space E. The convex hull $\operatorname{co} M$ is the smallest convex set (with respect to inclusion) containing M.

Theorem 1.2.6 (Mazur). *The closure of the convex hull of a compact set is compact.*

Let $F : M \to E_1$ be a mapping from some subset M of a Banach space E_0 to a Banach space E_1. In this situation, we say that F is an *operator* with *domain of definition* M and *range* E_1. The set

$$F(M) = \{y \in E_1 : y = F(x),\ x \in M\}$$

is the *image* of M under F.

The operator F is *continuous* at a point $x_0 \in M$ if, for every sequence (x_n) in M convergent to x_0 in the norm of E_1, the sequence $(F(x_n))$ converges to $F(x_0)$ in the norm of E_1, i.e., if

$$\lim_{n\to\infty} \|x_n - x_0\|_{E_0} = 0\ ,$$

implies

$$\lim_{n\to\infty} \|F(x_n) - F(x_0)\|_{E_1} = 0\ .$$

If F is continuous at each point of M, then F is *continuous* on M.

Theorem 1.2.7. *A continuous image of a compact set is compact.*

1.3 Linear Functionals and Dual Spaces

In this section, we introduce the notions of dual spaces, weak convergence and weak compactness.

1.3.1 Linear Functionals

Suppose that E is a normed vector space and $f : E \to \mathbb{R}$ is a function. Then f is called a *linear functional* if it is *additive* and *homogeneous*, i.e., for all $x, y \in E$ we have

$$f(x + y) = f(x) + f(y)$$

and for $x \in E, \alpha \in \mathbb{R}$ we have

$$f(\alpha x) = \alpha f(x)\ .$$

A linear functional f is *bounded* if

$$\sup_{\|x\|_E \leqslant 1} |f(x)| < \infty\ ;$$

then

$$\|f\| = \sup_{\|x\|_E \leqslant 1} |f(x)|$$

is the *norm* of f.

1.3.2 The Hahn–Banach Theorem

This is one of the fundamental results of functional analysis. It guarantees that a linear functional defined on some vector subspace can be extended to the whole space without increasing its norm.

Theorem 1.3.1 (Hahn–Banach). *For every bounded linear functional f_0 on a vector subspace E_0 of a normed vector space E, there exists a linear bounded functional f on E for which $f(x) = f_0(x)$ $(x \in E_0)$ and $\|f_0\|_{E_0} = \|f\|_E$.*

The Hahn–Banach theorem provides a tool for proving various *separability theorems.* We shall formulate the simplest of them.

Let M be a subset of a Banach space E and let $x_0 \notin M$. We say that a linear bounded functional f on E *separates* M from x_0 if

$$\sup_{x \in M} f(x) \leqslant f(x_0) \, .$$

If for some $\varepsilon > 0$ the inequality

$$\sup_{x \in M} f(x) \leqslant f(x_0) - \varepsilon$$

holds, then we say that f *strongly separates* M from x_0.

Theorem 1.3.2. *Suppose that M is a closed convex subset with $x_0 \notin M$. Then there exists a functional f that strongly separates M from x_0.*

1.3.3 The Dual Space

There are natural pointwise operations of addition and scalar multiplication for linear functionals on a vector space E: if f_1 and f_2 are linear functionals, then their *sum* $f_1 + f_2$ is the linear functional f defined by

$$f(x) = f_1(x) + f_2(x) \quad (x \in E) \, ,$$

and the *product* αf of a functional f by a scalar α is defined by

$$(\alpha f)(x) = \alpha(f(x)) \quad (x \in E).$$

With respect to these operations, the set of bounded linear functionals defined on E forms a vector space E^*, called the *dual space* of E. The value of a functional $f \in E^*$ at an element $x \in E$ is often denoted by $\langle f, x \rangle$.

The following result elucidates the structure of the dual of a Hilbert space H.

Theorem 1.3.3 (F. Riesz). *For every bounded linear functional $f \in H^*$ there exists a unique element $g \in H$ for which $\|f\|_{H^*} = \|g\|_H$ and*

$$f(x) = (x, g) \, . \tag{1.3.1}$$

The converse is obviously true: for every $g \in H$ the relation (1.3.1) defines a linear functional $f \in H^*$ for which $\|f\|_{H^*} = \|g\|_H$. Thus, (1.3.1) defines an isomorphism between the spaces H and H^*. For this reason, it is often convenient to identify a Hilbert space and its dual.

1.3.4 Reflexive Spaces

Let E be a Banach space and E^* its dual. Since E^* is also a Banach space, we can construct its dual $(E^*)^* = E^{**}$. Thus E^{**} is the space of bounded linear functionals φ on E^*. Consider a linear functional $f(x)$ on E. Here f is a fixed element from E^* and x is a variable element from E. However we can also consider the expression $f(x)$ from a different point of view, by fixing $x \in E$ and allowing $f \in E^*$ to vary. Then the expression $f(x)$ defines a functional on E^*; we denote it by φ_x. Thus, by definition,

$$\varphi_x(f) = f(x) .$$

It is easy to see that φ_x is a bounded linear functional on E^*, and consequently $\varphi_x \in E^{**}$. It follows immediately from the definition that the mapping $x \mapsto \varphi_x$ is injective and linear; this mapping $\pi : E \to E^{**}$ is called the *natural mapping from the space E into its double dual.*

If $\pi(E) = E^{**}$, then E is said to be a *reflexive* space.

For $1 < p < \infty$ the spaces $L_p(\Omega)$, $W_p^m(\Omega)$, $\overset{\circ}{W}{}_p^m(\Omega)$ are reflexive. The spaces $C(\overline{\Omega})$, $C^k(\overline{\Omega})$, $C^{k,\delta}(\overline{\Omega})$, $L_1(\Omega)$, $L_\infty(\Omega)$, $W_1^m(\Omega)$, and $W_\infty^m(\Omega)$ are not reflexive.

1.3.5 Weak Convergence

Let E be a normed space. A sequence (x_n) in E *weakly converges* to an element $x_0 \in E$ if

$$\lim_{n\to\infty} f(x_n) = f(x_0)$$

for all $f \in E^*$. The element x_0 is called the *weak limit* of the sequence (x_n).

Every strongly convergent sequence (x_n) in E is clearly weakly convergent, and in a finite-dimensional normed space strong convergence and weak convergence are equivalent.

Theorem 1.3.4. *Every weakly convergent sequence (x_n) in E is bounded.*

A subset M of E is said to be *weakly closed* if the weak limit of every weakly convergent sequence of elements of M belongs to M.

Theorem 1.3.5. *Every closed ball in a Banach space is weakly closed.*

1.3.6 Weak Compactness

A subset M of a Banach space E is *weakly compact* if each sequence in M has a subsequence that is weakly convergent to some element of M.

The notion of a weakly compact set in the dual space E^* of a space E is defined similarly; the role of the initial space and its dual are played by E^*

and E^{**}. However, we can define another type of convergence in E^* and thus yet another type of compactness.

A sequence (f_n) in E^* is *weak* convergent* to an element f_0 of E^* if

$$\lim_{n\to\infty} f_n(x) = f_0(x)$$

for all $x \in E$.

If E is a reflexive space, the notions of weak convergence and weak* convergence in E^* coincide.

A subset M of E^* is *weak* closed* if, for every sequence (f_n) in M which is weak* convergent to an element $f \in E^*$, we have $f \in M$.

Theorem 1.3.6. *Each closed ball in the dual space E^* is weak* closed.*

A subset M of E^* is *weak* compact* if every sequence in M has a subsequence that is weak* convergent to some element of M.

Theorem 1.3.7. *If E is a separable normed space then every closed ball in E^* is weak* compact.*

Theorem 1.3.8. *If E is a reflexive Banach space then every closed ball in E is weakly compact.*

1.4 Linear Operators

In this section we introduce linear operators; we describe their main properties and we consider some special classes of linear operators.

1.4.1 Definitions

A mapping A from a Banach space E_0 to a Banach space E_1 is called a *linear operator* if

$$A(x_1 + x_2) = Ax_1 + Ax_2 \tag{1.4.1}$$

for all $x_1, x_2 \in E_0$ and

$$A(\lambda x) = \lambda Ax \tag{1.4.2}$$

for all $x \in E_0$ and $\lambda \in \mathbb{R}$. Property (1.4.1) asserts the *additivity* of A and property (1.4.2) the *homogeneity* of A. The space E_0 is the *domain of definition* of A and E_1 is its *range*.

A linear operator A is *continuous* if it maps convergent sequences to convergent sequences, i.e., if whenever $x_0 \in E_0$, and (x_n) is a sequence in E_0 such that

$$\lim_{n\to\infty} \|x_n - x_0\|_{E_0} = 0 ,$$

then

$$\lim_{n\to\infty} \|Ax_n - Ax_0\|_{E_1} = 0 ,$$

and A is *bounded* if

$$\|Ax\|_{E_1} \leqslant C\,\|x\|_{E_0} \quad (x \in E_0) \tag{1.4.3}$$

for some constant C. The smallest constant C for which the inequality (1.4.3) holds is the *norm* of A and it is denoted by $\|A\|$. Thus

$$\|A\| = \sup_{\|x\|_{E_0}=1} \|Ax\|_{E_1} \,.$$

A linear operator $A : E_0 \to E_1$ is continuous if and only if it is bounded.

1.4.2 Convergence of Linear Operators

Let (A_n) be a sequence of bounded linear operators from E_0 to E_1. We say that (A_n) *converges uniformly* to a bounded linear operator $A : E_0 \to E_1$ if

$$\lim_{n\to\infty} \|A_n - A\| = 0 \,.$$

We say that (A_n) *strongly converges* to an operator $A : E_0 \to E_1$ if

$$\lim_{n\to\infty} \|A_n x - Ax\| = 0$$

for all $x \in E_0$.

Clearly uniform convergence implies strong convergence but the converse statement is not in general true.

Theorem 1.4.1 (Banach–Steinhaus). *If a sequence of linear bounded operators $A_n : E_0 \to E_1$ strongly converges to an operator $A : E_0 \to E_1$, then the sequence $(\|A_n\|)$ of norms is bounded.*

Corollary 1. *If a sequence of linear bounded operators $A_n : E_0 \to E_1$ strongly converges to an operator $A : E_0 \to E_1$, then A is a bounded linear operator.*

1.4.3 Inverse Operators

Let E_0 and E_1 be sets and suppose that $A : E_0 \to E_1$ is a bijective mapping. A mapping $B : E_1 \to E_0$ is an *inverse* of A if $BAx = x$ for all $x \in E_0$ and $ABy = y$ for all $y \in E_1$. If an inverse to A exists then it is unique, and it is usually denoted by A^{-1}.

If E_0 and E_1 are now normed vector spaces and A is a linear operator then A^{-1} is linear, and we can ask under what conditions A^{-1} is bounded. In general, the boundedness of A does not imply the boundedness of A^{-1}. The following result, one of the fundamental results of functional analysis, gives a sufficient condition for A^{-1} to be bounded.

Theorem 1.4.2 (Banach). *If A is a bijective linear operator from a Banach space E_0 to a Banach space E_1, then $A^{-1} : E_1 \to E_0$ is bounded.*

We formulate below two important corollaries of this theorem of Banach. The first is the well-known open mapping theorem.

A mapping $A : E_0 \to E_1$ from a Banach space E_0 to a Banach space E_1 is *open* if it maps every open subset of E_0 to an open subset of E_1.

Corollary 1 (open mapping theorem). *Each continuous surjective linear mapping $A : E_0 \to E_1$ from a Banach space E_0 to a Banach space E_1 is open.*

Corollary 2 (triple theorem). *Suppose that E_0, E_1 and E_2 are Banach spaces and $A_1 : E_0 \to E_1$, $A_2 : E_0 \to E_2$ are bounded linear operators, with $A_2E_0 = E_2$. Let*

$$\operatorname{Ker} A_1 \supset \operatorname{Ker} A_2 .$$

Then there exists a bounded linear operator $A_3 : E_2 \to E_1$ such that

$$A_1 = A_3A_2 .$$

1.4.4 Unbounded Operators

The domain of definition of a linear operator A may not be the whole Banach space E_0, but only some proper vector subspace $D(A)$. The image of this set is called the *range* $\operatorname{Im} A$ of the operator A, i.e.,

$$\operatorname{Im} A = \{Ax : \ x \in D(A)\} .$$

In applications, the domains of definition $D(A)$ of linear operators A are usually dense in E_0, i.e., $\overline{D(A)} = E_0$. If $\overline{D(A)} = E_0$ and A is bounded on $D(A)$, i.e.,

$$\|Ax\|_{E_1} \leqslant C\,\|x\|_{E_0} \quad (x \in D(A)) \tag{1.4.4}$$

for some constant $C > 0$, then A admits an extension from $D(A)$ to E_0 which is a linear bounded operator $\tilde{A} : E_0 \to E_1$. If there is no constant C for which (1.4.4) holds, then the operator A is *unbounded*.

1.4.5 The Spectrum of a Linear Operator

The *spectrum* $\sigma(A)$ of a linear operator $A : E \to E$ is another of the most important concepts in functional analysis. It is defined to be the complement in the complex plane $\mathbb{C}$ of the set $R(A)$ of regular values of A; a complex number λ is a *regular value* of A if the operator $A - \lambda I$ has a bounded inverse.

The *spectral radius* $r(A)$ of a bounded linear operator A is the radius of the smallest disc in $\mathbb{C}$ with center 0 which contains $\sigma(A)$. It can be determined using the *Gel'fand formula*

$$r(A) = \lim_{n\to\infty} \sqrt[n]{\|A^n\|} \; .$$

Two norms $\|\cdot\|$, $\|\cdot\|_*$ on a space E are called *equivalent* if there exist positive constants c, C such that

$$c\,\|x\| \leqslant \|x\|_* \leqslant C\,\|x\|$$

for all $x \in E$. When we pass from a norm to an equivalent norm, the convergent sequences, closed sets, open sets etc. remain unchanged.

1.4.6 Adjoint Operators

Let $A : E_0 \to E_1$ be a bounded linear operator, where E_0, E_1 are Banach spaces. Consider a functional $g \in E_1^*$ and set

$$f(x) = g(Ax) \; .$$

Clearly f is a bounded linear functional defined on E_0, i.e., $f \in E_0^*$. Thus from each functional $g \in E_1^*$ we obtain a functional $f \in E_0^*$. The resulting operator $A^* : E_1^* \to E_0^*$ is the *adjoint* of A; it is easy to check that it is a linear operator.

1.4.7 The Space of Linear Operators

The set of linear bounded operators from a Banach space E_0 to a Banach space E_1 is denoted by $\mathcal{L}(E_0, E_1)$. It is a vector space with respect to the natural pointwise operations of addition and scalar multiplication of bounded linear operators, defined by

$$(A+B)x = Ax + Bx, \quad (\lambda A)x = \lambda Ax,$$

and it becomes a Banach space with norm defined by

$$\|A\| = \sup_{\|x\|_{E_0}\leqslant 1} \|Ax\|_{E_1} \; .$$

1.4.8 Completely Continuous Operators

A linear operator $A : E_0 \to E_1$ is called *completely continuous* if it maps every bounded set in E_0 into a precompact subset of E_1. Clearly every completely continuous operator is bounded (and consequently continuous). In infinite-dimensional spaces the converse is in general not true.

1.4.9 Embedding Operators

We say that a Banach space E_0 is *embedded* into a Banach space E_1 if E_0 is a vector subspace of E_1. The corresponding operator $i : E_0 \to E_1$ mapping each element to itself is called the *embedding operator*; if it is continuous, then we say that E_0 is *continuously embedded* in E_1. If the embedding operator is completely continuous, then we say that the embedding $i : E_0 \to E_1$ is *compact*.

Conditions (1.1.2)–(1.1.4) determining embeddings of Sobolev spaces were given in Sec. 1.1.5. The embeddings defined by these conditions are continuous and the embeddings defined by (1.1.3), (1.1.4) are compact. If

$$\frac{1}{q} > \frac{1}{p} - \frac{m-k}{N} ,$$

then the embedding (1.1.2) is also compact.

1.5 Nonlinear Operators and Functionals

1.5.1 Continuity

Let E_0 and E_1 be Banach spaces. An operator $A : E_0 \to E_1$ is

(1) *continuous* if it maps every sequence that converges in the norm of E_0 to a sequence that converges in the norm of E_1;

(2) *bounded* if it maps every bounded set in E_0 to a bounded set in E_1;

(3) *weakly continuous* if it maps every weakly convergent sequence to a weakly convergent sequence;

(4) *strongly continuous* if it maps every weakly convergent sequence to a convergent sequence;

(5) *completely continuous* if it is continuous and maps every bounded set in E_0 into a precompact set in E_1;

(6) *demicontinuous* if it maps every strongly convergent sequence to a weakly convergent sequence;

(7) *hemicontinuous* if, for all x, $y \in E_0$ and every sequence of nonnegative numbers $(t_n) \to 0$, the sequence $(A(x + t_n y))$ weakly converges to $A(x)$.

These definitions extend easily to the case where the domain of definition of the operator A is a proper subset of E_0. Examples of operators having some of these properties will be given in Sec. 1.5.4. Other types of nonlinear operators (potential, monotonic, multivalued, etc.) will be discussed later on in this monograph.

1.5.2 Differentiability

An operator $A : E_0 \to E_1$ is *Fréchet differentiable* at a point $x \in E_0$ if there exists a linear bounded operator $B : E_0 \to E_1$ for which

$$A(x + h) - A(x) = Bh + \omega(x, h) ,$$

where

$$\lim_{\|h\|_{E_0}\to 0} \|h\|_{E_0}^{-1} \|\omega(x,h)\|_{E_1} = 0 .$$

The operator B is called the *Fréchet derivative* of A at x, and denoted by $A'(x)$. If A is Fréchet differentiable everywhere in some subset M of E_0 and the operator $A' : M \to \mathcal{L}(E_0, E_1)$ is continuous, then A is *Fréchet continuously differentiable* on M.

Theorem 1.5.1. *If $A : E_0 \to E_1$ is a completely continuous operator and A is Fréchet differentiable at x, then its Fréchet derivative $A'(x) : E_0 \to E_1$ is a completely continuous linear operator.*

An operator $A : E_0 \to E_1$ differentiable at each point of a subset M is *uniformly differentiable* on M if the remainder term $\omega(x,h)$ satisfies the condition

$$\lim_{\|h\|_{E_0}\to 0} \sup_{x\in M} \|h\|_{E_0}^{-1} \|\omega(x,h)\|_{E_1} = 0 .$$

An operator $A : E_0 \to E_1$ is *Gâteaux differentiable* at a point $x \in E_0$ if there exists a bounded linear operator $B : E_0 \to E_1$ for which

$$A(x+th) - A(x) = tBh + \omega(x,h,t) ,$$

where

$$\lim_{t\to 0} t^{-1}\|\omega(x,h,t)\|_{E_1} = 0$$

for all $h \in E_0$. Clearly a Fréchet differentiable operator is Gâteaux differentiable. The converse is not in general true. However, if an operator is Gâteaux differentiable and its Gâteaux derivative depends continuously (in the operator norm) on x, then this operator is Fréchet differentiable and the Gâteaux and Fréchet derivatives coincide. Just as for Fréchet derivatives, we denote the *Gâteaux derivative* of A by $A'(x)$.

If the operator $A' : E_0 \to \mathcal{L}(E_0, E_1)$ is Fréchet differentiable (Gâteaux differentiable) at a point x, then the operator A is said to be *twice Fréchet (Gâteaux) differentiable* at x, and higher derivatives of A are defined similarly.

The notion of Taylor differentiability, defined below, plays an important role in nonlinear analysis.

An operator

$$\mathcal{D}_k : \prod_{i=1}^{k} E_0 \to E_1$$

is said to be *multilinear* if it is linear and bounded with respect to each variable; its *norm* is then the number

$$\|\mathcal{D}_k\| = \sup_{\|x_1\|_{E_0},\ldots,\|x_k\|_{E_0}\leqslant 1} \|\mathcal{D}_k(x_1,\ldots,x_k)\|_{E_1} .$$

An operator B_k is a *homogeneous form of degree* k if there is a multilinear operator $\mathcal{D}_k(x_1, \ldots, x_k)$ such that

$$B_k(x) = \mathcal{D}_k(x, \ldots, x) ,$$

for all x; the *norm* of B_k is

$$\|B_k\| = \sup_{\|x\|_{E_0} \leqslant 1} \|B_k(x)\|_{E_1} .$$

We say that an operator $A : E_0 \to E_1$ is m-times *Taylor differentiable* at the point $x \in E_0$ if we can write

$$A(x+h) - A(x) = B_1(h) + \cdots + B_m(h) + \omega(x;h) ,$$

where B_k is a homogeneous form of degree k for $k = 1, \ldots, m$ and

$$\lim_{\|h\|_{E_0} \to 0} \|h\|_{E_0}^{-m} \|\omega(x;h)\|_{E_1} = 0 .$$

The operators $k!B_k$ $(k = 1, \ldots, m)$ are the *Taylor derivatives* of A at x and we write $A^{(k)}(x) = k!B_k$ for each k.

We say that A is *analytic* at x if, for all points $h \in E_0$ of sufficiently small norm, $A(x+h)$ can be represented by a uniformly convergent series

$$A(x+h) = A(x) + B_1(h) + \cdots + B_n(h) + \cdots ,$$

where B_n is a homogeneous form of order n for each n. In this case,

$$\overline{\lim_{n \to \infty}} \sqrt[n]{\|B_n\|} < \infty .$$

There is a close relation between Fréchet differentiability and Taylor differentiability. If an operator A is m-times Taylor differentiable in some neighborhood of a point x and its Taylor derivatives are continuous in norm in this neighborhood, then A is m-times Fréchet differentiable at x. Conversely, if A is m-times Fréchet differentiable in some neighborhood of x and the derivatives are continuous in the operator norm, then A is m-times Taylor differentiable at x.

The notions of continuity and differentiability described above for operators have special properties when applied to functionals. For instance, a functional is weakly continuous if and only if it is strongly continuous. While the Fréchet (or Gâteaux) derivative of a nonlinear operator $A : E_0 \to E_1$ is an operator $A' : E_0 \to \mathcal{L}(E_0, E_1)$, the Fréchet derivative of a nonlinear functional is an operator from E_0 to E_0^*. Indeed, by the definition given for operators, a functional $f : E \to \mathbb{R}$ is *Fréchet differentiable* at $x \in E$ if there exists a bounded linear functional l such that

$$f(x+h) - f(x) = \langle l, h \rangle + \omega(x;h) ,$$

where

$$\lim_{\|h\|_E \to 0} \|h\|_E^{-1} \, \omega(x; h) = 0 \, .$$

This functional $l \in E^*$ is called the *Fréchet gradient* of f at the point x and is denoted by $\nabla f(x)$. In particular, if E is a Hilbert space, then ∇f is an operator from E to E. The gradient of a weakly continuous uniformly differentiable functional has the following important property.

Theorem 1.5.2. *The gradient $\nabla f : E \to E^*$ of a weakly continuous uniformly Fréchet differentiable functional f is a completely continuous operator.*

If $f : E \to \mathbb{R}$ is twice Fréchet differentiable at a point x, then its second derivative is called the *Hessian* of f at x and is denoted by $\nabla^2 f(x)$; it is an operator from E to $\mathcal{L}(E, E^*)$.

A functional $f : E \to \mathbb{R}$ is *Gâteaux differentiable* at a point x if there exists a bounded linear functional $l \in E^*$ such that

$$f(x + th) - f(x) = t\langle l, h\rangle + \omega(x, h, t)$$

for all $h \in E$, where for each $h \in E$

$$\lim_{t \to 0} t^{-1} \, \omega(x, h, t) = 0 \, .$$

The functional l is the *Gâteaux gradient* of f at x. Like the Fréchet gradient, the Gâteaux gradient will be denoted by $\nabla f(x)$.

The *Lagrange formula* (or the *finite increments formula*) holds for continuously differentiable functionals: if x_0, $x_1 \in E$, then

$$f(x_0) - f(x_1) = \langle \nabla f((1 - \tau)x_0 + \tau x_1), x_0 - x_1\rangle \, , \qquad (1.5.1)$$

for some $\tau \in (0, 1)$.

The Lagrange formula cannot be generalized to operators. However, if M is a convex subset of E, and A is an operator that is Fréchet differentiable on M and satisfies

$$\sup_{x \in M} \|A'(x)\| \leqslant q \, ,$$

then

$$\|A(x_0) - A(x_1)\| \leqslant q \, \|x_0 - x_1\|$$

for all x_0, $x_1 \in M$.

1.5.3 Lipschitzian and Convex Functionals

Let E be a real Banach space. We say that a functional $f : E \to \mathbb{R}$ satisfies a *Lipschitz condition* with the constant L on a subset M of E if

$$|f(x_0) - f(x_1)| \leqslant L \, \|x_0 - x_1\| \quad (x_0, x_1 \in M) \, .$$

If f satisfies a Lipschitz condition on each ball of E, then f is said to be *locally Lipschitzian* on E. (It should be emphasized that a locally Lipschitzian functional may not be Lipschitzian on E.)

Suppose that f satisfies a Lipschitz condition with constant L in a neighborhood of a point $x \in E$ and that $v \in E$. The *generalized Clarke derivative* of f at x in the direction of v is the number

$$f^0(x;v) = \overline{\lim_{\substack{y \to x \\ t \to +0}}} \, t^{-1}(f(y+tv) - f(y)) \, .$$

It follows from the definition of the generalized Clarke derivative that $f^0(x;v)$ is upper semicontinuous as a function of $(x;v)$ and satisfies a Lipschitz condition with respect to v with constant L. In particular,

$$\left| f^0(x;v) \right| \leqslant L \left\| v \right\| \, .$$

The *generalized gradient* $\partial f(x)$ at x of a Lipschitzian functional f is the set of continuous linear functionals $\xi \in E^*$ satisfying the inequality

$$f^0(x;v) \geqslant \langle \xi, v \rangle \quad (v \in E) \, .$$

The following result holds.

Theorem 1.5.3. *Let the functional f satisfy a Lipschitz condition in a neighborhood of a point $x \in E$. Then*

(a) *$\partial f(x)$ is a nonempty convex weak* compact set in E^*,*

(b) *for every $v \in E$ we have*

$$f^0(x;v) = \max_{\xi \in \partial f(x)} \langle \xi, v \rangle \, ,$$

(c) *the equality*

$$\partial f(x) = \bigcap_{\delta > 0} \bigcup_{y \in B(\delta, x)} \partial f(y)$$

holds, and

(d) *if E is finite-dimensional, then the multivalued mapping $x \mapsto \partial f(x)$ is upper semicontinuous.*

Regular and *convex* functionals form important classes of Lipschitzian functionals. Here are the definitions.

Let f be a functional on a space E and let v be an element of E. If the limit

$$\lim_{t \to +0} t^{-1}(f(x+tv) - f(x))$$

exists, then it is called the *derivative of f at x in the direction of v*. The directional derivative is denoted by $f'(x;v)$.

The functional f is *regular* at a point x if for every v it has a directional derivative $f'(x;v)$ and

$$f'(x;v) = f^0(x;v) .$$

It follows from this definition that any finite linear combination with nonnegative coefficients of functionals that are regular at x is again regular at x. In addition, if f is a regular functional and is Gâteaux differentiable at x, then

$$\partial f(x) = \{\nabla f(x)\} .$$

The following results greatly simplify the calculation of generalized gradients of Lipschitzian functionals.

Proposition 1.5.1. *If $f : E \to \mathbb{R}$ is a Lipschitzian functional, then*

$$\partial(sf)(x) = s\partial f(x)$$

for every $s \in \mathbb{R}$.

Proposition 1.5.2. *Let $f_i : E \to \mathbb{R}$ $(i = 1, \dots, n)$ be Lipschitzian functionals. Then*

$$\partial\left(\sum_{i=1}^{n} f_i\right)(x) \subset \sum_{i=1}^{n} \partial f_i(x).$$

If each f_i is regular at x, then

$$\partial\left(\sum_{i=1}^{n} f_i\right)(x) = \sum_{i=1}^{n} \partial f_i(x) .$$

Proposition 1.5.3. *Let $f_i : E \to \mathbb{R}$ $(i = 1, \dots, n)$ be Lipschitzian functionals. Then*

$$\partial\left(\sum_{i=1}^{n} s_i f_i\right)(x) \subset \sum_{i=1}^{n} s_i \partial f_i(x)$$

for all $s_1, \dots, s_n \in \mathbb{R}$. If each f_i is regular at x and $s_i \geqslant 0$ for each i, then

$$\partial\left(\sum_{i=1}^{n} s_i f_i\right)(x) = \sum_{i=1}^{n} s_i \partial f_i(x) .$$

Consider a family $f_i : E \to \mathbb{R}$ $(i = 1, \dots, n)$ of Lipschitzian functionals. We set

$$f(x) = \max\{f_i(x) : \; i = 1, \dots, n\} .$$

Then the functional f is also Lipschitzian. Writing $I(x)$ for the set of indices i such that $f_i(x) = f(x)$, we have the following result.

Proposition 1.5.4. *The inclusion*

$$\partial f(x) \subset \operatorname{co}\{\partial f_i(x) : \ i \in I(x)\}$$

holds. If f_i is regular at x for each $i \in I(x)$, then f is regular at x and

$$\partial f(x) = \operatorname{co}\{\partial f_i(x) : \ i \in I(x)\} .$$

We shall now give the definition and some properties of convex functionals.

A functional $f : U \to \mathbb{R}$ on a convex set $U \subset E$ is *convex* if, for all points $x_0, x_1 \in U$ and all $\lambda \in [0, 1]$, we have

$$f\left((1-\lambda)x_0 + \lambda x_1\right) \leqslant (1-\lambda)f(x_0) + \lambda f(x_1) .$$

If this inequality is strict for all distinct x_0, $x_1 \in U$ and $\lambda \in (0, 1)$, then f is *strictly convex*. The *Jensen inequality* holds for convex functionals, namely, if $x_i \in U$ $(i = 1, \ldots, n)$, $\alpha_i \geqslant 0$ $(i = 1, \ldots, n)$, and $\sum_{i=1}^{n} \alpha_i = 1$, then

$$f\left(\sum_{i=1}^{n} \alpha_i x_i\right) \leqslant \sum_{i=1}^{n} \alpha_i f(x_i) .$$

The following result shows that convex functionals are, in general, Lipschitzian.

Theorem 1.5.4. *Let $f : U \to \mathbb{R}$ be a convex functional on a convex set U in E containing a δ-neighborhood of a set V, and suppose that*

$$|f(x)| \leqslant C \quad (x \in U)$$

for a constant $C \geqslant 0$. Then

$$|f(x_0) - f(x_1)| \leqslant \frac{2C}{\delta} \|x_0 - x_1\| \quad (x_0, x_1 \in V) .$$

The *subdifferential* of a convex functional $f : E \to \mathbb{R}$ at $x \in E$ is the set $\partial f(x)$ of all points $\xi \in E^*$ such that

$$f(x+v) - f(x) \geqslant \langle \xi, v \rangle$$

for all $v \in E$.

We use the same notation for the subdifferential of a convex functional f and the generalized gradient of f, because the generalized gradient of f coincides with its subdifferential and the generalized derivative $f^0(x; v)$ coincides with the directional derivative $f'(x; v)$ for all $v \in E$.

If a convex functional $f : E \to \mathbb{R}$ is bounded in some neighborhood of a point x, then f is regular at x. Therefore, except for pathological cases, convex functionals are regular.

1.5.4 Some Special Nonlinear Operators and their Properties

The superposition operator. Let Ω be a bounded set in $\mathbb{R}^N$ and let

$$f(x; u_1, \ldots, u_k) \quad (x \in \overline{\Omega}, \quad u_i \in \mathbb{R}, \quad i = 1, \ldots, k)$$

be a function that is continuous with respect to the collection of variables $u_1, \ldots, u_k$ for almost all $x \in \overline{\Omega}$ and measurable with respect to x for all $u_1, \ldots, u_k$. We define an operator $\mathfrak{f}$, called the *superposition operator* or *Nemytskii operator*, by setting

$$\mathfrak{f}(u_1, \ldots, u_k) = f(x, u_1(x), \ldots, u_k(x)).$$

If $f(x; u_1, \ldots, u_k)$ is continuous with respect to its collection of variables, then $\mathfrak{f}$ is an operator from $C(\overline{\Omega}) \times \cdots \times C(\overline{\Omega})$ to $C(\overline{\Omega})$.

If we have

$$|f(x; u_1, \ldots, u_k)| \leqslant c \sum_{i=1}^{k} |u_i|^{p_i/p} + \varphi(x) \,, \tag{1.5.2}$$

with $p_i, p \geqslant 1$ and $\varphi \in L_p(\Omega)$, then $\mathfrak{f}$ maps $L_{p_1}(\Omega) \times \cdots \times L_{p_k}(\Omega)$ to $L_p(\Omega)$ and $\mathfrak{f}$ is bounded and continuous. One should bear in mind that $\mathfrak{f}$ is not completely continuous (except in the trivial case when $f(x; u_1, \ldots, u_k)$ is independent of $u_1, \ldots, u_k$).

If $f(x; u_1, \ldots, u_k)$ is continuous with respect to its collection of variables and the partial derivatives $f'_{u_i}(x; u_1, \ldots, u_k)$ $(i = 1, \ldots, k)$ exist, then $\mathfrak{f}$, regarded as an operator from $C(\overline{\Omega}) \times \cdots \times C(\overline{\Omega})$ to $C(\overline{\Omega})$, is Fréchet differentiable, and its Fréchet derivative at the point $\{u_1, \ldots, u_k\} = \{u_1(x), \ldots, u_k(x)\}$ is defined by

$$\mathfrak{f}'(u_1, \ldots, u_k)\{h_1, \ldots, h_k\} = \sum_{i=1}^{k} f'_{u_i}(x; u_1(x), \ldots, u_k(x)) h_i(x) \,. \tag{1.5.3}$$

Suppose that $\mathfrak{f}$ maps $L_{p_1}(\Omega) \times \cdots \times L_{p_k}(\Omega)$ to $L_p(\Omega)$ and that $p_i > p$ $(i = 1, \ldots, k)$. If $f(x; u_1, \ldots, u_k)$ is differentiable with respect to $u_1, \ldots, u_k$ and the inequalities

$$\left| f'_{u_i}(x; u_1, \ldots, u_k) \right| \leqslant c_i \sum_{i=1}^{k} |u_i|^{p_i/p-1} + \varphi_i(x) \quad (i = 1, \ldots, k) \tag{1.5.4}$$

hold, where $\varphi_i(x) \in L_{p_i p/(p_i - p)}$, then $\mathfrak{f} : L_{p_1}(\Omega) \times \cdots \times L_{p_k}(\Omega) \to L_p(\Omega)$ is Fréchet differentiable and its derivative is given by (1.5.3).

The Hammerstein operator. This is the nonlinear integral operator

$$A(u) = \int_{\Omega} K(x, y) f(y, u(y)) \, dy \,.$$

It is the composite of the linear integral operator

$$\mathcal{K}u = \int_{\Omega} K(x,y)u(y)\,dy$$

and the superposition operator $\mathfrak{f}$ defined by

$$\mathfrak{f}(u) = f(x, u(x)) .$$

For the Hammerstein operator to be completely continuous on the space $L_p(\Omega)$, it is sufficient for $\mathfrak{f}$ to be an operator from $L_p(\Omega)$ to some space $L_q(\Omega)$ and for the linear operator $\mathcal{K}$ to be a completely continuous mapping from $L_q(\Omega)$ to $L_p(\Omega)$.

If the superposition operator $\mathfrak{f}$, regarded as an operator from $L_p(\Omega)$ to $L_q(\Omega)$, is differentiable at some point $u = u(x)$ and $\mathcal{K}$ is a mapping from $L_q(\Omega)$ to $L_p(\Omega)$, then the Hammerstein operator $A = \mathcal{K} \circ \mathfrak{f}$ is also differentiable at u, and we have

$$A'(u)h = \mathcal{K} \circ \mathfrak{f}'(u)h = \int_{\Omega} K(x,y) f'_u(y, u(y))h(y)\,dy .$$

1.5.5 Extensions of Mappings and the Partition of Unity Theorem

In many problems of nonlinear analysis, we need to extend mappings from their domain of definition to larger sets, while preserving some important properties. The theorem of Tietze and Urysohn is the main tool for doing this.

Theorem 1.5.5 (Tietze–Urysohn). *Suppose that M_0 is a closed subset of a complete metric space M and $f_0 : M_0 \to \mathbb{R}$ is a continuous function for which*

$$|f_0(x)| \leqslant k \quad (x \in M_0) .$$

Then there exists a continuous function $f : M \to \mathbb{R}$, coinciding with f_0 on M_0, for which

$$|f(x)| \leqslant k \quad (x \in M) .$$

This theorem implies

Theorem 1.5.6. *If M_0 is a closed subset of a complete metric space M then every continuous mapping $F : M_0 \to \mathbb{R}^N$ can be extended to a continuous mapping with domain of definition M.*

The following result is often useful.

Theorem 1.5.7. *Suppose that K is a closed cube in $\mathbb{R}^N$, K_0 is a compact subset of K, $\varphi_0 : K_0 \to \mathbb{R}^M \setminus \{0\}$ is a continuous mapping, and $M > N$. Then φ_0 can be extended to a continuous mapping $\varphi : K \to \mathbb{R}^M \setminus \{0\}$.*

Let Ω be a domain in $\mathbb{R}^N$ that is covered by a family Ω_i $(i \in I)$ of open subsets of Ω , i.e.

$$\Omega = \bigcup_{i \in I} \Omega_i \, .$$

Theorem 1.5.8 (partition of unity). *There exists a family of infinitely differentiable functions* $\alpha_i : \mathbb{R} \to \mathbb{R}$ $(i \in I)$ *such that* $\operatorname{supp} \alpha_i \subset \Omega_i$ *for every* $i \in I$*, and with the following properties:*

$$0 \leqslant \alpha_i(x) \leqslant 1 \quad (x \in \mathbb{R}^N, \ i \in I) \, ;$$

$$\sum_{i \in I} \alpha_i(x) = 1 \quad (x \in \Omega) \, .$$

(Here, as usual, the support $\operatorname{supp} \beta$ of a mapping β is defined by $\operatorname{supp} \beta = \{x \in \mathbb{R}^N : \beta(x) \neq 0\}$.)

1.6 Contraction Mappings, the Implicit Function Theorem, and Differential Equations in a Banach Space

1.6.1 Contraction Mappings

Let M be a subset of a Banach space E. A mapping $A : M \to E$ is said to be a *contraction mapping* on M if for some constant $q < 1$ we have

$$\|A(x_0) - A(x_1)\| \leqslant q \, \|x_0 - x_1\|$$

for all x_0, $x_1 \in M$. A point $x_* \in M$ is a *fixed point* of A if

$$x_* = A(x_*) \, .$$

Thus A has a fixed point if and only if the operator equation

$$x = A(x)$$

is solvable.

The following result is one of the main tools in nonlinear analysis for proving the solvability of operator equations.

Theorem 1.6.1 (Banach). *Suppose that* M *is a closed subset of a Banach space and that* $A : M \to M$ *is a contraction mapping. Then* A *has a unique fixed point* x_* *in* M*.*

Theorem 1.6.1 is called the *contraction mapping theorem.* The fixed point x_* can be constructed by the *method of successive approximations*, as the limit of the sequence

$$x_{n+1} = A(x_n) \quad (n = 0, 1, \ldots) \, .$$

If $x_0 \in M$ is the initial approximation, then the successive approximations x_n converge in norm to x_* and an estimate for the rate of convergence is

$$\|x_n - x_*\| \leqslant \frac{q^n}{1-q} \|x_0 - A(x_0)\| \, .$$

1.6.2 The Implicit Function Theorem

Suppose that E, F, G are Banach spaces and that $A : E \times F \to G$ is an operator. Consider the equation

$$A(x, y) = 0 \, , \tag{1.6.1}$$

which we regard as an equation defining an element $y(x) \in F$ from an element $x \in E$. If such an element $y(x)$ can be found for all x in some subset M of E, then $y : M \to F$ is called the *implicit function* defined by Eq. (1.6.1).

Suppose that

$$A(x_0, y_0) = 0$$

for some $x_0 \in E$ and $y_0 \in F$ and that A is continuous in some neighborhood U in $E \times F$ of the pair (x_0, y_0). The following result holds.

Theorem 1.6.2. *Suppose that for every pair $(x, y) \in U$ the operator A has a Fréchet derivative $A'_y(x, y) : F \to G$ which is continuous in the operator norm, and suppose that the operator $A'_y(x_0, y_0) : F \to G$ has a bounded inverse defined on G. Then, on some ball $B(r, x_0)$ with $r > 0$, Eq. (1.6.1) defines a unique function $y : B(r, x_0) \to F$ satisfying the condition*

$$y(x_0) = y_0 \, .$$

In addition, the function y is continuously differentiable on $B(r, x_0)$ and its Fréchet derivative satisfies

$$y'(x) = -A'_y(x, y(x))^{-1} \circ A'_x(x, y(x)) \, .$$

Note that if, with the hypothesis of Theorem 1.6.2, the operator A is m-times continuously differentiable on U, then the implicit function $y : E \to F$ is m-times differentiable on the ball $B(r, x_0)$. In particular, it follows that if A is infinitely differentiable, then so is the implicit function. If A is an analytic operator, then the implicit function $y(x)$ is analytic.

1.6.3 The Cauchy Problem

Suppose that E is a real Banach space and $f : E \to E$ is a continuous mapping. Consider the differential equation

$$\frac{dx}{dt} = f(x) . \tag{1.6.2}$$

A continuously differentiable mapping $x : [0, T] \to E$ is called a *solution* of Eq. (1.6.1) on the interval $[0, T]$ if

$$\frac{dx(t)}{dt} \equiv f(x(t)) \quad (0 \leqslant t \leqslant T) .$$

As in the case when E is finite-dimensional, we seek solutions of Eq. (1.6.2) satisfying initial conditions. The problem of finding solutions of Eq. (1.6.2) with the initial condition

$$x(0) = x_0 \tag{1.6.3}$$

is known as the *Cauchy problem*; in general, it need not have a solution if f is merely continuous. However, if f satisfies a local Lipschitz condition, then, for sufficiently small T, the Cauchy problem (1.6.2), (1.6.3) has a unique solution on $[0, T]$. If the solution $x(t)$ is defined for all $t \geqslant 0$, then this solution is said to be *nonlocally extendable.* For example, the solutions of Eq. (1.6.2) are certainly nonlocally extendable if for some constant C we have

$$\|f(x)\| \leqslant C(1 + \|x\|) .$$

A point x_0 is a *state of equilibrium* of Eq. (1.6.2) if

$$f(x_0) = 0 .$$

It is then clear that the constant function $x(t) \equiv x_0$ $(0 \leqslant t < \infty)$ is a solution of Eq. (1.6.2).

A state of equilibrium x_0 is *Lyapunov stable* if, for all $\varepsilon > 0$, there exists some $\delta > 0$ such that whenever x satisfies

$$\|x - x_0\| < \delta ,$$

the Cauchy problem for Eq. (1.6.2) has a solution $p(t, x)$ with the initial condition $p(0, x) = x$, defined for all $t > 0$ and satisfying

$$\|x_0 - p(t, x)\| < \varepsilon \quad (0 \leqslant t < \infty) .$$

If a state of equilibrium x_0 of Eq. (1.6.2) is Lyapunov stable and

$$\lim_{t \to \infty} \|x_0 - p(t, x)\| = 0$$

for all x in a sufficiently small neighborhood of x_0, then x_0 is said to be an *asymptotically stable* state of equilibrium.

The *method of Lyapunov functionals* is frequently used to investigate the stability and asymptotic stability of states of equilibrium. Here are two basic results concerning this method that will be used in this monograph.

Theorem 1.6.3. *Suppose that there exists a continuously differentiable functional V on a ball $B(r, x_0)$ for which $V(x_0) = 0$ and such that*

$$\inf_{\|x-x_0\|=\rho} V(x) \geqslant \mu(\rho) > 0 \quad (0 < \rho \leqslant r) \tag{1.6.4}$$

and

$$\langle \nabla V(x), f(x) \rangle \leqslant 0 \quad (x \in B(r, x_0)) \, . \tag{1.6.5}$$

Then x_0 is a Lyapunov stable state of equilibrium.

Theorem 1.6.4. *Suppose that x_0 is a state of equilibrium of Eq.* (1.6.2) *and that there exists a continuously differentiable functional V in some neighborhood $B(r, x_0)$ of x_0 for which $V(x_0) = 0$ and such that*

$$\inf_{\|x-x_0\|=\rho} V(x) \geqslant \mu(\rho) > 0 \quad (0 < \rho \leqslant r) \tag{1.6.6}$$

and

$$\langle \nabla V(x), f(x) \rangle \leqslant \nu(\|x - x_0\|) < 0 \quad (x \neq x_0) \, , \tag{1.6.7}$$

where $\nu(\rho)$ $(0 < \rho \leqslant r)$ is a continuous function. Then x_0 is an asymptotically stable state of equilibrium.

The functionals V appearing in Theorems 1.6.3 and 1.6.4 are called *Lyapunov functionals.*

A Lyapunov stable state of equilibrium of Eq. (1.6.2) is said to be *stable in the large* if

$$\lim_{t \to \infty} \|x_0 - p(t, x)\| = 0$$

for all $x \in E$. Clearly every state of equilibrium of Eq. (1.6.2) that is stable in the large is also asymptotically stable. The converse statement is not true even if x_0 is the unique state of equilibrium of Eq. (1.6.2) and the space E is finite-dimensional.

1.7 Minimizers of Nonlinear Functionals

1.7.1 Critical Points

Let $f : E \to \mathbb{R}$ be a Gâteaux differentiable functional. A point $x_0 \in E$ is called a *critical point* of f if

$$\nabla f(x_0) = 0 \, .$$

A point $x_0 \in E$ is a *local minimizer* for a functional f if in some neighborhood $U(x_0)$ of x_0 we have

$$f(x_0) \leqslant f(x) \quad (x \in U(x_0)). \tag{1.7.1}$$

If the inequality (1.7.1) is strict for $x \neq x_0$, then x_0 is a *strict local minimizer*.

If the inequality

$$f(x_0) \leqslant f(x) \quad (x \in M) \tag{1.7.2}$$

holds on a subset M of E containing x_0, then x_0 is an *absolute minimizer* for f *on* M, and the notion of a *strict absolute minimizer* for f on M is defined similarly. In particular, if $M = E$, then the minimizer is called a *global minimizer* (*strict global minimizer*) for f.

If (1.7.2) is satisfied on the set $M \cap U$, where U is a neighborhood of x_0, then x_0 is called a *local minimizer for f on M* or a *conditional local minimizer* for f.

The following theorem explains the importance of critical points.

Theorem 1.7.1. *If $x_0 \in E$ is local minimizer of a differentiable functional f, then x_0 is a critical point of f.*

Theorem 1.7.1 gives a necessary condition for minimizers: local minimizers of a smooth functional are necessarily critical points. The condition in Theorem 1.7.1 is called the *first-order necessary condition for a minimizer.*

1.7.2 Second-Order Necessary and Sufficient Conditions for a Minimizer

Suppose that f is a functional that is twice Fréchet differentiable at a point $x_0 \in E$ and that $\nabla^2 f(x_0) : E \to E^*$ is its Jacobian at x_0. Then for $x_0 + h$ in some neighborhood of x_0 we can write

$$f(x_0 + h) - f(x_0) = \langle \nabla f(x_0), h \rangle + \tfrac{1}{2} \langle \nabla^2 f(x_0) h, h \rangle + \omega(h) \;, \tag{1.7.3}$$

where

$$\lim_{\|h\| \to 0} \|h\|^{-2} \omega(h) = 0 \;. \tag{1.7.4}$$

If x_0 is a critical point of f then $\nabla f(x_0) = 0$, and (1.7.3) assumes the form

$$f(x_0 + h) - f(x_0) = \langle \nabla^2 f(x_0) h, h \rangle + \omega(h) \;. \tag{1.7.5}$$

The representation (1.7.5) implies the following results.

Theorem 1.7.2. *Let x_0 be a local minimizer of a functional f. Then*

$$\langle \nabla^2 f(x_0) h, h \rangle \geqslant 0 \quad (h \in E) \;. \tag{1.7.6}$$

Theorem 1.7.3. *Let x_0 be a critical point of a functional f and suppose that*

$$\langle \nabla^2 f(x_0) h, h \rangle \geqslant a \, \|h\|^2 \quad (h \in E) \tag{1.7.7}$$

for some $a > 0$. Then x_0 is a local minimizer for f.

The quadratic functional

$$b(h) = \tfrac{1}{2}\langle \nabla^2 f(x_0)h, h\rangle \tag{1.7.8}$$

is the *second variation* of f at x_0. Conditions (1.7.6), (1.7.7) assert respectively the *nonnegative definiteness* and the *strong positive definiteness* of the second variation of f. The conditions given in Theorem 1.7.2 and Theorem 1.7.3 are respectively the *second-order necessary condition for a minimizer* and the *second-order sufficient condition for a minimizer*.

Every critical point of a convex functional is a minimizer:

Theorem 1.7.4. *Suppose that M is a convex domain in E and that $x_0 \in M$ is a critical point of a convex differentiable functional f. Then x_0 is an absolute minimum for f on M.*

If $x_0 \in M$ is a minimizer for a strictly convex functional f, then clearly x_0 is a strict absolute minimizer for f on M. Therefore Theorem 1.7.4 implies the following results.

Theorem 1.7.5. *Suppose that M is a convex domain in E and $x_0 \in M$ is a critical point of a differentiable strictly convex functional f. Then x_0 is an absolute strict minimizer for f on M.*

Theorem 1.7.6. *Suppose that $f : E \to \mathbb{R}$ is a differentiable convex (resp. strictly convex) functional and x_0 is a critical point of f. Then x_0 is a global (resp. strict global) minimizer for f.*

1.7.3 Conditional Extrema

So far we have considered results concerning minimization of functionals without any constraints on their domain of definition. Such problems are known as problems of *unconditional minimization.* In important parts of mathematics and its applications (classical calculus of variations, optimal control theory, linear and convex programming, etc.) we encounter extremal problems with constraints. In this section, we give a necessary condition for an extremum in such problems.

Let E_0 and E_1 be Banach spaces and $F : E_0 \to E_1$ be an operator. We set

$$M = \{x \in E_0 : F(x) = 0\} \tag{1.7.9}$$

and consider the problem of minimization of a functional $f : E_0 \to \mathbb{R}$ on the set M. This problem is denoted by

$$\begin{cases} f(x) \to \min\,, \\ F(x) = 0\,. \end{cases} \tag{1.7.10}$$

Let x_0 be a solution of problem (1.7.10) and suppose that both f and F are continuously differentiable in some neighborhood of x_0. Then the following result holds.

Theorem 1.7.7. *If the image of E_0 under the mapping $F'(x_0) : E_0 \to E_1$ is closed, then there exist a number $\lambda_0 \neq 0$ and a nonzero functional $l_0 \in E_1^*$ for which*

$$\lambda_0 \nabla f(x_0) + (F'(x_0))^* l_0 = 0 . \tag{1.7.11}$$

Theorem 1.7.7 is an infinite-dimensional analog of the well-known method of *Lagrange multipliers* from classical analysis for locating conditional extrema. Indeed, if we seek a minimizer for a function $f_0(x_1, \dots, x_n)$ under the conditions

$$f_i(x_1, \dots, x_n) = 0 \quad (i = 1, \dots, m) , \tag{1.7.12}$$

then we take $E_0 = \mathbb{R}^n$, $E_1 = \mathbb{R}^m$, we define the mapping $F : E_0 \to E_1$ by

$$F(x) = \{f_1(x_1, \dots, x_n), \dots, f_m(x_1, \dots, x_n)\} ,$$

and we take as the functional $l_0 \in E_1^*$ the vector $l_0 = \{\lambda_1, \dots, \lambda_m\}$. Equation (1.7.11) becomes

$$\lambda_0 \nabla f(x_1, \dots, x_n) + \sum_{i=1}^{m} \lambda_i \nabla f_i(x_1, \dots, x_n) = 0 , \tag{1.7.13}$$

and the points which must be investigated for a conditional extremum can be determined from the system of equations (1.7.12), (1.7.13). In this way we recover the method of Lagrange multipliers for finding a conditional extremum.

Now we consider a special case of (1.7.10):

$$\begin{cases} f(x) \to \min , \\ g(x) = 0 , \end{cases} \tag{1.7.14}$$

where $g : E_0 \to \mathbb{R}$ is a continuously differentiable functional. Theorem 1.7.7 implies

Theorem 1.7.8. *If x_0 is a solution of* (1.7.14) *and $\nabla g(x_0) \neq 0$, then*

$$\nabla f(x_0) = \lambda_0 \nabla g(x_0)$$

for some $\lambda_0 \in \mathbb{R}$.

1.7.4 Extrema of Lipschitzian and Convex Functionals

In this section we consider critical and conditionally critical points of convex and Lipschitzian functionals and give necessary conditions for an extremum.

Let M be a nonempty subset of a Banach space E. The *distance functional* $d_M : E \to \mathbb{R}$ is defined by

$$d_M(x) = \inf_{y \in M} \|x - y\| .$$

If M is closed, then $x \in M$ if and only if $d_M(x) = 0$. Although d_M is not differentiable at all points $x \in E$ it satisfies a global Lipschitz condition.

Theorem 1.7.9. *The inequality*

$$|d_M(x_0) - d_M(x_1)| \leqslant \|x_0 - x_1\|$$

holds for all x_0, $x_1 \in E$.

Using the distance functional we may now define the tangent cone and normal cone to a subset M. These will be used to formulate further necessary conditions for a minimizer.

Let x be a point of M. A vector $v \in E$ is a *tangent* to M at x if the generalized derivative of d_M in the direction of v at x is zero:

$$d_M^0(x; v) = 0 .$$

The set of all tangent vectors to M at x is denoted by $T_M(x)$. It follows immediately from the properties of the generalized derivative (see 1.5.3) that $T_M(x)$ is a closed convex cone in E. It is called the *tangent cone* to M at x.

The *normal cone* to M at x is the cone $N_M(x)$ dual to the tangent cone:

$$N_M(x) = \{\xi \in E^* : \langle \xi, v \rangle \leqslant 0,\ v \in T_M(x)\} .$$

The normal cone is often equivalently defined in terms of the generalized gradient as follows:

$$N_M(x) = \operatorname{cl}\left(\bigcup_{\lambda \geqslant 0} \lambda \partial d_M(x)\right) ,$$

where cl denotes weak* closure.

The critical points of smooth functionals are the zeros of their gradients. If $f : E \to \mathbb{R}$ is a Lipschitzian functional, then x_0 is a *critical point* of f if

$$0 \in \partial f(x_0) . \tag{1.7.15}$$

If f is also regular and differentiable at x_0, then x_0 is also a critical point of f in the classical sense. However this assertion is not in general true without the regularity hypothesis.

The following result explains why it is natural to take (1.7.15) as the definition of a critical point of a Lipschitzian functional f.

Theorem 1.7.10. *If x_0 is a point of local extremum of a Lipschitzian functional f then x_0 is a critical point of f.*

Suppose that f is a functional defined on a subset M of E and that f is Lipschitzian on M. Then a point $x_0 \in M$ is a *conditionally critical point* of f if

$$0 \in \partial f(x_0) + N_M(x_0) , \tag{1.7.16}$$

where $N_M(x_0)$ is the normal cone to M at x_0. In particular, if $M = E$, then

$$N_E(x) = \{0\}$$

for each point x, and this definition reduces to the definition of a critical point given in (1.7.15). The following fundamental result holds.

Theorem 1.7.11. *If x_0 is a local minimizer of a Lipschitzian functional f on a set M, then x_0 is a conditionally critical point of f on M.*

1.7.5 Weierstrass Theorems

A continuous function on a closed and bounded set in a finite-dimensional space attains its extrema. This fundamental result of classical analysis, known as the Weierstrass theorem, does not extend directly to infinite-dimensional spaces. In this section, we present some extensions to various classes of functionals on subsets of infinite-dimensional spaces.

Theorem 1.7.12. *Suppose that K is a compact set in a Banach space E and that $f : K \to \mathbb{R}$ is a continuous functional. Then f is bounded on K and attains in K its least upper bound and greatest lower bound.*

Theorem 1.7.12 also is known as the Weierstrass theorem. The hypothesis that K is compact cannot be omitted. Thus, for instance, for every ball in an infinite-dimensional Banach space there is a continuous functional unbounded on this ball.

The Weierstrass theorem can be generalized to semicontinuous functionals. A functional $f : M \to \mathbb{R}$ is said to be *lower semicontinuous* on M if the condition

$$\lim_{n\to\infty} \|x_n - x_0\| = 0 \quad (x_n, x_0 \in M,\ n = 1, 2, \ldots) \tag{1.7.17}$$

implies the inequality

$$f(x_0) \leqslant \varliminf_{n\to\infty} f(x_n)\ .$$

Similarly, $f : M \to \mathbb{R}$ is *upper semicontinuous* on M if condition (1.7.17) implies the inequality

$$f(x_0) \geqslant \varlimsup_{n\to\infty} f(x_n)\ .$$

Theorem 1.7.13. *Suppose that K is a compact set and that the functional $f : K \to \mathbb{R}$ is lower semicontinuous. Then f is bounded from below on K and attains its greatest lower bound.*

Theorem 1.7.14. *Suppose that K is a compact set and that the functional $f : K \to \mathbb{R}$ is upper semicontinuous. Then f is bounded from above on K and attains its least upper bound.*

In the investigation of extrema of functionals defined on noncompact sets Theorems 1.7.12–1.7.14 are inapplicable, and the property of *weak semicontinuity* of functionals often proves useful.

A functional $f : M \to \mathbb{R}$ on a subset M of a Banach space E is *lower* (resp. *upper*) *weakly semicontinuous* at a point $x_0 \in M$ if, whenever (x_n) is a sequence in M that is weakly convergent to a point x_0, the inequality

$$f(x_0) \leqslant \varliminf_{n\to\infty} f(x_n) \quad (\text{resp. } f(x_0) \geqslant \varlimsup_{n\to\infty} f(x_n))$$

holds.

Theorem 1.7.15. *Each lower weakly semicontinuous functional on a weakly compact subset M is bounded on M and attains its greatest lower bound.*

In a reflexive Banach space every bounded weakly closed set is weakly compact, and so Theorem 1.7.15 implies

Theorem 1.7.16. *If M is a bounded weakly closed subset of a reflexive Banach space then every lower weakly semicontinuous functional on M is bounded on M and attains its greatest lower bound.*

Similar results hold for upper weakly semicontinuous functionals on weakly compact and bounded weakly closed subsets.

Criteria for weak semicontinuity of functionals are important in connection with Theorems 1.7.15 and 1.7.16. Below we give conditions for weak semicontinuity.

Theorem 1.7.17. *A functional $f : E \to \mathbb{R}$ is lower weakly semicontinuous if and only if for each $c \in \mathbb{R}$ the Lebesgue set*

$$\mathcal{L}_c = \{x \in E : f(x) \leqslant c\}$$

is weakly closed.

A functional $f : E \to \mathbb{R}$ on a Banach space E will be said to be *growing* if

$$\lim_{\|x\| \to \infty} f(x) = \infty .$$

Theorem 1.7.18. *Let $f : E \to \mathbb{R}$ be a functional on a reflexive Banach space E and suppose that f is lower weakly semicontinuous and growing. Then f attains its greatest lower bound on E.*

For convex functionals, the above theorems on minimizers can be refined.

Theorem 1.7.19. *Let M be a convex set in a Banach space E. Then a convex functional $f : M \to \mathbb{R}$ is lower weakly semicontinuous on M if and only if f is lower semicontinuous on M.*

Theorems 1.7.15 and 1.7.19 imply

Theorem 1.7.20. *If M is a convex closed bounded subset of a reflexive Banach space E, then every lower semicontinuous convex functional $f : M \to \mathbb{R}$ attains its greatest lower bound on M.*

Simple examples show that the requirement that M is bounded cannot be omitted in Theorem 1.7.20. However, this requirement is not needed for strongly convex functionals.

A functional $f : M \to \mathbb{R}$ on a convex set M of a Banach space E is *strongly convex* if there exists a constant $\varkappa > 0$ such that

$$f((1-\lambda)x_0 + \lambda x_1) \leqslant (1-\lambda)f(x_0) + \lambda f(x_1) - \lambda(1-\lambda)\varkappa \|x_0 - x_1\|^2$$

for all x_0, $x_1 \in M$ and $\lambda \in [0,1]$. Such a constant $\varkappa$ is called a *constant of strong convexity* of f. Clearly a strongly convex functional is also convex.

Theorem 1.7.21. *Each strongly convex functional $f : E \to \mathbb{R}$ on a Banach space E is growing.*

Theorems 1.7.20 and 1.7.21 imply

Theorem 1.7.22. *If M is a convex closed subset of a reflexive Banach space E then every strongly convex lower semicontinuous functional $f : M \to \mathbb{R}$ attains its greatest lower bound in M.*

1.8 Monotonicity

In this section we introduce monotonic operators and potential operators and give conditions for potential operators to be monotonic.

1.8.1 Definitions

An operator $A : E \to E^*$ from a Banach space E to its dual is *monotonic* if

$$\langle A(x_1) - A(x_2), x_1 - x_2 \rangle \geqslant 0 \tag{1.8.1}$$

for all x_1, $x_2 \in E$, *strictly monotonic* if

$$\langle A(x_1) - A(x_2), x_1 - x_2 \rangle > 0 \tag{1.8.2}$$

for $x_1 \neq x_2$, and *strongly monotonic* if there is some $\varkappa > 0$ such that

$$\langle A(x_1) - A(x_2), x_1 - x_2 \rangle \geqslant \varkappa \|x_1 - x_2\|^2 \tag{1.8.3}$$

for all $x_1, x_2 \in E$. An operator $A : E \to E^*$ is sometimes said to be strongly monotonic if it satisfies the condition

$$\langle A(x_1) - A(x_2), x_1 - x_2 \rangle \geqslant \gamma \left(\|x_1 - x_2\|\right) \|x_1 - x_2\| \,, \tag{1.8.4}$$

where $\gamma(t)$ $(0 \leqslant t < \infty)$ is a continuous function satisfying the conditions

$$\gamma(0) = 0, \quad \gamma(t) > 0 \quad (0 < t < \infty)$$

and

$$\lim_{t \to \infty} \gamma(t) = \infty \,.$$

1.8.2 Potential Operators

An operator $A : E \to E^*$ is said to be a *potential* operator if there exists a differentiable functional $f : E \to \mathbb{R}$ for which

$$A(x) = \nabla f(x) .$$

The functional f is the *potential* of the operator A.

The Nemytskii operators or superposition operators (see Sect. 1.5.4) form an important class of potential operators.

Let Ω be a bounded domain in $\mathbb{R}^N$ and suppose that $g(x, u)$ ($x \in \Omega$, $u \in \mathbb{R}$) is a function continuous in u for almost all $x \in \Omega$ and x-measurable for all u. Then (see Sect. 1.5.4), if the inequality

$$|g(x,u)| \leqslant a(x) + b\,|u|^{p-1} , \tag{1.8.5}$$

where $a(x) \in L_q(\Omega)$, $p > 1$, $p^{-1} + q^{-1} = 1$, $b > 0$, is satisfied, then the Nemytskii operator

$$\mathfrak{f}(u) = g(x, u(x)) \tag{1.8.6}$$

from $L_p(\Omega)$ to $L_q(\Omega)$ is a continuous potential operator. Direct computations show that the functional

$$f(u) = \int\limits_{\Omega} \left(\int\limits_{0}^{u(x)} g(x,v)dv \right) dx \tag{1.8.7}$$

is its potential.

The potential of the Nemytskii operator is called the *Hammerstein–Goloumb functional.*

The condition of monotonicity of the operator $\mathfrak{f} : L_p(\Omega) \to L_q(\Omega)$ (where $p + q = 1$) assumes the form

$$\int\limits_{\Omega} (g(x, u_1(x)) - g(x, u_2(x)))(u_1(x) - u_2(x))\, dx \geqslant 0,$$

for all u_1, $u_2 \in L_p(\Omega)$. This inequality holds if and only if for every fixed $x \in \Omega$ the function $g(x, u)$ is non-decreasing as a function of u.

1.8.3 Monotonicity and Convexity

The monotonicity of a gradient is closely connected with the convexity of the corresponding functional.

Theorem 1.8.1. *The gradient $\nabla f : E \to E^*$ of a functional $f : E \to \mathbb{R}$ is monotonic if and only if the functional f is convex.*

Theorem 1.8.2. *The gradient* $\nabla f : E \to E^*$ *of a functional* $f : E \to \mathbb{R}$ *is strictly monotonic if and only if the functional* f *is strictly convex.*

Theorem 1.8.3. *The gradient* $\nabla f : E \to E^*$ *of a functional* $f : E \to \mathbb{R}$ *is strongly monotonic if and only if the functional* f *is strongly convex.*

The following criteria for convexity, strict convexity and strong convexity are often useful.

Theorem 1.8.4. *Let* $f : E \to \mathbb{R}$ *be a Fréchet differentiable functional on* E. *Then*

(a) f *is convex if and only if*

$$f(x+h) \geqslant f(x) + \langle \nabla f(x), h \rangle \quad (x, h \in E) \; ; \tag{1.8.8}$$

(b) f *is strictly convex if and only if*

$$f(x+h) > f(x) + \langle \nabla f(x), h \rangle \quad (x, h \in E, h \neq 0) \; ; \tag{1.8.9}$$

(c) f *is strongly convex if and only if for some* $\varkappa > 0$ *we have*

$$f(x+h) \geqslant f(x) + \langle \nabla f(x), h \rangle + \tfrac{1}{2}\varkappa \|h\|^2 \quad (x, h \in E) \; . \tag{1.8.10}$$

If the functional f is twice Fréchet differentiable, then the conditions for it to be convex, strictly convex, and strongly convex can be conveniently formulated in terms of its Hessian $\nabla^2 f(x)$:

Theorem 1.8.5. *The functional* $f : E \to \mathbb{R}$ *is (a) convex, (b) strictly convex, or (c) strongly convex if and only if the corresponding inequality below holds:*

$$\langle \nabla^2 f(x)h, h \rangle \geqslant 0 \quad (h \in E) \; , \tag{1.8.11}$$

$$\langle \nabla^2 f(x)h, h \rangle > 0 \quad (h \in E, h \neq 0) \; , \tag{1.8.12}$$

$$\textit{for some } \varkappa > 0, \quad \langle \nabla^2 f(x)h, h \rangle \geqslant \varkappa \|h\|^2 \quad (h \in E) \; . \tag{1.8.13}$$

2 Finite-Dimensional Problems

In this chapter we shall use the homotopy method to investigate critical points of various types of functions of finitely many variables.

2.1 Nondegenerate Deformations of Smooth Functions

2.1.1 Invariance of Local Minimizers

Let G be a domain in $\mathbb{R}^N$, let $\lambda \in [0,1]$ be a real parameter, and suppose that

$$f : G \times [0,1] \to \mathbb{R} \tag{2.1.1}$$

is a one-parameter family of continuously differentiable functions. We denote by ∇ the gradient operator of $f(x;\lambda)$ with respect x. We assume that both the family (2.1.1) and the gradient $\nabla f(x;\lambda)$ are continuous maps on $G\times[0,1]$.

Suppose that for every $\lambda \in [0,1]$ the function $f(\cdot;\lambda)$ has an extremum $x(\lambda) \in G$ which continuously depends on λ. The one-parameter family of functions (2.1.1) is said to be a *nondegenerate deformation on the domain G* of the function

$$f_0(x) = f(x;0) \quad (x \in G)$$

into the function

$$f_1(x) = f(x;1) \quad (x \in G)$$

if, for every $\lambda \in [0,1]$, the point $x(\lambda)$ is the only critical point of $f(x;\lambda)$ in G.

The following theorem is the main result of this section.

Theorem 2.1.1 (deformation principle for a minimizer). *Suppose that there exists a nondegenerate deformation of the function $f_0 : G \to \mathbb{R}$ into the function $f_1 : G \to \mathbb{R}$. If the critical point $x_0 = x(0)$ of f_0 is a local minimizer, then $x_1 = x(1)$ is a local minimizer for f_1.*

For the proof of the theorem, we require an auxiliary result.

Let $\mathcal{D}$ be a domain in $\mathbb{R}^N$ and $f : \mathcal{D} \to \mathbb{R}$ be a continuously differentiable function. Consider the ordinary differential equation

$$\frac{dx}{dt} = -\nabla f(x) \tag{2.1.2}$$

on $\mathcal{D}$ and let $p(t,x)$ be a solution satisfying the initial condition

$$p(0,x) = x \,. \tag{2.1.3}$$

Lemma 2.1.1. *Suppose that*

$$p(t,x_0) \in \mathcal{D} \quad (0 \leqslant t \leqslant t_0) \tag{2.1.4}$$

for some $x_0 \in \mathcal{D}$ *and* $t_0 > 0$, *and that*

$$|\nabla f(x)| \geqslant \alpha > 0 \quad (x \in \mathcal{D}) \,. \tag{2.1.5}$$

Then

$$f(x_0) - f(p(t_0,x_0)) \geqslant \alpha |x_0 - p(t_0,x_0)| \,. \tag{2.1.6}$$

Proof. We have

$$f(x_0) - f(p(t_0,x_0)) = \int_0^{t_0} |\nabla f(p(t,x_0))|^2 dt \,.$$

We change the variable of integration in the right-hand side to the length s of the curve $\{p(t,x_0) : 0 \leqslant t \leqslant t_0\}$. Then

$$s(t) = \int_0^t |p'_\tau(\tau,x)| d\tau$$

and

$$f(x_0) - f(p(t_0,x_0)) = \int_0^l |\nabla f(p(t(s),x_0))| ds \geqslant \alpha l \,,$$

where $l = s(t_0)$ is the length of the curve $p(t,x_0)$. Since

$$l \geqslant |x_0 - p(t_0,x_0)| \,,$$

the inequality (2.1.6) follows. □

Now we prove Theorem 2.1.1. We only need to consider the case when there is a deformation $f(x;\lambda)$ which is twice continuously differentiable with respect to x, since we can pass to the general case by standard smoothing. We may also assume, for simplicity and without loss of generality, that

$$\overline{G} = B, \quad x(\lambda) \equiv 0 \quad (0 \leqslant \lambda \leqslant 1), \quad f(0;\lambda) \equiv 0 \quad (0 \leqslant \lambda \leqslant 1) \,.$$

We denote by Λ the set of $\lambda \in [0,1]$ such that 0 is a local minimizer for $f(\cdot;\lambda)$. Clearly Λ is open in $[0,1]$ and nonempty; if we establish that it is closed, then it follows that $\Lambda = [0,1]$, since the interval $[0,1]$ is connected.

Consequently, by the definition of Λ, the point 0 is a local minimizer for f_1, and the theorem will be proved.

In order to prove that Λ is closed, it suffices to construct a ball $B(\rho)$ such that 0 is an absolute minimizer for $f(\cdot;\lambda)$ on $B(\rho)$ for every $\lambda \in \Lambda$. Indeed, if such a ball exists, then the inequalities

$$f(x;\lambda_n) \geqslant 0 \quad (x \in B(\rho),\ n = 1, 2, \ldots)$$

hold for any sequence (λ_n) in Λ that converges to some point $\lambda_0 \in [0,1]$. Passing to the limit as $n \to \infty$, we obtain

$$f(x;\lambda_0) \geqslant 0 \quad (x \in B(\rho)) \,,$$

i.e., 0 is local minimizer for $f(\cdot;\lambda_0)$ and, consequently, $\lambda_0 \in \Lambda$.

We set

$$M = \max_{x \in B,\ \lambda \in [0,1]} |\nabla f(x;\lambda)| \,, \tag{2.1.7}$$

$$m = \min_{\frac{1}{2} \leqslant |x| \leqslant 1,\ \lambda \in [0,1]} |\nabla f(x;\lambda)| \,, \tag{2.1.8}$$

and show that we can take

$$\rho = \frac{m}{4M} \,. \tag{2.1.9}$$

Let $\lambda_0 \in \Lambda$. We shall show that 0 is an absolute minimizer for $f(\cdot;\lambda_0)$ on $B(\rho)$, where ρ is defined by (2.1.9). Since the gradient of $f(\cdot;\lambda_0)$ does not vanish on $B \setminus 0$, it follows that 0 is a strict local minimizer for $f(\cdot;\lambda_0)$ and so is an asymptotically stable state of equilibrium of the gradient differential equation

$$\frac{dx}{dt} = -\nabla f(x;\lambda_0) \quad (x \in \mathring{B}) \,. \tag{2.1.10}$$

In order to prove this fact, it suffices to note that $f(\cdot;\lambda_0)$ itself is a Lyapunov function for Eq. (2.1.10).

Write ω for the *attraction set* of the zero state of equilibrium of Eq. (2.1.10); it consists of those $x \in \mathring{B}$ for which the solutions $p(t,x)$ ($p(0,x) = x$) of Eq. (2.1.10) for $t \geqslant 0$ lie in $\mathring{B}$ and for which

$$\lim_{t \to \infty} p(t,x) = 0 \,. \tag{2.1.11}$$

The set ω is open.

Let $\mathring{B}(r)$ the largest open ball lying in ω. Since (2.1.11) holds for the points $x \in \mathring{B}(r)$ and the function $f(\cdot;\lambda_0)$ decreases monotonically on the trajectories of Eq. (2.1.10), in order to prove the theorem it suffices to show that $r \geqslant \rho$. This inequality is clear if $r \geqslant \frac{1}{2}$. Suppose that $r < \frac{1}{2}$.

Consider a point x_0 which lies on the sphere $S(r)$ and does not belong to ω, and the solution $p(t,x_0)$ of Eq. (2.1.10) with the initial condition x_0.

Since $x_0 \notin \omega$, there exists some $t_0 > 0$ such that

$$\tfrac{3}{4} < |p(t_0, x_0)| < 1 .$$

Since the solutions of Eq. (2.1.10) depend continuously on the initial conditions, this equation has a solution $p(t, x_1)$ with the initial condition $x_1 \in \mathring{B}(r)$ for which

$$\tfrac{3}{4} < |p(t_0, x_1)| < 1 .$$

By construction, $x_1 \in \omega$, i.e.,

$$\lim_{t\to\infty} p(t, x_1) = 0 .$$

Therefore there exist t_1 and t_2 such that

$$t_0 < t_1 < t_2 ,$$

$$|p(t_1, x_1)| = \tfrac{3}{4}, \quad |p(t_2, x_1)| = \tfrac{1}{2}$$

and

$$\tfrac{1}{2} \leqslant |p(t, x_1)| \leqslant \tfrac{3}{4} \quad (t_1 \leqslant t \leqslant t_2) .$$

By (2.1.7), we have

$$f(x_1; \lambda_0) \leqslant M|x_1| < Mr . \tag{2.1.12}$$

On the other hand,

$$f(x_1; \lambda_0) > f(p(t_1, x_1); \lambda_0) - f(p(t_2, x_1); \lambda_0)$$

and, from (2.1.8) and Lemma 2.1.1,

$$f(p(t_1, x_1); \lambda_0) - f(p(t_2, x_1); \lambda_0) \geqslant m|p(t_1, x_1) - p(t_2, x_1)| \geqslant \tfrac{1}{4}m .$$

These two inequalities imply that

$$f(x_1; \lambda_0) \geqslant \tfrac{1}{4}m .$$

Comparing this with (2.1.12), we obtain

$$r \geqslant \frac{m}{4M} ,$$

and the theorem is proved. □

Theorem 2.1.1 implies

Theorem 2.1.2. *Let f_0, $f_1 : B \to \mathbb{R}$ be continuously differentiable functions which have a unique critical point in B, located at the point 0. Suppose that for all nonzero $x \in B$ the gradients $\nabla f_0(x)$ and $\nabla f_1(x)$ of f_0 and f_1 are not oppositely directed. If 0 is a local minimizer for f_0 then 0 is a local minimizer for f_1.*

Proof. A nondegenerate deformation of f_0 into f_1 on $\mathring{B}$ can be defined by

$$f(x; \lambda) = (1 - \lambda) f_0(x) + \lambda f_1(x) \quad (0 \leqslant \lambda \leqslant 1) .$$

The theorem now follows from Theorem 2.1.1. □

2.1.2 Invariance of a Global Minimizer

Assume that under the conditions of Theorem 2.1.1 the point x_0 is an absolute minimizer for the function f_0 on G, i.e.,

$$f_0(x) \geqslant f_0(x_0) \quad (x \in G) .$$

Must x_1 then be an absolute minimizer for f_1 on G? In general this is not true: a counter-example illustrating this will be given later. Thus, under the conditions of Theorem 2.1.1, we only discuss the invariance of local minimizers. The non-preservation of global minimizers under nondegenerate deformations is related to the following fact, which is surprising at first glance: for $N \geqslant 2$ there exist smooth functions on $\mathbb{R}^N$ with a unique critical point, which is a local minimizer but not a global minimizer. Later we shall give an example of such a function.

In this section, we study conditions for the homotopy invariance of a global minimizer. Recall that x_* is a global minimizer for a function $f : \mathbb{R}^N \to \mathbb{R}$ if

$$f(x) \geqslant f(x_*) \quad (x \in \mathbb{R}^N) .$$

Suppose that $f : \mathbb{R}^N \to \mathbb{R}$ is a continuously differentiable function and that f has a unique critical point x_* which is a local minimizer for f. Without additional assumptions x_* need not be a global minimizer. Consider, for example, the function of two real variables

$$f(x_1, x_2) = \frac{3x_1^2 - 2x_1^3 - 1}{1 + x_2^2} + (3x_1^2 - 2x_1^3)e^{-x_2} .$$

The only critical point of this function is 0, which is a local minimizer since the matrix

$$\nabla^2 f(0,0) = \begin{bmatrix} 12 & 0 \\ 0 & 2 \end{bmatrix}$$

is positive definite. However, $f(0,0) = -1,\ f(2,0) = -9$.

One of the effective criteria for identifying a global minimizer is connected with the notion of a growing function. We say that $f : \mathbb{R}^N \to \mathbb{R}$ is a *growing* function if

$$\lim_{|x| \to \infty} f(x) = \infty . \tag{2.1.13}$$

The following statement is clear.

Theorem 2.1.3. *If the unique critical point x_0 of a continuously differentiable growing function f is a local minimizer, then x_0 is a global minimizer.*

In applications, we sometimes have to establish that a function f is growing just from information about the behavior of its gradient ∇f. Below we give a useful criterion for f to be growing, formulated in terms of the rate of increase of its gradient at infinity.

Let $f : \mathbb{R}^N \to \mathbb{R}$ be a differentiable function with a unique critical point. We assume for simplicity that the critical point is the point 0 of $\mathbb{R}^N$, that $f(0) = 0$, and that the gradient ∇f is locally Lipschitzian.

Theorem 2.1.4. *Suppose that* 0 *is a local minimizer for the function* f *and that*

$$|\nabla f(x)| \geqslant \alpha(|x|) \quad (x \in \mathbb{R}^N) , \tag{2.1.14}$$

where $\alpha(s)$ $(0 < s < \infty)$ *is a positive continuous function for which*

$$\int_0^\infty \alpha(s)ds = \infty . \tag{2.1.15}$$

Then f *is growing.*

Proof. Since 0 is a local minimizer for f and the gradient of f does not vanish at nonzero points, it follows that 0 is an absolute strict minimizer for f in some ball $B(r)$ with $r > 0$. Therefore f is a Lyapunov function in a neighborhood of the zero state of equilibrium of the gradient differential equation

$$\frac{dx}{dt} = -\nabla f(x) \quad (x \in \mathbb{R}^N) . \tag{2.1.16}$$

Consequently, the zero state of equilibrium of this equation is asymptotically stable. Let ω be the attraction set of the zero state of equilibrium of Eq. (2.1.16), i.e., the set of $x \in \mathbb{R}^N$ for which the solutions $p(t,x)$ of Eq. (2.1.16) with the initial condition x satisfy

$$\lim_{t\to\infty} |p(t,x)| = 0 .$$

Clearly ω nonempty and open; we shall show that it is closed.

Consider an arbitrary sequence (x_n) in ω that converges to a point x_0. We shall show that

$$x_0 \in \omega . \tag{2.1.17}$$

We assume the contrary, and denote by $p(t,x_0)$ a solution of Eq. (2.1.16) with the initial condition x_0 and by $[0,a)$ the interval of continuity of this solution. Then we have

$$\inf_{0\leqslant t<a} |p(t,x_0)| > r \tag{2.1.18}$$

for some $r > 0$. We set

$$R = \sup_{0\leqslant t<a} |p(t,x_0)|$$

and suppose first that $R < \infty$. In this case $a = \infty$. The inequality (2.1.14) gives

$$\inf_{0\leqslant t<a} |\nabla f\,(p(t,x_0))\,| \geqslant \beta > 0 ,$$

where

$$\beta = \min_{r \leqslant s \leqslant R} \alpha(s) .$$

Therefore

$$\frac{d}{dt} f(p(t, x_0)) = (\nabla f(p(t, x_0)), \frac{d}{dt} p(t, x_0)) = -|\nabla f(p(t, x_0))|^2 \leqslant -\beta^2$$

and, consequently,

$$\lim_{t \to \infty} f(p(t, x_0)) = -\infty ,$$

which contradicts the fact that f is bounded on the ball $B(R)$.

Thus we have $R = \infty$. We set

$$M(t) = \sup_{0 \leqslant s \leqslant t} |p(s, x_0)| .$$

Then

$$\lim_{t \to a} M(t) = \infty . \tag{2.1.19}$$

Consider the function f on the half-trajectory $p(t, x_0)$ $(0 \leqslant t < a)$. We have

$$\begin{aligned} f(p(t, x_0)) &= f(x_0) + \int_0^t \frac{d}{ds} f(p(s, x_0))\, ds \\ &= f(x_0) - \int_0^t |\nabla f(p(s, x_0))|^2\, ds . \end{aligned} \tag{2.1.20}$$

In order to estimate the integral

$$J(t) = \int_0^t |\nabla f(p(s, x_0))|^2\, ds ,$$

we consider the set

$$\chi = \left\{ s \in [0, t] : \frac{d}{ds} |p(s, x_0)| > 0 \right\} .$$

Then

$$J(t) \geqslant \int_\chi |\nabla f(p(s, x_0))|^2\, ds . \tag{2.1.21}$$

Changing the variable of integration by setting

$$\tau = |p(s, x_0)| , \tag{2.1.22}$$

we have

$$\begin{aligned}\int\limits_{\chi} |\nabla f(p(s,x_0))|^2\, ds &= -\int\limits_{\psi} \frac{|\nabla f(p(s(\tau),x_0))|^2 \tau}{(p(s(\tau),x_0), \nabla f(p(s(\tau),x_0)))}\, d\tau \\ &\geqslant \int\limits_{\psi} \frac{|\nabla f(p(s(\tau),x_0))|^2 \tau}{|p(s(\tau),x_0)| \cdot |\nabla f(p(s(\tau),x_0))|}\, d\tau \\ &= \int\limits_{\psi} |\nabla f(p(s(\tau),x_0))|\, d\tau\,,\end{aligned} \tag{2.1.23}$$

where ψ is the image of χ under the mapping (2.1.22). From (2.1.14) we have

$$\begin{aligned}\int\limits_{\psi} |\nabla f\,(p(s(\tau),x_0))\,|\, d\tau &\geqslant \int\limits_{|x_0|}^{M(t)} |\nabla f\,(p(s(\tau),x_0))\,|\, d\tau \\ &\geqslant \int\limits_{|x_0|}^{M(t)} \alpha\,(|p(s(\tau),x_0)|)\, d\tau = \int\limits_{|x_0|}^{M(t)} \alpha(\tau)\, d\tau\,.\end{aligned} \tag{2.1.24}$$

The inequalities (2.1.21), (2.1.23), and (2.1.24) yield

$$J(t) \geqslant \int\limits_{|x_0|}^{M(t)} \alpha(\tau)\, d\tau\,. \tag{2.1.25}$$

It follows from (2.1.15), (2.1.19), and (2.1.25) that

$$\lim_{t\to a} J(t) = \infty\,. \tag{2.1.26}$$

Therefore, by (2.1.20), we have

$$\lim_{t\to a} f\,(p(t,x_0)) = -\infty$$

so that

$$f\,(p(t_0,x_0)) < 0$$

for some $t_0 > 0$, and since the solutions $p(t,x)$ of Eq. (2.1.16) depend continuously on the initial conditions, we have

$$f\,(p(t_0,x_n)) < 0 \tag{2.1.27}$$

for large n. On the other hand, since $x_n \in \omega$, the function

$$\varphi(t) = f\,(p(t,x_n))$$

decreases monotonically, and

$$\lim_{t\to\infty} \varphi(t) = 0 ,$$

it follows that

$$f\left(p(t_0, x_n)\right) > 0 . \tag{2.1.28}$$

The inequalities (2.1.27) and (2.1.28) contradict each other. Consequently, the assertion of (2.1.17) holds, and therefore the set ω is closed. On the other hand, ω is nonempty and open, and so $\omega = \mathbb{R}^N$.

We have thus shown that the zero state of equilibrium of Eq. (2.1.16) is stable in the large. Thus, fixing a point $x \in \mathbb{R}^N$ and considering the trajectory $p(t,x)$, we have

$$\lim_{t\to\infty} |p(t,x)| = 0 . \tag{2.1.29}$$

Therefore

$$f(x) = f(0) - \int_0^\infty \frac{d}{ds} f\left(p(s,x)\right) ds = -\int_0^\infty |\nabla f\left(p(s,x)\right)|^2 \, ds . \tag{2.1.30}$$

Estimating this integral in the same way as we estimated the integral $J(t)$, we obtain

$$\int_0^\infty |\nabla f\left(p(s,x)\right)|^2 \, ds \geqslant \int_0^{|x|} \alpha(s) \, ds . \tag{2.1.31}$$

The relations (2.1.15), (2.1.30), and (2.1.31) now give

$$\lim_{|x|\to\infty} f(x) = \infty ,$$

and the theorem is proved. □

Next we study the invariance of global minimizers under smooth deformations.

Let f_0, $f_1 : \mathbb{R}^N \to \mathbb{R}$ be continuously differentiable functions with unique critical points x_0, x_1. The one-parameter family of functions

$$f : \mathbb{R}^N \times [0,1] \to \mathbb{R} \tag{2.1.32}$$

is called a *global nondegenerate deformation* of f_0 into f_1 if both the family (2.1.32) and the gradient $\nabla f(x;\lambda)$ are continuous on $\mathbb{R}^N \times [0,1]$, if $f(\cdot;\lambda)$ has a unique critical point $x(\lambda)$ continuously depending on λ for $\lambda \in [0,1]$, and if

$$f(\cdot;0) = f_0, \quad f(\cdot;1) = f_1 .$$

Theorems 2.1.1 and 2.1.4 imply

Theorem 2.1.5. *Suppose that there exists a global nondegenerate deformation* $f : \mathbb{R}^N \times [0,1] \to \mathbb{R}$ *of* f_0 *into* f_1. *Let* x_0 *be a local minimizer for* f_0 *and* x_1 *a critical point of* f_1, *and suppose that the gradient* ∇f_1 *of* f_1 *satisfies*

$$|\nabla f_1(x)| \geqslant \alpha(|x - x_1|) , \tag{2.1.33}$$

where $\alpha(s)$ $(0 < s < \infty)$ *is a continuous positive function for which*

$$\int_0^\infty \alpha(s)\, ds = \infty . \tag{2.1.34}$$

Then x_1 *is a global minimizer for* f_1, *and*

$$\lim_{|x| \to \infty} f_1(x) = \infty .$$

Under the conditions of Theorem 2.1.5, if (2.1.33) and (2.1.34) do not hold, then x_1 need not be a global minimizer for f_1, even if f_0 is growing.

2.1.3 Linear Deformations

We saw in the preceding section that the existence of a global nondegenerate deformation of a function f_0 into a function f_1 does not guarantee the preservation of a global minimizer, even if f_0 is growing. However, we show in this section that if the deformation is *linear*, i.e., has the form

$$f(x;\lambda) = (1-\lambda) f_0(x) + \lambda f_1(x) \quad (x \in \mathbb{R}^N,\ 0 \leqslant \lambda \leqslant 1) , \tag{2.1.35}$$

and f_0 is growing, then a global minimizer is preserved.

Theorem 2.1.6. *Let* $x = 0$ *be the unique critical point of the continuously differentiable functions* f_0 *and* f_1, *and suppose that for* $x \neq 0$ *the gradients* $\nabla f_0(x)$ *and* $\nabla f_1(x)$ *of* f_0 *and* f_1 *are not oppositely directed. Suppose, finally, that*

$$\lim_{|x| \to \infty} f_0(x) = \infty . \tag{2.1.36}$$

Then $x = 0$ *is a global minimizer for* f_1.

Proof. We may suppose for simplicity that

$$f_0(0) = f_1(0) = 0 .$$

We consider an arbitrary point $x_0 \in \mathbb{R}^N$ and show that

$$f_1(0) \leqslant f_1(x_0) .$$

Indeed, we can find $c > 0$ such that

$$x_0 \in \Omega , \tag{2.1.37}$$

where

$$\Omega = \{x \in \mathbb{R}^N : f_0(x) < c\} .$$

By (2.1.36) the set $\overline{\Omega}$ is compact. Therefore f_1 has minimizer x_1 in $\overline{\Omega}$. We shall show that $x_1 \in \Omega$. Indeed, if $x_1 \in \partial\Omega$, then, by Theorem 1.7.8, we have

$$\nabla f_1(x_1) = -\lambda_0 \nabla f_0(x_1) \quad (\lambda_0 > 0).$$

Since $x_1 \neq 0$, it follows that the vectors $\nabla f_0(x)$ and $\nabla f_1(x)$ are oppositely directed at x_1, and this contradicts our hypothesis. Thus $x_1 \in \Omega$. We conclude that

$$\nabla f_1(x_1) = 0 ,$$

and consequently $x_1 = 0$. Therefore, by (2.1.37),

$$f_1(0) = \min_{x\in\Omega} f_1(x) \leqslant f_1(x_0) ,$$

and the theorem is proved. □

Note that the conditions of Theorem 2.1.6 do not require f_1 to be growing: an example is given by the functions

$$f_0(x) = x^2, \quad f_1(x) = \frac{x^2}{1+x^2} .$$

Suppose that f_0 and f_1 are continuously differentiable functions. Let $\mathfrak{M}$ be the set of critical points of f_1, and assume that $\mathfrak{M}$ is bounded. The following theorem holds.

Theorem 2.1.7. *Suppose that for all x outside of some ball the gradients $\nabla f_0(x)$ and $\nabla f_1(x)$ of f_0 and f_1 are not oppositely directed. Suppose also that f_0 is growing, i.e.,*

$$\lim_{|x|\to\infty} f_0(x) = \infty .$$

Then f_1 has a global minimizer on $\mathfrak{M}$.

This theorem can be proved using the same method as for Theorem 2.1.6. It has the following corollary.

Corollary 1. *Suppose that f is a continuously differentiable function all of whose critical points lie in the ball $\mathring{B}(r)$, and that the vectors $\nabla f(x)$ are not opposite in direction to the vectors x on the sphere $\partial B(r)$. Then f has a critical point which is an absolute minimizer on $B(r)$.*

2.2 Nondegenerate Deformations of Nonsmooth Functions

2.2.1 Generalized Derivatives and Generalized Gradients

In this section we describe the deformation principle for minimizers of Lipschitzian and continuous functions. We recall the following definitions.

Let G be a domain in $\mathbb{R}^N$. A function $f : G \to \mathbb{R}$ is *locally Lipschitzian* if, for every $R > 0$, this function satisfies a Lipschitz condition on the set $G \cap B(R)$. If f is locally Lipschitzian on G and $x \in G$ then

$$f^0(x; v) = \overline{\lim_{y \to x,\ t \to +0}} \frac{f(y + tv) - f(y)}{t} \tag{2.2.1}$$

is the *generalized derivative* in the direction of v of f at x, and the *generalized gradient* of f at x is the set

$$\partial f(x) = \{y \in \mathbb{R}^N : (y, v) \leqslant f^0(x; v), v \in \mathbb{R}^N\} . \tag{2.2.2}$$

For every $x \in G$, the set $\partial f(x)$ is a nonempty convex compact set. The generalized gradient can also be defined by

$$\partial f(x) = \mathrm{co}\{y \in \mathbb{R}^N : \nabla f(x_i) \to y, x_i \to x, x_i \in \Lambda_f \cap G\} .$$

Here, Λ_f is the set of points of differentiability of f and $\nabla f(x_i)$ is the classical gradient of f at the point $x_i \in \Lambda_f$.

A point $x_* \in G$ is a *critical point* of f if $0 \in \partial f(x_*)$.

2.2.2 The Deformation Theorem

Theorem 2.1.1 on the invariance of a minimizer under nondegenerate smooth deformations has an analog for Lipschitzian functions.

Let $f_0 : G \to \mathbb{R}$, $f_1 : G \to \mathbb{R}$ be locally Lipschitzian functions which have unique critical points x_0 and x_1 respectively. We say that a one-parameter family of functions $f(\cdot; \lambda) : G \to \mathbb{R}$ $(0 \leqslant \lambda \leqslant 1)$ is a *nondegenerate deformation* of f_0 into f_1 in G if

(1) f is continuous on $G \times [0, 1]$ and is locally Lipschitzian with respect to x for every $\lambda \in [0, 1]$,

(2) the multivalued mapping

$$\partial_x f(\cdot; \cdot) : G \times [0, 1] \to \mathbb{R}^N$$

is upper semicontinuous,

(3) for every $\lambda \in [0, 1]$, the function $f(\cdot; \lambda)$ has a unique critical point $x(\lambda)$ which depends continuously on the parameter λ, with

$$x(0) = x_0, \quad x(1) = x_1 ,$$

(4) $f(\cdot; 0) = f_0, \quad f(\cdot; 1) = f_1$.

Theorem 2.2.1. *Suppose that there exists a nondegenerate deformation* $f(\cdot;\lambda)$ $(0 \leqslant \lambda \leqslant 1)$ *of the function* f_0 *into the function* f_1. *If* x_0 *is a local minimizer for* f_0 *then* x_1 *is a local minimizer for* f_1.

First we prove two auxiliary results.

Let D be a domain in $\mathbb{R}^N$ and g be a continuously differentiable function defined on D. We consider the differential equation

$$\dot{x} = -\nabla g(x) \tag{2.2.3}$$

and let $p(t,x)$ be a solution satisfying the initial condition $p(0,x) = x$. Suppose that D contains the spherical layer

$$T(r,R) = \{x \in \mathbb{R}^N : r \leqslant |x| \leqslant R\} ,$$

and that for all $s \in [r,R]$ we have

$$\min_{s \leqslant |x| \leqslant R} |\nabla g(x)| \geqslant m(s) , \tag{2.2.4}$$

where m is a positive continuous function. Lemma 2.1.1 implies

Lemma 2.2.1. *Let* $p(t,x) \in T(r,R)$ *for* $0 \leqslant t \leqslant T$ *and*

$$|p(0,x)| = R, \quad |p(T,x)| = r . \tag{2.2.5}$$

Then

$$g(x) - g(p(T,x)) \geqslant \int\limits_r^R m(s)\,ds . \tag{2.2.6}$$

Suppose that D contains the ball $B(R)$. We consider a one-parameter family of functions $g(x,\lambda)$ $(x \in D,\ 0 \leqslant \lambda \leqslant 1)$ such that they and the gradients $\nabla f(x;\lambda)$ are continuous in both of the variables and satisfy, for some $r \in (0,R]$, the inequalities

$$\min_{0 \leqslant \lambda \leqslant 1} |\nabla g(x;\lambda)| > m(|x|) > 0 \quad (r \leqslant |x| \leqslant R) , \tag{2.2.7}$$

$$\max_{0 \leqslant \lambda \leqslant 1} |g(x;\lambda)| < M(|x|) \quad (0 \leqslant |x| \leqslant R) , \tag{2.2.8}$$

where $m(s)$ $(r \leqslant s \leqslant R)$ and $M(s)$ $(0 \leqslant s \leqslant R)$ are continuous monotonically increasing functions and $M(0) = 0$.

Lemma 2.2.2. *Suppose that* r *is such that*

$$\int\limits_r^\rho m(s)\,ds - 3M(r) > 0 , \tag{2.2.9}$$

where

$$\rho = M^{-1}\left(\int\limits_r^R m(s)\,ds - 2M(r)\right). \tag{2.2.10}$$

Let

$$\min_{|x|=\rho} g(x;0) > M(r)\,. \tag{2.2.11}$$

Then

$$\min_{|x|=\rho} g(x;1) > M(r)\,. \tag{2.2.12}$$

Proof. Let Λ be the set of $\lambda \in [0,1]$ for which

$$\min_{|x|=\rho} g(x;\lambda) > M(r)\,. \tag{2.2.13}$$

Then Λ is nonempty and open; we establish that Λ is closed. For this purpose, we first prove the inequality

$$\min_{|x|=\rho} g(x;\lambda) \geqslant \int\limits_r^\rho m(s)\,ds - M(r) \quad (\lambda \in \Lambda)\,. \tag{2.2.14}$$

In order to prove (2.2.14), we consider the differential equation

$$\dot{x} = -\nabla g(x;\lambda) \quad (\lambda \in \Lambda)$$

on the ball $\mathring{B}(R)$ and let $p(t,x)$ be a solution satisfying the initial condition $p(0,x) = x$. We denote by Ω the set of x for which

$$p(t,x) \in \mathring{B}(R) \quad (0 \leqslant t < \infty)\,.$$

By (2.2.13), this set is open and contains the ball $B(R)$. If $x \in T(r,R) \cap \Omega$, then, for some $t = t(x)$, we have

$$|p(t(x),x)| = r\,. \tag{2.2.15}$$

Indeed, if

$$r < |p(t,x)| \leqslant R \quad (0 \leqslant t < \infty)\,,$$

then, by (2.2.8),

$$\begin{aligned}
\lim_{t\to\infty} g(p(t,x);\lambda) &= \lim_{t\to\infty} \int\limits_0^t (\nabla g(p(s,x),\lambda), p_s'(s,x))\,ds + g(x;\lambda) \\
&= -\lim_{t\to\infty} \int\limits_0^t |\nabla g(p(s,x);\lambda)|^2\,ds + g(x;\lambda) \\
&\leqslant -\lim_{t\to\infty} t m^2(r) + g(x;\lambda) = -\infty\,,
\end{aligned}$$

and this contradicts the boundedness of g on $T(r, R)$.

Now consider the maximal open ball $\mathring{B}(\rho_0)$ that lies in Ω; we show that

$$\rho_0 \geqslant \rho .$$

If $\rho_0 = R$, this is clear. Suppose therefore that $\rho_0 < R$. Let

$$y \in \partial B(\rho_0) \cap \partial \Omega .$$

Since $y \notin \Omega$, we have

$$|p(t, y)| < R \quad (0 \leqslant t < T), \quad |p(T, y)| = R \tag{2.2.16}$$

for some $T > 0$. We fix a small $\varepsilon > 0$ and choose a point $z \in \mathring{B}(\rho_0)$ for which

$$|p(T, z)| \geqslant R - \varepsilon . \tag{2.2.17}$$

Since $z \in \Omega$, we have

$$|p(t(z), z)| = r$$

for some $t(z) > T$. Hence, by Lemma 2.2.1, we have

$$g(z; \lambda) - g(p(t(z), z); \lambda) \geqslant g(p(T, z); \lambda) - g(p(t(z), z); \lambda) \geqslant \int_r^{R-\varepsilon} m(s)\, ds . \tag{2.2.18}$$

On the other hand,

$$g(z; \lambda) - g(p(t(z), z); \lambda) \leqslant M(\rho_0) + M(r) . \tag{2.2.19}$$

The inequalities (2.2.18) and (2.2.19) give

$$M(\rho_0) \geqslant \int_r^R m(s)\, ds - M(r)$$

and hence

$$\rho_0 \geqslant M^{-1} \left(\int_r^R m(s)\, ds - M(r) \right) .$$

Thus $\rho_0 > \rho$.

Consider an arbitrary point $x \in \partial B(\rho)$. We have

$$x \in T(r, R) \cap \Omega ,$$

and consequently

$$|p(t(x), x)| = r$$

for some $t = t(x)$. Then, by Lemma 2.2.1,

$$g(x;\lambda) \geqslant \int\limits_r^\rho m(s)\,ds + g(p(t(x),x);\lambda) \geqslant \int\limits_r^\rho m(s)\,ds - M(r)\,. \qquad (2.2.20)$$

This gives the inequality (2.2.14).

Now we prove that the set Λ is closed. Let $\lambda_n \in \Lambda$ for each $n \in \mathbb{N}$ and $\lambda_n \to \lambda_0$. By (2.2.14) we have

$$\min_{|x|=\rho} g(x;\lambda_n) \geqslant \int\limits_r^R m(s)\,ds - M(r)\,. \qquad (2.2.21)$$

It follows from (2.2.9) and (2.2.21) that

$$\min_{|x|=\rho} g(x;\lambda_0) > M(r)\,, \qquad (2.2.22)$$

i.e., $\lambda_0 \in \Lambda$. Thus Λ is closed. Consequently $\Lambda = [0,1]$, and the lemma follows.
□

Now we prove Theorem 2.2.1. We may assume for simplicity that the domain G contains the unit ball B, that

$$x(\lambda) \equiv 0 \quad (0 \leqslant \lambda \leqslant 1)\,,$$

$$f(0;\lambda) \equiv 0 \quad (0 \leqslant \lambda \leqslant 1)\,,$$

and that 0 is an absolute strict minimizer for f_0 on B.

We shall conduct the proof in three steps. In the first, we construct real functions m and M from the given nondegenerate deformation and in the second, we use m and M to construct a smooth approximation of the nondegenerate deformation. In the third step, we complete the proof of the theorem using constructions from the second step and Lemma 2.2.2.

Step 1. Consider an extension of the function f from the set $G \times [0,1]$ to a set $G \times [-\varepsilon, 1+\varepsilon]$ for some $\varepsilon > 0$, which preserves continuity, the Lipschitzian property with respect to x, and the upper semicontinuity of the mapping

$$\partial_x f : G \times [-\varepsilon, 1+\varepsilon] \to \mathbb{R}^N\,.$$

We maintain the notation f for the extended function and assume that for (the extended) f and every $\lambda \in [-\varepsilon, 1+\varepsilon]$, the only critical point of $f(\cdot;\lambda)$ on G is the point 0, and we also assume that ε is chosen such that the ball $B(1+\varepsilon)$ lies in G.

Consider the function

$$\varphi(x;\lambda) = \min_{y \in \partial_x f(x;\lambda)} |y| \quad (x \in G, -\varepsilon \leqslant \lambda \leqslant 1+\varepsilon)\,.$$

Since the mapping $\partial_x f(\cdot;\lambda)$ is upper semicontinuous, $\varphi(\cdot;\lambda)$ is lower semicontinuous. Therefore there exists a continuous monotonically increasing function $m(s)$ $(0 \leqslant s \leqslant 1+\varepsilon)$ for which

$$m(0) = 0$$

and

$$\min_{\substack{-\varepsilon \leqslant \lambda \leqslant 1+\varepsilon \\ s \leqslant |x| \leqslant 1+\varepsilon}} \varphi(x;\lambda) > m(s) \quad (0 < s \leqslant 1+\varepsilon)\ .$$

Next, we define M to be a continuous monotonically increasing function for which

$$M(0) = 0$$

and

$$\max_{\substack{-\varepsilon \leqslant \lambda \leqslant 1+\varepsilon \\ |x| \leqslant s}} |f(x;\lambda)| < M(s) \quad (0 < s \leqslant 1+\varepsilon)\ .$$

Step 2. Let Λ_f be the set of points of the cylinder

$$U = B(1+\varepsilon) \times [-\varepsilon, 1+\varepsilon]$$

at which f has a classical gradient $\nabla f(x;\lambda)$ with respect to the variable x. For each point

$$z = (x;\lambda) \in (B \setminus 0) \times [0,1]$$

we can find a ball $B(\rho(z), z)$ (in $\mathbb{R}^{N+1}$) of nonzero radius such that for some $\delta(x;\lambda) > 0$ we have

$$\nabla f(y;\mu) \in \partial_x f(x;\lambda) + \delta(x;\lambda) B\ , \tag{2.2.23}$$

whenever $(y;\mu) \in B(\rho(z), z) \cap \Lambda_f$, and

$$\varphi(x;\lambda) - \delta(x;\lambda) > m(|x| + \rho(z))\ . \tag{2.2.24}$$

Let $r \in (0,1)$ and denote by $T(r)$ the spherical layer

$$\{x \in \mathbb{R}^N : r \leqslant |x| \leqslant 1\}\ .$$

We consider the cylinder $T(r) \times [0,1]$ and choose a finite subcovering from its covering by the balls $B(\frac{1}{2}\rho(z), z)$. Let $\rho(r)$ be the smallest radius of a ball in this subcovering. Thus ρ is a positive function defined on the interval $(0,1)$.

We set

$$\omega(t) = \begin{cases} \exp(-(1-t^2)^{-1}) & \text{for } |t| < 1\ , \\ 0 & \text{for } |t| \geqslant 1 \end{cases}$$

and, using the function $\omega(t)$, construct a sequence of averaging kernels

$$\omega_n(x,\lambda) = n^{N+1} \omega\left(\sqrt{|x|^2 + \lambda^2}\right) \left(\int_{-\infty}^{\infty} \omega(s)\, ds \right)^{-1}\ .$$

Consider the sequence of functions

$$f_n(x;\lambda) = \int\limits_{\mathbb{R}^{N+1}} \omega_n(y,\mu) f(x+y;\lambda+\mu)\,dyd\mu\,.$$

The functions f_n are infinitely differentiable and converge uniformly to f on the cylinder $B \times [0,1]$. Therefore for all sufficiently large n we have

$$\max_{\substack{0\leqslant\lambda\leqslant 1\\ r\leqslant|x|\leqslant s}} |f_n(x;\lambda)| < M(s) \quad (r \leqslant s \leqslant 1) \tag{2.2.25}$$

for every $r \in (0,1)$. We show now that for all sufficiently large n we have

$$\min_{\substack{0\leqslant\lambda\leqslant 1\\ s\leqslant|x|\leqslant 1}} |\nabla_x f_n(x,\lambda)| > m(|s|) \quad (r \leqslant s \leqslant 1) \tag{2.2.26}$$

for every $r \in (0,1)$. To do this, we fix a point

$$z_0 = (x_0;\lambda_0) \in T(r) \times [0,1]\,.$$

Then z_0 lies in some ball $B(\frac{1}{2}\rho(z_1), z_1)$ (with $z_1 = (x_1;\lambda_1)$) belonging to our finite covering of $T(r) \times [0,1]$. By the definition of $\rho(r)$, we have

$$B(\rho(r), z_0) \subset B(\rho(z_1), z_1)\,.$$

Consequently, by (2.2.23), we have

$$\nabla_y f(y;\mu) \in \partial_x f(x_1;\lambda_1) + \delta(x_1;\lambda_1)B$$

for

$$(y;\mu) \in B(\rho(r), z_0) \cap \Lambda_f.$$

Thus, by (2.2.24), there exists a linear functional $l : \mathbb{R}^N \to \mathbb{R}$ for which $|l| = 1$ and

$$l(\nabla_y f(y;\mu)) > m(|x_1| + \rho(z_1))\,.$$

However

$$|x_1| + \rho(z_1) \geqslant |x_0|\,,$$

and therefore

$$l(\nabla_y f(y;\mu)) > m(|x_0|) \quad ((y;\mu) \in B(\rho(r), z_0) \cap \Lambda_f))\,.$$

Let

$$n \geqslant \rho(r)^{-1}\,.$$

Then

$$\begin{aligned}|\nabla f_n(x_0;\lambda_0)| &\geqslant l(\nabla f_n(x_0;\lambda_0))\\ &= \int\limits_{\mathbb{R}^{N+1}} \omega_n(y,\mu) l(\nabla f_n(x_0+y;\lambda_0+\mu))\, dy d\mu\\ &= \int\limits_{B(n^{-1})} \omega_n(y,\mu) l(\nabla f_n(x_0+y;\lambda_0+\mu)\, dy d\mu > m(|x_0|)\,.\end{aligned} \tag{2.2.27}$$

This gives the inequality (2.2.26).

We have thus shown that for every $r \in (0,1)$ there exists a number $n(r)$ such that for $n \geqslant n(r)$ the functions f_n satisfy (2.2.25) and (2.2.26). We may now construct from the functions f_n a sequence

$$g_n : B \times [0,1] \to \mathbb{R} \quad (n \geqslant n(r))$$

of continuously differentiable functions such that for all $n \geqslant n(r)$ the function g_n coincides with f_n on $T(r) \times [0,1]$ and satisfies

$$\max_{\substack{0\leqslant\lambda\leqslant 1\\ |x|\leqslant s}} |g_n(x;\lambda)| < M(s) \quad (0 < s \leqslant 1,\ n \geqslant n(r))\,,$$

and such that

$$\lim_{n\to\infty} g_n(x;\lambda) = f(x,\lambda)$$

uniformly with respect to $x \in B$, $\lambda \in [0,1]$.

Step 3. We choose $r \in (0,1)$ so small that for all $n \geqslant n(r)$ the one-parameter family of functions

$$g_n : B \times [0,1] \to \mathbb{R}$$

satisfies the conditions of Lemma 2.2.2. We fix $n \geqslant n(r)$ and consider the function $g_n(\cdot;1)$ on the ball $B(\rho)$, where ρ is defined by (2.2.10). Since

$$\min_{|x|=\rho} g_n(x;1) > M(r)\,,$$

the absolute minimizer x_n of $g_n(x,1)$ on the ball $B(\rho)$ is in the interior of this ball. Thus

$$\nabla g_n(x_n;1) = 0$$

and, by (2.2.7) (with $f(x;\lambda)$ replaced by $g_n(x;\lambda)$), we have

$$|x_n| < r\,.$$

Without loss of generality, we can assume that $x_n \to x_*$ as $n \to \infty$, where $x_* \in B(r)$. Passing to the limit in the inequalities

$$g_n(x;1) \geqslant g_n(x_n;1) \quad (x \in B(\rho))\,,$$

we obtain

$$f_1(x) \geqslant f_1(x_*) \quad (x \in B(\rho)) .$$

Therefore x_* is a minimizer for f_1 on $B(\rho)$. However

$$B(r) \subset \mathring{B}(\rho) ,$$

and so x_* is in the interior of $B(\rho)$, and hence x_* is a critical point of f_1. Thus $x_* = 0$, i.e., the point 0 is a minimizer of f_1 on $B(\rho)$. □

2.2.3 Global Deformations of Lipschitzian Functions

Our aim in this section is to generalize to Lipschitzian functions the results of Sect. 2.1.2 on the invariance of global minimizers of smooth functions. We begin with some auxiliary results.

Lemma 2.2.3. *Let $\varphi : \mathbb{R}^N \to \mathbb{R}$ be a twice continuously differentiable function. Suppose that there exist positive numbers $r_0 < r_1$ and a continuous positive function $\alpha : [r_0, \infty) \to \mathbb{R}$ for which*

$$\min_{x \in \partial B(r_1)} \varphi(x) > \max_{x \in B(r_0)} |\varphi(x)| , \tag{2.2.28}$$

$$|\nabla\varphi(x)| \geqslant \alpha(|x|) \quad (|x| \geqslant r_0) , \tag{2.2.29}$$

$$\int_{r_0}^{\infty} \alpha(s)\, ds = \infty . \tag{2.2.30}$$

Then

$$\lim_{|x| \to \infty} \varphi(x) = \infty . \tag{2.2.31}$$

Proof. We set

$$c = \min_{x \in \partial B(r_1)} \varphi(x)$$

and

$$\mathfrak{N} = \{x \in B(r_1) : \varphi(x) < c\} .$$

By (2.2.28), the set $\mathfrak{N}$ is nonempty. Consider the differential equation

$$\dot{x} = -\nabla\varphi(x) \quad (x \in \mathbb{R}^N) \tag{2.2.32}$$

and let $p(t, x)$ be a solution satisfying the initial condition $p(0, x) = x$.

Let $\mathfrak{M}$ be the set of points x in $\mathbb{R}^N$ for which we have

$$p(t(x), x) \in \mathfrak{N} \tag{2.2.33}$$

for some $t = t(x)$. Then $\mathfrak{M}$ is nonempty (since $\mathfrak{N} \subset \mathfrak{M}$) and open. We shall show that $\mathfrak{M}$ is also closed.

If this is false, then there exists a sequence (x_n) in $\mathfrak{M}$ that converges to a point $x_0 \notin \mathfrak{M}$. Consider a solution $p(t, x_0)$ and let $[0, T)$ be the interval of extendability of this solution. We shall show that

$$\lim_{k\to\infty} |p(t_k, x_0)| = \infty \tag{2.2.34}$$

for some sequence (t_k) in $[0, T)$. Indeed, if (2.2.34) does not hold, then

$$\sup_{0\leqslant t\leqslant T} |p(t, x_0)| = r_2 < \infty \,. \tag{2.2.35}$$

It follows from (2.2.35) that $T = \infty$. However, then the inequalities

$$r_0 \leqslant |p(t, x)| \leqslant r_2 \quad (0 \leqslant t < \infty)$$

and (2.2.29) yield

$$\begin{aligned} \varphi(p(t, x_0)) &= \varphi(x_0) + \int_0^t \frac{d}{ds}\varphi(p(s, x_0))\, ds \\ &= \varphi(x_0) - \int_0^t |\nabla\varphi(p(s, x_0))|^2\, ds \leqslant \varphi(x_0) - t\alpha_0^2 \,, \end{aligned}$$

where

$$\alpha_0 = \min_{r_0\leqslant s\leqslant r_2} \alpha(s) \,.$$

Therefore

$$\lim_{t\to\infty} \varphi(p(t, x_0)) = -\infty \,. \tag{2.2.36}$$

On the other hand,

$$\sup_{0\leqslant t\leqslant\infty} |\varphi(p(t, x_0))| \leqslant \max_{r_0\leqslant |x|\leqslant r_2} |\varphi(x)| < \infty \,,$$

and this contradicts (2.2.36).

Thus (2.2.34) holds for some sequence (t_k). Since the solutions of (2.2.32) depend continuously on the initial conditions, there exists a sequence (y_k) in $\mathfrak{M}$ for which

$$\lim_{k\to\infty} |y_k - x_0| = 0 \,, \tag{2.2.37}$$

$$\lim_{k\to\infty} |p(t_k, y_k)| = \infty \,. \tag{2.2.38}$$

Since $y_k \in \mathfrak{M}$, for all sufficiently large k we can choose $\tau_k > t_k$ such that

$$|p(\tau_k, y_k)| = r_1$$

and

$$r_1 \leqslant |p(t, y_k)| \leqslant |p(t_k, y_k)| \quad (t_k \leqslant t \leqslant \tau_k) \,.$$

By Lemma 2.2.1,

$$\varphi(p(t_k, y_k)) - \varphi(p(\tau_k, x_k)) \geqslant \int_{r_1}^{|p(t_k,y_k)|} \alpha(s)\, ds \,. \tag{2.2.39}$$

However,

$$\varphi(y_k) \geqslant \varphi(p(t_k, y_k)) \,.$$

Therefore, by (2.2.39), we have

$$\varphi(y_k) \geqslant \varphi(p(\tau_k, y_k)) + \int_{r_1}^{|p(t_k,y_k)|} \alpha(s)\, ds \,. \tag{2.2.40}$$

It follows from (2.2.30), (2.2.38), (2.2.40) that

$$\lim_{k\to\infty} \varphi(y_k) = \infty \,.$$

On the other hand, by (2.2.37), we have

$$\lim_{k\to\infty} \varphi(y_k) = \varphi(x_0) \,.$$

This contradiction shows that $\mathfrak{M}$ must be closed, and hence that $\mathfrak{M} = \mathbb{R}^N$.

Consider an arbitrary point x with $|x| \geqslant r_1$ and a solution $p(t,x)$ of (2.2.32). Since $\mathfrak{M} = \mathbb{R}^N$, we have

$$p(t_0, x) \in \mathfrak{N}$$

for some t_0, and since $\mathfrak{N} \subset B(r_1)$ we can choose $t_1 < t_0$ such that

$$|p(t_1, x)| = r_1 \tag{2.2.41}$$

and

$$|p(t, x)| \geqslant r_1 \quad (0 \leqslant t \leqslant t_1) \,.$$

Thus by Lemma 2.2.1 we have

$$\varphi(x) - \varphi(p(t_1, x)) \geqslant \int_{r_1}^{|x|} \alpha(s)\, ds \,.$$

This inequality and (2.2.41) give

$$\varphi(x) \geqslant \int_{r_1}^{|x|} \alpha(s)\, ds + \min_{y\in\partial B(r_1)} \varphi(y) \,.$$

Therefore

$$\lim_{|x|\to\infty} \varphi(x) = \infty \,,$$

as required. □

Lemma 2.2.4. *Suppose that $f : \mathbb{R}^N \to \mathbb{R}$ is a locally Lipschitzian function whose only critical point is* 0, *and that* 0 *is a local minimizer for f. Let*

$$\min_{y \in \partial f(x)} |y| \geqslant \beta(|x|) , \tag{2.2.42}$$

where $\beta : (0, \infty) \to \mathbb{R}$ is a positive continuous function for which

$$\int_0^\infty \beta(s) = \infty .$$

Then 0 *is a global minimizer for f and*

$$\lim_{|x| \to \infty} f(x) = \infty .$$

Proof. We may assume for simplicity that $f(0) = 0$. Since 0 is the only critical point of f, it is a strict local minimizer. Therefore there exist positive numbers $r_0 < r_1$ and ε such that

$$\min_{x \in \partial B(r_1)} f(x) - \varepsilon \geqslant \max_{x \in B(r_0)} |f(x)| .$$

Let φ be a twice continuously differentiable function on $\mathbb{R}^N$ for which

$$\sup_{x \in \mathbb{R}^N} |f(x) - \varphi(x)| < \tfrac{1}{2}\varepsilon ,$$

and

$$\min_{y \in \partial f(x)} |\nabla\varphi(x) - y)| < \tfrac{1}{2}\beta(|x|) \quad (|x| \geqslant r_0) .$$

Then φ satisfies the conditions of Lemma 2.2.3 and

$$\alpha(s) = \tfrac{1}{2}\beta(s) \quad (r_0 \leqslant s < \infty) .$$

Consequently

$$\lim_{|x| \to \infty} \varphi(x) = \infty .$$

Therefore

$$\lim_{|x| \to \infty} f(x) = \infty .$$

It follows that f has a global minimizer x_* which is a critical point. Thus $x_* = 0$, and the lemma is proved. □

We say that a nondegenerate deformation

$$f : G \times [0, 1] \to \mathbb{R}$$

of a locally Lipschitzian function f_0 into a locally Lipschitzian function f_1 is a *global nondegenerate deformation* if $G = \mathbb{R}^N$.

Theorem 2.2.2. *Suppose that f_0, f_1 are locally Lipschitzian functions and that there exists a global nondegenerate deformation $f : \mathbb{R}^N \times [0,1] \to \mathbb{R}$ of f_0 into f_1. Let x_0 be a minimizer for f_0 and suppose that for some $r > 0$ we have*

$$\min_{y \in \partial f_1(x)} |y| \geqslant \alpha(|x|) \quad (|x| \geqslant r) , \tag{2.2.43}$$

where $\alpha : [r, \infty) \to \mathbb{R}$ is a continuous positive function for which

$$\int_r^\infty \alpha(s)\, ds = \infty . \tag{2.2.44}$$

Then x_1 is a global minimizer of f_1, and

$$\lim_{|x| \to \infty} f_1(x) = \infty . \tag{2.2.45}$$

Proof. By Theorem 2.2.1, the point x_1 is a local minimizer for f_1. Thus by Lemma 2.2.4, it is a global minimizer for f_1 and (2.2.45) holds. □

2.2.4 Linear Deformations of Lipschitzian Functions

Recall that a locally Lipschitzian function f is *regular* if for each point x and for each $v \neq 0$ it has an ordinary derivative $f'(x, v)$ at x in the direction of v and

$$f'(x, v) = f^0(x, v) .$$

If there is a global nondegenerate deformation $f(\cdot, \lambda)$ of such a function f_0 into a function f_1 that is *linear*, i.e., has the form

$$f(x, \lambda) = (1 - \lambda) f_0(x) + \lambda f_1(x) ,$$

and if f_0 is regular at each point and growing, then a global minimizer is preserved even without conditions (2.2.43), (2.2.44):

Theorem 2.2.3. *Let $x = 0$ be the only critical point of the locally Lipschitzian functions $f_0 : \mathbb{R}^N \to \mathbb{R}$ and $f_1 : \mathbb{R}^N \to \mathbb{R}$ and suppose that for $x \neq 0$ the convex hull*

$$\mathrm{co}\{\partial f_0(x), \partial f_1(x)\}$$

of the generalized gradients $\partial f_0(x)$ and $\partial f_1(x)$ of f_0 and f_1 does not contain 0. Suppose, finally, that f_0 is regular at each point and that

$$\lim_{|x| \to \infty} f_0(x) = \infty . \tag{2.2.46}$$

Then $x = 0$ is a global minimizer of f_1.

Proof. We may assume for simplicity that

$$f_0(0) = f_1(0) = 0 .$$

Consider an arbitrary point $x \in \mathbb{R}^N$ and choose a number $c > 0$ such that $f_0(x) < c$. We set

$$Q = \{x \in \mathbb{R}^N : f_0(x) \leqslant c\} .$$

By (2.2.46), the set Q is compact. Therefore f_1 attains its minimum on Q. Let x_* be a minimizer for f_1 on Q. We shall show that $x_* \in \operatorname{int} Q$.

Indeed, if $x_* \in \partial Q$, then (by Theorem 1.7.10)

$$0 \in \partial f_1(x_*) + N_Q(x_*) , \tag{2.2.47}$$

where $N_Q(x_*)$ is the normal cone to Q at x_*. Since

$$f_0(x_*) = c > 0 ,$$

it follows that $x_* \neq 0$. Thus

$$0 \notin \partial f_0(x_*)$$

and, since f_0 is regular, we have

$$N_Q(x_*) = \bigcup_{\lambda \geqslant 0} \lambda \partial f_0(x_*) .$$

Therefore from (2.2.47) we have $v + \lambda u = 0$ for some vectors $u \in \partial f_0(x_*)$ and $v \in \partial f_1(x_*)$ and for some $\lambda \geqslant 0$. It follows that

$$0 \in \operatorname{co}\{\partial f_0(x_*), \partial f_1(x_*)\} ,$$

and this contradicts the hypothesis of the theorem.

Thus $x_* \in \operatorname{int} Q$. It follows that x_* is a critical point of f_1, regarded as a functional on the open set $\operatorname{int} Q$. Therefore

$$0 \in \partial f_1(x_*)$$

and, consequently, $x_* = 0$. Thus

$$f_1(x) > f_1(0) ,$$

and the theorem is proved. □

2.2.5 Regular and Critical Points of Continuous Functions

The most widely used technique for determining critical points of nonsmooth functions depends on properties of generalized gradients. However, these are only available for special types of functions (convex, Lipschitzian, quasi-differentiable, etc.). We present here a different definition of critical points which covers all continuous functions.

Let U be a domain in $\mathbb{R}^N$ and α and β be positive numbers. We denote by $H(\alpha,\beta,U)$ the class of continuous mappings $h : [-\alpha,\alpha] \times U \to \mathbb{R}^N$ for which

$$|x - h(t,x)| \leqslant \beta\,|t| \quad (|t| \leqslant \alpha, \quad x \in U)\,. \tag{2.2.48}$$

We say that a subset M of $\mathbb{R}^N$ is a *regular set* of a continuous function $f : \mathbb{R}^N \to \mathbb{R}$ if there exists a sequence of mappings h_n from some class $H(\alpha,\beta,U)$ such that $\overline{M} \subset U$ and

$$\lim_{n\to\infty} \sup_{|t|\leqslant\alpha, x\in U} |f(x) - f(h_n(t,x)) - t| = 0\,. \tag{2.2.49}$$

Points x such that $\{x\}$ is regular will be called *regular points* and nonregular points will be called *critical points.* It follows from the definition that the set $R(f)$ of regular points of a continuous function f is open and the set $C(f)$ of critical points is closed.

The following result gives a criterion for a compact set to be regular.

Lemma 2.2.5. *Let M be a compact set that lies in a bounded domain W in $\mathbb{R}^N$. Suppose that there exists a sequence of continuously differentiable functions f_n, uniformly convergent to f in W, such that*

$$|\nabla f_n(x)| > a > 0 \quad (x \in \overline{W};\ n = 1,2,\ldots)\,. \tag{2.2.50}$$

Then M is a regular set of f.

Proof. Choose $\rho > 0$ such that the neighborhood

$$U(\rho, M) = \{x \in \mathbb{R}^N : |y - x| < \rho,\ y \in M\}$$

of M is contained in $\overline{W}$. We associate with each f_n a twice continuously differentiable function g_n for which

$$|\nabla g_n(x)| > a \quad (x \in \overline{W};\ n = 1,2,\ldots)\,,$$

$$|g_n(x) - f_n(x)| < \frac{1}{n} \quad (x \in \overline{W};\ n = 1,2,\ldots)\,.$$

Let $p_n(t;x)$ be a solution of the Cauchy problem for the differential equation

$$\frac{dx}{dt} = -|\nabla g_n(x)|^{-2}\nabla g_n(x)$$

with the initial condition

$$p_n(0;x) = x\,.$$

Then, for

$$U = U\left(\tfrac{1}{2}\rho; M\right), \quad \alpha = \tfrac{1}{2}a\rho, \quad \beta = a^{-1}\,,$$

the mappings

$$h_n(t,x) = p_n(t;x)$$

belong to $H(\alpha,\beta,U)$ and satisfy condition (2.2.49), as required. □

We could use Lemma 2.2.5 as the basis for a general definition of regular sets of continuous functions f, taking a subset M to be regular for f if there exist a neighborhood W of the closure $\overline{M}$ and a sequence (f_n) of smooth functions convergent uniformly to f on $\overline{W}$ and satisfying (2.2.50). It would be interesting to construct a theory of critical points of continuous functions f based on this definition if the regularity for f of each point of a compact set M implies the regularity of M. However, we do not know whether this is true in general.

Direct verification shows that, for a continuously differentiable function f, the regular points are precisely the points at which the gradient of f is nonzero and the critical points are the points at which the gradient is zero. For a locally Lipschitzian function f, a point x_0 is regular if $0 \notin \partial f(x_0)$, and so if $0 \in \partial f(x_0)$, then x_0 is a critical point of f.

Lemma 2.2.6. *If $\mathcal{D}_0$ and $\mathcal{D}_1$ are compact sets that are regular sets of a continuous function f then $\mathcal{D}_0 \cup \mathcal{D}_1$ is a regular set of f.*

Before proving the lemma, we consider the following construction.

Let $h \in H(\alpha,\beta,U)$. Consider a smooth function $\mu : \mathbb{R}^N \to [0,1]$ whose support M lies in U. We set

$$g(t,x) = \begin{cases} h(\mu(x)t,x) & \text{for } |t| \leqslant \alpha,\ x \in U\ , \\ x & \text{for } |t| \leqslant \alpha,\ x \in \mathbb{R}^N \setminus U\ , \end{cases}$$

and

$$s(t,x,y) = \begin{cases} h(\mu(x)t,y) & \text{for } |t| \leqslant \alpha,\ x \in \mathbb{R}^N, \qquad y \in U\ , \\ y & \text{for } |t| \leqslant \alpha,\ x \in \mathbb{R}^N \setminus M,\ y \in \mathbb{R}^N\ . \end{cases}$$

Then

$$g \in H(\alpha,\beta,\mathbb{R}^N)$$

and s maps $[-\alpha,\alpha] \times ((\mathbb{R}^N \times U) \cup (M \times \mathbb{R}^N))$ continuously into $\mathbb{R}^N$.

Let us now prove the lemma. We can assume that $\mathcal{D}_0 \cap \mathcal{D}_1 \neq \varnothing$. By hypothesis, there exist sequences of mappings

$$h_n^0(t,x) \in H(\alpha_0,\beta_0,U_0) \quad (\mathcal{D}_0 \subset U_0)$$

and

$$h_n^1(t,x) \in H(\alpha_1,\beta_1,U_1) \quad (\mathcal{D}_1 \subset U_1)$$

for which

$$\lim_{n\to\infty} \sup_{|t|\leqslant \alpha_0, x\in U_0} \left| f(x) - f(h_n^0(t,x)) - t \right| = 0 \tag{2.2.51}$$

and

$$\lim_{n\to\infty} \sup_{|t|\leqslant\alpha_1, x\in U_1} \left|f(x) - f(h_n^1(t,x)) - t\right| = 0 \,. \tag{2.2.52}$$

We choose bounded neighborhoods U_0^1, U_0^2; U_1^1, U_1^2 of $\mathcal{D}_0$ and $\mathcal{D}_1$ such that $\overline{U}_0^2 \subset U_0^1$, $\overline{U}_0^1 \subset U_0$, $\overline{U}_1^2 \subset U_1^1$, $\overline{U}_1^1 \subset U_1$. By Theorem 1.5.8 on the partition of unity, there exist smooth functions μ_0, $\mu_1 : \mathbb{R}^N \to \mathbb{R}$ for which

$$0 \leqslant \mu_0(x) \leqslant 1, \quad 0 \leqslant \mu_1(x) \leqslant 1 \quad (x \in \mathbb{R}^N)\,,$$

$$\operatorname{supp} \mu_0 \subset U_0^1, \quad \operatorname{supp} \mu_1 \subset U_1^1\,,$$

$$\mu_0(x) + \mu_1(x) = 1 \quad (x \in U_0^2 \cup U_1^2)\,.$$

We set

$$W = (U_0^1 \cap U_1^2) \cup (U_0^2 \cap U_1^1)\,.$$

Since $\overline{W} \subset U_0$, there exists some $\alpha_2 > 0$ such that for $|t| \leqslant \alpha_2$ and $x \in W$ we have

$$h_n^1(t,x) \in U_0\,.$$

Let

$$U = U_0^2 \cup U_1^2, \quad \alpha = \min(1, \alpha_0, \alpha_1, \alpha_2), \quad \beta = \beta_0 + \beta_1$$

and

$$h_n(t,x) = h_n^0(\mu_0(x)t, h_n^1(\mu_1(x)t, x))\,.$$

Each mapping $h_n(t,x)$ is defined and continuous on $[-\alpha,\alpha] \times U$, and for $|t| \leqslant \alpha$, $x \in U$ we have

$$\begin{aligned}
|x - h_n(t,x)| &= \left|x - h_n^0(\mu_0(x)t, h_n^1(\mu_1(x)t,x))\right| \\
&\leqslant \left|x - h_n^1(\mu_1(x)t,x)\right| \\
&\quad + \left|h_n^1(\mu_1(x)t,x) - h_n^0(\mu_0(x)t, h_n^1(\mu_1(x)t,x))\right| \\
&\leqslant \beta_1\mu_1(x)\,|t| + \beta_0\mu_0(x)\,|t| = \beta\,|t|\,,
\end{aligned}$$

i.e., $h_n(t,x) \in H(\alpha,\beta,U)$.

By (2.2.51) and (2.2.52), we have the following sequence of inequalities:

$$\begin{aligned}
&\lim_{n\to\infty} \sup_{|t|\leqslant\alpha, x\in U} |f(x) - f(h_n(t,x)) - t| \\
&\quad\leqslant \lim_{n\to\infty} \sup_{|t|\leqslant\alpha, x\in U} \left|f(x) - f(h_n^1(\mu_1(x)t,x)) - \mu_1(x)t\right| \\
&\qquad + \lim_{n\to\infty} \sup_{|t|\leqslant\alpha, x\in U} \left|f(h_n^1(\mu_1(x)t,x)) - f(h_n^0(\mu_0(x)t, h_n^1(\mu_1(x)t,x))) - \mu_0(x)t\right| \\
&\quad\leqslant \lim_{n\to\infty} \sup_{|t|\leqslant\alpha, x\in U_1} \left|f(x) - f(h_n^1(t,x)) - t\right| \\
&\qquad + \lim_{n\to\infty} \sup_{|t|\leqslant\alpha, x\in U_0} \left|f(x) - f(h_n^0(t,x)) - t\right| = 0\,.
\end{aligned}$$

Hence the mappings h_n satisfy condition (2.2.49). □

Lemma 2.2.7. *If each point of a compact set $\mathcal{D} \subset \mathbb{R}^N$ is a regular point of the continuous function f, then $\mathcal{D}$ is a regular set of f.*

The points of local extremum of smooth and Lipschitzian functions are critical points. A similar statement holds for continuous functions.

Theorem 2.2.4. *The points of local extremum of a continuous function f are critical points of f.*

Proof. Suppose, for definiteness, that x_* is a local minimizer for f, and assume that x_* is a regular point of f. Then there exist a ball $B(\rho, x_*)$ whose points x satisfy

$$f(x) - f(x_*) \geqslant 0\,, \tag{2.2.53}$$

and a sequence (h_n) of mappings from some class $H(\alpha, \beta, U)$, where $B(\rho, x_*) \subset U$, which satisfy condition (2.2.49). We can assume without loss of generality that $\alpha < \rho/\beta$ and that

$$|f(x) - f(h_m(t,x)) - t| < \alpha \quad (|t| \leqslant \alpha,\ x \in U) \tag{2.2.54}$$

for some m. Since the operator $\mathcal{D}$ defined by

$$\mathcal{D}(x) = x - h_m(\alpha, x) - x_*$$

maps $B(\rho, x_*)$ into itself, the equation

$$h_m(\alpha, x) = x_*$$

has a solution $x_0 \in B(\rho, x_*)$, by Brouwer's classical fixed-point theorem. Thus by (2.2.54) we have

$$f(x_*) - f(x_1) > 0\,,$$

which contradicts (2.2.53). □

2.2.6 Deformations of Continuous Functions

Consider a one-parameter family of continuous functions $f(\cdot;\lambda) : \mathbb{R}^N \to \mathbb{R}$ $(0 \leqslant \lambda \leqslant 1)$. We say that the family $f(\cdot;\lambda)$ is a *nondegenerate* deformation of the function

$$f_0(\cdot) = f(\cdot;0)$$

into the function

$$f_1(\cdot) = f(\cdot;1)$$

on a ball B if

(1) the function $f : \mathbb{R}^N \times [0,1] \to \mathbb{R}$ is continuous,

(2) for all $\lambda \in [0,1]$ and $\rho \in (0,1)$, there exists a sequence of mappings

$$h_n(\cdot,\cdot;\lambda,\rho) : [-\alpha(\rho), \alpha(\rho)] \times T(\rho) \to \mathbb{R}^N\,,$$

from the class $H(\alpha(\rho), \beta(\rho), T(\rho))$, where

$$T(\rho) = \{x \in \mathbb{R}^N : \rho < |x| < 1\} ,$$

for which

$$\lim_{n\to\infty} \sup_{x\in T(\rho),\ |t|\leqslant\alpha(\rho),\ 0\leqslant\lambda\leqslant1} |f(x;\lambda) - f(h_n(t,x;\lambda,\rho);\lambda) - t)| = 0 , \quad (2.2.55)$$

(3) for each $\lambda \in [0,1]$ the point 0 is the unique critical point of $f(\cdot;\lambda)$ in B.

Theorem 2.2.5. *Suppose that f_1, f_2 are continuous functions and that there exists a nondegenerate deformation f of f_0 into f_1. If 0 is a critical point and a local minimizer for f_0, then 0 is a local minimizer for f_1.*

Proof. We can assume without loss of generality that

$$f(0;\lambda) \equiv 0 \quad (0 \leqslant \lambda \leqslant 1) .$$

Let Λ be the set of $\lambda \in [0,1]$ for which 0 is a local minimizer for $f(\cdot;\lambda)$. Since $f(\cdot;\lambda)$ has no nonzero critical points in B, it follows by Theorem 2.2.4 that 0 is a strict local minimizer of $f(\cdot;\lambda)$. Hence Λ is open.

Now we show that Λ is closed. It suffices to construct some $r > 0$ such that for all $\lambda \in \Lambda$ the point 0 is an absolute minimizer of $f(\cdot;\lambda)$ on $B(r)$.

We set

$$\varphi(\rho) = \rho + \max_{x\in B(\rho),\ 0\leqslant\lambda\leqslant1} |f(x;\lambda)| ,$$

$$r = \min\left\{\tfrac{1}{4}, \varphi^{-1}\left(\alpha\left(\tfrac{1}{4}\right)\right)\right\} . \quad (2.2.56)$$

We fix some $\lambda_0 \in \Lambda$ and denote by $l(a)$ the connected component of the Lebesgue set

$$\mathcal{L}(a) = \{x \in \mathbb{R}^N : f(x;\lambda_0) < a\}$$

containing 0. Let A be the set of all $a > 0$ for which

$$l(a) \subset B\left(\tfrac{1}{2}\right).$$

Then A is a nonempty interval; let a_0 be its right-hand endpoint and set

$$l = \bigcup_{a\in A} l(a) .$$

Since $f(\cdot;\lambda_0)$ assumes a positive value a_0 on the boundary ∂l of l and the point 0 is the only critical point of f in $\bar{l}$, the point 0 in $\bar{l}$ is an absolute minimizer of this function. Therefore, to prove the theorem, it suffices to show that

$$B(r) \subset \bar{l} ,$$

where r is defined by (2.2.56).

We shall first show that

$$\partial l \cap S(\tfrac{1}{2}) \neq \varnothing .$$

If this is false, then

$$\min_{x \in \partial l,\ y \in S(1/2)} |x - y| = \delta > 0 .$$

Since ∂l consists of regular points of $f(\cdot;\lambda_0)$, by Lemma 2.2.7 the compact set ∂l is a regular set of $f(\cdot;\lambda_0)$. Therefore, for some α, $\beta > 0$ and some domain $V \supset \partial l$, there exists a sequence (h_n) of mappings in $H(\alpha, \beta, V)$ for which

$$\lim_{n\to\infty} \sup_{|t| \leqslant \alpha, \alpha \in V} |f(x;\lambda_0) - f(h_n(t,x);\lambda_0) - t| = 0 . \tag{2.2.57}$$

Let

$$\gamma = \min_{x \in \partial l} |x|, \quad \alpha_0 = \min\{\alpha, \gamma/\beta, \delta/\beta\} .$$

We set

$$\Gamma_n = \{y \in \mathbb{R}^N : y = h_n(\alpha_0, x),\ x \in \partial l\} .$$

Since each domain $l(a_0 + n^{-1})$ has points in common with the sphere $S(\frac{1}{2})$, there exist continuous curves

$$s_n : [0,1] \to \mathbb{R}^N$$

for which

$$s_n(0) = 0, \quad |s_n(1)| = \tfrac{1}{2}$$

and

$$f(s_n(\mu);\lambda_0) \leqslant a_0 + n^{-1} \quad (0 \leqslant \mu \leqslant 1) . \tag{2.2.58}$$

Since

$$s_n([0,1]) \cap \Gamma_n \neq \varnothing ,$$

we have

$$h_n(\alpha_0, x_n) = s_n(\mu_n) \quad (n = 1, 2, \ldots)$$

for some $\mu_n \in [0,1]$ and $x_n \in \partial l$. Therefore

$$\lim_{n\to\infty} |a_0 - f(s_n(\mu_n);\lambda_0) - \alpha_0| = 0 ,$$

and this contradicts the inequalities (2.2.58).

Thus

$$\partial l \cap S(\tfrac{1}{2}) = \varnothing .$$

This implies that there is a point $x_0 \in l$ such that

$$|x_0| \geqslant \tfrac{1}{2}$$

and

$$f(x_0;\lambda_0) < a_0 \,. \tag{2.2.59}$$

Let $B(\rho_0)$ be the ball of maximal radius contained in l. Then 0 is an absolute minimizer for $f(\cdot;\lambda_0)$ on this ball. If $\rho_0 > \frac{1}{4}$ then $B(r) \subset B(\rho_0)$, and 0 is an absolute minimizer for $f(\cdot;\lambda_0)$ on $B(r)$.

Suppose that

$$\rho_0 < \tfrac{1}{4} \,.$$

Since $f(\cdot;\lambda)$ is a nondegenerate deformation of f_0 into f_1, it follows that

$$\lim_{n\to\infty} \sup_{x\in T(\frac{1}{4}),|t|\leqslant\alpha(\frac{1}{4})} \left|f(x;\lambda_0) - f\left(h_n\left(t,x;\lambda_0,\tfrac{1}{4}\right);\lambda_0\right) - t\right| = 0 \,. \tag{2.2.60}$$

Consider the mappings

$$g_n(x) = h_n\left(\alpha(\tfrac{1}{4}),x;\lambda_0,\tfrac{1}{4}\right).$$

Since $g_n(x_0) \in l$ for all large n, and since 0 is an absolute minimizer for $f(\cdot;\lambda_0)$ on l, we have

$$f(g_n(x_0);\lambda_0) \geqslant 0 \tag{2.2.61}$$

for large n. On the other hand, from (2.2.53) and (2.2.60) we have

$$f(g_n(x_0);\lambda_0) < a_0 + \alpha(\tfrac{1}{4}) \tag{2.2.62}$$

for large n. The inequalities (2.2.61) and (2.2.62) imply that

$$a_0 \geqslant \alpha(\tfrac{1}{4}).$$

However,

$$a_0 \leqslant \varphi(\rho_0)\,,$$

and therefore

$$\rho_0 \geqslant \varphi^{-1}(\alpha(\tfrac{1}{4})).$$

Thus Λ is closed. It follows that

$$\Lambda = [0,1]\,,$$

and the theorem is proved. □

2.3 Converses of Deformation Theorems

2.3.1 Real Polynomials

Suppose that x_* is an isolated critical point of both $f_0 : \mathbb{R}^N \to \mathbb{R}$ and $f_1 : \mathbb{R}^N \to \mathbb{R}$ and that x_* is a local minimizer for f_0 and f_1. Can we find a nondegenerate deformation of f_0 into f_1 in some neighborhood of x_*? Although we do not have a definitive answer to this question, there are satisfactory results, described below, for some types of functions. In this section, we investigate the existence of nondegenerate deformations for real polynomials in N variables. By a *real polynomial* $P(x)$ (where $x = (x_1, \dots, x_N) \in \mathbb{R}^N$) we understand a function of the form

$$P(x) = \sum a_{\alpha_1 \dots \alpha_N} x_1^{\alpha_1} \dots x_N^{\alpha_N} , \tag{2.3.1}$$

where $a_{\alpha_1 \dots \alpha_N}$ are real coefficients, $\alpha_1, \dots, \alpha_N$ are nonnegative integers, and the sum contains a finite number of terms.

Theorem 2.3.1. *Suppose that* 0 *is an isolated critical point of the real polynomials* P_0 *and* P_1 *and that* 0 *is a local minimizer of these polynomials. Then for some* $\rho > 0$ *the linear deformation*

$$P(x; \lambda) = (1 - \lambda) P_0(x) + \lambda P_1(x) \quad (0 \leqslant \lambda \leqslant 1) \tag{2.3.2}$$

is nondegenerate on the ball $B(\rho)$ *in* $\mathbb{R}^N$.

For the proof we need an auxiliary result. A subset V of $\mathbb{R}^N$ is a *real algebraic set* if it is an intersection of the sets of zeros of a finite family of real polynomials.

Lemma 2.3.1 (on the choice of curves [171]). *Suppose that* V *is a real algebraic set in* $\mathbb{R}^N$, *and that* U *is an open set in* $\mathbb{R}^N$ *defined by finitely many polynomial inequalities:*

$$U = \{x \in \mathbb{R}^N : g_1(x) > 0, \dots, g_k(x) > 0\} .$$

If

$$0 \in \overline{U \cap V} ,$$

then there exists an analytic curve

$$p : [0, \varepsilon) \to \mathbb{R}^N$$

such that

$$p(0) = 0 \quad \textit{and} \quad p(t) \in U \cap V \quad (0 < t < \varepsilon) .$$

Now we prove Theorem 2.3.1. We may assume that

$$P_0(0) = P_1(0) = 0 .$$

In order to prove the theorem, it suffices to show that there is a neighborhood of 0 in which for $x \neq 0$ the gradients $\nabla P_0(x)$ and $\nabla P_1(x)$ are not oppositely directed.

Let us assume the contrary. Then the intersection $U \cap V$ of the sets

$$U = \{x \in \mathbb{R}^N : (\nabla P_0(x), \nabla P_1(x)) < 0\} ,$$

$$V = \{x \in \mathbb{R}^N : (\nabla P_0(x), \nabla P_1(x))^2 - |\nabla P_0(x)|^2 |\nabla P_1(x)|^2 = 0\}$$

(consisting of points at which the gradients $\nabla P_0(x)$ and $\nabla P_1(x)$ are oppositely directed) is nonempty, and

$$0 \in \overline{U \cap V} .$$

By the lemma on the choice of curves, there exists an analytic curve

$$x = p(t) \quad (0 \leqslant t < \varepsilon)$$

all of whose points (except for the point $0 = p(0)$) lie in $U \cap V$. Since we have

$$f_0(p(t)) > 0 \quad \text{and} \quad f_1(p(t)) > 0$$

for small $t > 0$, it follows, by the analyticity of the functions $f_0(p(\cdot))$ and $f_1(p(\cdot))$ for small $t > 0$, that for small $t > 0$ we have

$$f_0'(p(t)) = (\nabla f_0(p(t)), p'(t)) > 0 ,$$

$$f_1'(p(t)) = (\nabla f_1(p(t)), p'(t)) > 0 ,$$

i.e., for small $t > 0$ the vectors $\nabla f_0(p(t))$ and $\nabla f_1(p(t))$ form an acute angle with $p'(t)$, and this contradicts the inclusion $p(t) \in U \cap V$ $(0 < t < \varepsilon)$. □

2.3.2 Real Analytic Functions

A function $f : \mathbb{R}^N \to \mathbb{R}$ is *real analytic* at a point x if, in some neighborhood of this point, f can be represented by a power series

$$f(x+h) = f(x) + \sum a_{\alpha_1 \dots \alpha_N} h_1^{\alpha_1} \dots h_N^{\alpha_N} ,$$

and f is analytic in a domain Ω in $\mathbb{R}^N$ if it is analytic at every point of Ω.

There is a result corresponding to Theorem 2.3.1 for real analytic functions.

Theorem 2.3.2. *Suppose that 0 is an isolated critical point of the real analytic functions f_0 and f_1, and that 0 is a local minimizer for f_0, f_1. Then, on some ball $B(\rho)$, there exists a nondegenerate deformation of f_0 into f_1.*

The proof depends on an important result, the so-called *Lojasiewicz inequality.*

Suppose that Ω is an open set in $\mathbb{R}^N$ and $f : \Omega \to \mathbb{R}$ is a real analytic function. Let E be the set of zeros of f lying in Ω, and denote by $d_E(x)$ the distance from the point x to E.

Theorem 2.3.3 (Lojasiewicz; see [166]). *For every compact set K of Ω there exist constants C, $m > 0$ such that*

$$|f(x)| \geqslant C(d_E(x))^m$$

for all $x \in K$.

Let us now prove Theorem 2.3.2. Since 0 is an isolated critical point of f_0 and f_1, by the Lojasiewicz inequality there are positive numbers c_0, c_1, m_0, m_1 such that, for small $\rho > 0$, we have

$$|\nabla f_0(x)| > c_0 |x|^{m_0} \quad (x \in B(\rho)) ,$$

$$|\nabla f_1(x)| > c_1 |x|^{m_1} \quad (x \in B(\rho)) .$$

These inequalities imply that we can obtain linear deformations of f_0 and f_1 into polynomials g_0 and g_1 which are truncations of the power series expansions of f_0 and f_1 in a neighborhood of 0; we can ensure that the linear deformations are nondegenerate in some neighborhood of 0 by retaining a sufficient number of terms of the power series. Now 0 is both an isolated critical point and a local minimizer for g_0 and g_1, and by Theorem 2.3.1 we can connect g_0 and g_1 by a linear deformation which is nondegenerate in some neighborhood of 0. Therefore there exists a nondegenerate deformation of f_0 into f_1 in some neighborhood of 0. □

2.3.3 Smooth Functions

Considerable difficulties are encountered when trying to prove analogs of Theorems 2.3.1 and 2.3.2 for smooth functions, and we have only been able to prove a converse to the deformation principle for a minimizer in dimensions $N = 2$ and $N \geqslant 6$. We do not know whether there is a converse of Theorem 2.1.1 in dimensions 3, 4, and 5.

The proof of the converse of Theorem 2.1.1 in dimensions $N \geqslant 6$ is based largely on Smale's h-cobordism theorem (see [170]). In order to formulate this theorem, we need some definitions and auxiliary results.

Let U and V be domains in $\mathbb{R}^N$ and $\mathbb{R}^M$ respectively. Each mapping $f : U \to V$ is defined by a collection of M functions $f_i : U \to \mathbb{R}$, called the *components* of f. We say that f is *smooth* if all partial derivatives

$$\frac{\partial^n f_i(x_1, \ldots, x_N)}{(\partial x_{i_1})^{\alpha_1} \ldots (\partial x_{i_N})^{\alpha_N}}$$

$$(i = 1, \ldots, M; \ \alpha_i \geqslant 0, \ \alpha_1 + \cdots + \alpha_N = n; \ n = 1, 2, \ldots)$$

of the components f_i exist and are continuous on U.

If X and Y are subsets of $\mathbb{R}^N$ and $\mathbb{R}^M$ respectively, then a mapping $f : X \to Y$ is *smooth* if, for every point $x \in X$, there exist a neighborhood U of x in $\mathbb{R}^N$ and a smooth mapping $F : U \to \mathbb{R}^M$ coinciding with f on $U \cap X$.

A continuous one-to-one mapping $f : X \to Y$ (where $X \subset \mathbb{R}^N$, $Y \subset \mathbb{R}^M$) is a *homeomorphism* from X to Y if $f(X) = Y$ and the inverse mapping $f^{-1} : Y \to X$ is continuous; f is a *diffeomorphism* if in addition both f and f^{-1} are smooth. If there exists a diffeomorphism $f : X \to Y$, the sets X and Y are said to be *diffeomorphic*.

A subset $\mathfrak{M}$ of $\mathbb{R}^N$ is a *smooth* m-dimensional manifold without boundary if, for every point $x \in \mathfrak{M}$, there exists a neighborhood U of x in $\mathbb{R}^N$ such that $\mathfrak{M} \cap U$ is diffeomorphic to some open set V in $\mathbb{R}^m$. The set $\mathfrak{M} \cap U$ is a neighborhood of x in $\mathfrak{M}$. A subset $\mathfrak{M}$ of $\mathbb{R}^N$ is a *zero-dimensional manifold* if every point $x \in \mathfrak{M}$ has a neighborhood U in $\mathbb{R}^N$ such that $\mathfrak{M} \cap U = \{x\}$.

A diffeomorphism $g : V \to \mathfrak{M} \cap U$ is called a *parametrization* of the neighborhood $\mathfrak{M} \cap U$. The inverse diffeomorphism $g^{-1} : \mathfrak{M} \cap U \to V$ is called a *coordinate system* on $\mathfrak{M} \cap U$.

Let U be a domain in $\mathbb{R}^N$ and $V \subset \mathbb{R}^M$. The *derivative* f'_x at $x \in U$ of a smooth mapping $f : U \to V$ is defined by

$$f'_x(h) = \lim_{t \to 0} \frac{f(x+th) - f(x)}{t} ,$$

for $h \in \mathbb{R}^N$; it is a linear mapping from $\mathbb{R}^N$ to $\mathbb{R}^M$. If $f_1, \ldots, f_M$ are the components of f, then f'_x is specified by the $M \times N$ matrix

$$\begin{bmatrix} \dfrac{\partial f_1}{\partial x_1} & \cdots & \dfrac{\partial f_1}{\partial x_N} \\ \cdots & \cdots & \cdots \\ \dfrac{\partial f_M}{\partial x_1} & \cdots & \dfrac{\partial f_M}{\partial x_N} \end{bmatrix} .$$

We often identify this matrix with f'_x.

Next we define the tangent space to a smooth m-dimensional manifold $\mathfrak{M}$ in $\mathbb{R}^N$.

Let $x \in \mathfrak{M}$. We choose a parametrization

$$g : V \to \mathfrak{M}$$

of a neighborhood $g(V)$ of x, where V is an open subset of $\mathbb{R}^m$, and we write $v = g^{-1}(x)$. Consider the derivative $g'_v : \mathbb{R}^m \to \mathbb{R}^N$ of g. The *tangent space* $T\mathfrak{M}_x$ to $\mathfrak{M}$ at x is the image of $\mathbb{R}^m$ under g'_v:

$$T\mathfrak{M}_x = g'_v(\mathbb{R}^m) .$$

This definition is independent of the particular parametrization $g : V \to \mathfrak{M}$ of the chosen neighborhood of x and therefore the tangent space is well defined.

Suppose that $\mathfrak{M} \subset \mathbb{R}^N$ and $\mathfrak{N} \subset \mathbb{R}^M$ are smooth manifolds of dimensions m and n respectively and $f : \mathfrak{M} \to \mathfrak{N}$ is a smooth mapping of $\mathfrak{M}$ to $\mathfrak{N}$. We fix a point $x \in \mathfrak{M}$ and consider the derivative f'_x of f at x; it is a linear operator from $T\mathfrak{M}_x$ to $T\mathfrak{N}_y$, where $y = f(x)$.

Since f is a smooth mapping, we can find an open neighborhood U of x in $\mathbb{R}^N$ and a smooth mapping $F : U \to \mathbb{R}^M$ that agrees with f on $\mathfrak{M} \cap U$. We set

$$f'_x(h) = F'_x(h) \quad (h \in T\mathfrak{M}_x) .$$

This definition is independent of the choice of the mapping F and so f'_x is well defined.

We set

$$H^m = \{x = (x_1, \ldots, x_m) \in \mathbb{R}^m : x_m \geqslant 0\} .$$

Thus H^m is a closed half-space in $\mathbb{R}^m$, with *boundary* ∂H^m the hyperplane

$$\partial H^m = \{x = (x_1, \ldots, x_m) \in \mathbb{R}^m : x_m = 0\} ,$$

and so $\partial H^m = \mathbb{R}^{m-1} \times 0$.

A subset X of $\mathbb{R}^N$ is called a *smooth m-dimensional manifold with boundary* if, for every point $x \in X$, there exists a neighborhood $X \cap U$ diffeomorphic to a neighborhood $H^m \cap V$ (where U and V are open sets in $\mathbb{R}^N$ and $\mathbb{R}^m$ respectively). The *boundary* ∂X of X is the set of all points of X that are mapped by the above diffeomorphisms to points of ∂H^m. The boundary ∂X of a smooth m-dimensional manifold X is a smooth $(m-1)$-dimensional manifold. The set $X \setminus \partial X$ is the *interior* of X; it is a smooth manifold of dimension m.

The tangent space TX_x to a manifold with boundary is defined in the same way as for closed manifolds. Note that TX_x has the same dimension as X. This tangent space should be distinguished from the tangent space $T\partial X_x$ to ∂X at x. There is a natural embedding $T\partial X_x \subset TX_x$, and we have

$$\dim T\partial X_x = \dim TX_x - 1 .$$

Suppose that X is a smooth m-dimensional manifold, $f : X \to \mathbb{R}$ is a smooth function, and p is an interior point of X. Then p is said to be a *critical point* of f if

$$\frac{\partial f(p)}{\partial x_1} = \cdots = \frac{\partial f(p)}{\partial x_m} = 0$$

in some coordinate system.

A critical point p is *nondegenerate* if

$$\det \left(\frac{\partial^2 f(p)}{\partial x_i \partial x_j} \right)_{i,j=1,\ldots,m} \neq 0 .$$

If p is a nondegenerate critical point of a smooth function $f : X \to \mathbb{R}$, then we may write

$$f(x_1, \ldots, x_m) = f(p) - x_1^2 - \cdots - x_\lambda^2 + x_{\lambda+1}^2 + \cdots + x_m^2$$

in some coordinate system in a neighborhood of p (see, e.g., [169]). The number λ is called the *Morse index* of p.

We shall call a triple $(W; V_0, V_1)$ of smooth manifolds a *triad of smooth manifolds* if W is a compact smooth N-dimensional manifold and its boundary ∂W is the disjoint union of V_0 and V_1. A *Morse function* of this triad of smooth manifolds is a smooth function $f : W \to [0, 1]$ for which

(1) $f^{-1}(0) = V_0$, $f^{-1}(1) = V_1$,

(2) all critical points of f are nondegenerate.

It turns out (see, e.g., [170]) that every triad $(W; V_0, V_1)$ of smooth manifolds possesses Morse functions. The *Morse number* μ of a triad $(W; V_0, V_1)$ is the minimum of the number of critical points taken over all Morse functions $f : W \to \mathbb{R}$.

Let A be a subset of a topological space X. A continuous mapping

$$h : X \times [0,1] \to X$$

is a *deformation retraction* from X to A if

$$h(x;0) = x \quad (x \in X)\,,$$

$$h(x;1) \in A \quad (x \in X)\,,$$

$$h(x;1) = x \quad (x \in A)\,.$$

If there exists a deformation retraction from X to a subset A, then A is called a *deformation retract* of X. A deformation retraction h is *strict* if

$$h(x;\lambda) = x \quad (x \in A)$$

for every $\lambda \in [0,1]$.

A topological space X is *path-connected* if, for any two points $x_0, x_1 \in X$, there exists a continuous mapping

$$g : [0,1] \to X$$

for which

$$g(0) = x_0, \quad g(1) = x_1\,.$$

Such a mapping g is called a *path* from x_0 to x_1. A path-connected space X is *simply connected* if any two paths $g_0 : [0,1] \to X$ and $g_1 : [0,1] \to X$ with the same endpoints x_0 and x_1, i.e., such that

$$g_0(0) = g_1(0) = x_0, \quad g_0(1) = g_1(1) = x_1\,,$$

are homotopic, i.e., if there exists a continuous mapping

$$g : [0,1] \times [0,1] \to X$$

for which

$$g(\cdot;0) = g_0, \quad g(\cdot;1) = g_1\,.$$

Clearly X is simply connected if and only if any one-dimensional cycle in X can be contracted to a point, i.e., any continuous mapping $g : S^1 \to X$ is homotopic to a constant mapping.

The following remarkable statement, the so-called h-cobordism theorem, is the basic tool of this section. We formulate it in a form convenient for our application.

Theorem 2.3.4 (Smale; see [170]). *Let $(W; V_0, V_1)$ be a triad of smooth manifolds, with* $\dim W \geqslant 6$. *Suppose that the manifolds V_0 and V_1 are simply connected and are deformation retracts of W. Then the Morse number of the triad $(W; V_0, V_1)$ is 0.*

In other words, under the conditions of the h-cobordism theorem, there exists a smooth function $f : W \to [0,1]$ with no critical points such that

$$V_0 = \{x \in W : f(x) = 0\}, \quad V_1 = \{x \in W : f(x) = 1\} .$$

The following theorem is the main result of this section.

Theorem 2.3.5. *Let $N \geqslant 6$ and let f_0, $f_1 : B^N \to \mathbb{R}$ be smooth functions. Suppose that 0 is the only critical point on B^N which is a local minimizer for f_0, f_1. Then there exists a nondegenerate deformation $f(\cdot;\lambda)$ of f_0 into f_1 in some neighborhood of 0.*

Proof. We can assume without loss of generality that

$$f_0(x) = |x|^2, \quad f_1(0) = 0 .$$

We choose positive numbers $r_0 < r_1 < r_2 < 1$ and $c_0 < c_1$ such that

(1) the connected component l_0 of the Lebesgue set

$$L(c_0) = \{x \in \mathbb{R}^N : f_1(x) \leqslant c_0\}$$

containing 0 lies in the ball $B(r_1)$ and contains the ball $B(r_0)$,

(2) the connected component l_1 of the Lebesgue set

$$L(c_1) = \{x \in \mathbb{R}^N : f_1(x) \leqslant c_1\}$$

containing 0 lies in the ball B^N and contains the ball $B(r_2)$.

Let

$$W = B^N \setminus \operatorname{int} l_0 .$$

This set is a smooth N-dimensional manifold with boundary

$$\partial W = V_0 \cap V_1 ,$$

where

$$V_0 = \partial l_0, \quad V_1 = S^{N-1} .$$

The triple $(W; V_0, V_1)$ forms a triad of smooth manifolds. We shall show that all conditions of Theorem 2.3.4 are satisfied for this triad.

Consider the differential equation

$$\frac{dx}{dt} = -\nabla f_1(x) \quad (x \in B) . \tag{2.3.3}$$

Let $p(t,x)$ be a solution of the Cauchy problem for this equation with the initial condition $p(0,x) = x$.

We begin by showing that V_0 is path-connected. Let x_0, $x_1 \in V_0$. We choose t_0, $t_1 \leqslant 0$ such that

$$p(t_0, x_0) \in T(r_1, r_2), \quad p(t_1, x_1) \in T(r_1, r_2)\,,$$

where $T(r_1, r_2)$ is the spherical layer

$$T(r_1, r_2) = \{x \in \mathbb{R}^N : r_1 \leqslant |x| \leqslant r_2\}\,.$$

We connect the points $y_0 = p(t_0, x_0)$ and $y_1 = p(t_1, x_1)$ by a continuous path

$$\gamma : [0,1] \to T(r_1, r_2), \quad \gamma(0) = y_0, \quad \gamma(1) = y_1\,.$$

Let $\tau(y)$ $(y \in l_1)$ be the time for which

$$p(\tau(y), y) \in V_0\,.$$

Since the trajectories of Eq. (2.3.3) are orthogonal to V_0, the function τ is continuous. Then the continuous mapping

$$p(\tau(\gamma(\cdot)), \gamma(\cdot)) : [0,1] \to V_0$$

defines a path in V_0 connecting x_0 and x_1.

Next we establish that V_0 is simply connected. It suffices to show that every smooth mapping

$$q : S^1 \to V_0$$

is homotopic to a constant mapping.

We choose a smooth function t defined on $q(S^1)$ such that

$$p(t(x), x) \in T(r_1, r_2) \quad (x \in q(S^1))\,.$$

Then we choose a neighborhood U of the set

$$Q = p(t(q(S^1)), q(S^1))$$

which lies in $T(r_1, r_2)$ and is homeomorphic to the ball $\mathring{B}^N$, and we let

$$h : U \times [0,1] \to U$$

be a deformation retraction of U to some point $y_0 \in U$. For every $\lambda \in [0,1]$ the mapping

$$h(\cdot; \lambda) : U \to U$$

is a homeomorphism. Then the family of mappings

$$\varphi(\cdot;\cdot) : S^1 \times [0,1] \to V_0$$

defined by

$$\varphi(s;\lambda) = p(\tau(h(p(t(q(s)),q(s));\lambda),h(p(t(q(s)),q(s));\lambda))$$

is a homotopy of the mapping

$$q : S^1 \to V_0$$

into the constant mapping

$$\varphi(s;1) \equiv p(\tau(y_0), y_0)$$

with values in V_0. In a similar way, we can establish that W is simply connected.

Now we show that V_0 and V_1 are deformation retracts of W. We set

$$\psi(x;\lambda) = \begin{cases} x(1-2\lambda+2\lambda r_2/|x|) & \text{for } x \in T(1,r_2),\ 0 \leqslant \lambda \leqslant \frac{1}{2}, \\ x & \text{for } x \in W \setminus T(1,r_2),\ 0 \leqslant \lambda \leqslant \frac{1}{2}. \end{cases}$$

Then the mapping

$$h(x;\lambda) = \begin{cases} \psi(x;\lambda) & \text{for } x \in W,\ 0 \leqslant \lambda \leqslant \frac{1}{2}, \\ p\left((2\lambda-1)\tau\left(\psi\left(x;\frac{1}{2}\right)\right),\ \psi\left(x;\frac{1}{2}\right)\right) & \text{for } x \in W,\ \frac{1}{2} \leqslant \lambda \leqslant 1 \end{cases}$$

is a deformation retraction of W to V_0. Similarly, we can construct a deformation retraction of W to V_1.

Thus all conditions of Theorem 2.3.4 are satisfied, and consequently there exists a smooth function

$$w : W \to \mathbb{R},$$

without critical points, for which

$$w(x) = c_0 \quad (x \in V_0),$$
$$w(x) = 1 \quad (x \in S^{N-1}).$$

We set

$$v(x) = \begin{cases} f_1(x) & \text{for } x \in B \setminus W, \\ w(x) & \text{for } x \in W, \\ |x|^2 & \text{for } x \in \mathbb{R}^N \setminus B. \end{cases}$$

The function $v : \mathbb{R}^N \to \mathbb{R}$ is continuous. It is smooth everywhere except possibly the manifold V_0 and the unit sphere S^{N-1}. Since the vectors $\nabla f_1(x)$, $\nabla w(x)$ (and $\nabla w(x)$, x respectively) are collinear for $x \in V_0 \cup S^{N-1}$, there exists a smooth function $g : \mathbb{R}^N \to \mathbb{R}$, having only 0 as a critical point, which agrees with f_1 on $B(r_0)$ and with $|x|^2$ outside of $B(2)$. We set

$$f(x;\lambda) = \begin{cases} \lambda^2 g(\lambda^{-1}x) & \text{for } x \in \mathbb{R}^N \setminus 0,\ 0 \leqslant \lambda \leqslant 1, \\ 0 & \text{for } x = 0,\ 0 < \lambda \leqslant 1, \\ |x|^2 & \text{for } x \in \mathbb{R}^N,\ \lambda = 0. \end{cases}$$

The family $f(\cdot;\cdot) : B(r_0) \times [0,1] \to \mathbb{R}$ is a deformation of $|x^2|$ into f, and it is nondegenerate on $B(r_0)$. This concludes the proof of the theorem. □

2.4 Theorems of Hopf and Parusinski

2.4.1 Statement of the Problem

The main result of this chapter, Theorem 2.1.1, stated that the property of being a local minimizer is a homotopy invariant of critical points under nondegenerate smooth deformations; i.e., if a critical point is a local minimizer for one value of the parameter, then it is a local minimizer for all values of the parameter. Therefore, for instance, a nondegenerate deformation cannot transform a minimizer into a maximizer: a homotopy deformation of a function whose critical point is a minimizer into one whose critical point is a maximizer inevitably results in a bifurcation of the critical point. (We shall see later that in infinite-dimensional spaces such a bifurcation does not always occur: there are one-parameter families of smooth functionals having a unique critical point which is a minimizer for some values of the parameter but not a minimizer for other values.)

If we generalize the notion of nondegenerate deformation, the situation changes radically.

Let f_0, $f_1 : B^N \to \mathbb{R}$ be two smooth functions for which 0 is the only critical point. We shall call a one-parameter family of smooth functions

$$f(\cdot;\cdot) : B^N \times [0,1] \to \mathbb{R}$$

a *generalized nondegenerate deformation* of f_0 into f_1 if both $f(\cdot;\cdot)$ and the gradient $\nabla f(\cdot;\cdot)$ are continuous on $B^N \times [0,1]$ and in addition

$$f(x;0) = f_0(x) \quad (x \in B^N) ,$$

$$f(x;\lambda) = f_1(x) \quad (x \in B^N) ,$$

and

$$\nabla f(x;\lambda) \neq 0 \quad (x \in S^{N-1},\ 0 \leqslant \lambda \leqslant 1) .$$

Thus, in contrast to the definition of a nondegenerate deformation given in Sect. 2.1.1, we have weakened the requirement of uniqueness of the critical point to the requirement that there are no critical points on the boundary of the domain of definition of the functions under consideration. We shall show later that under generalized nondegenerate deformations of smooth functions, the topological type of critical points may change. Thus, for instance, in even-dimensional spaces, generalized nondegenerate deformations can homotopically transform minimizers of smooth functions into maximizers.

The question that now arises is how minimizers behave under generalized nondegenerate deformations. Section 2.4 is devoted to an investigation of this question.

2.4.2 The Degree of a Mapping and Rotation of a Vector Field

The notion of the degree of a mapping, and the equivalent notion of the rotation of a vector field, are important tools in nonlinear analysis. A detailed account of their construction and properties are beyond the scope of this monograph. Therefore we shall describe this theory only in outline and quote without proof the results concerning rotations of vector fields that we require. We shall only prove Hopf's famous classification theorem, the result which lies at the heart of the constructions used in this section.

Consider the set of bases in $\mathbb{R}^N$, i.e., the set of ordered N-tuples of linearly independent vectors. We divide this set into two equivalence classes, by letting two bases $(a_1, \dots, a_N)$ and $(b_1, \dots, b_N)$ belong to the same class if the determinant $\det C$ of the transition matrix $C = (c_{ij})$ from the basis $(a_1, \dots, a_N)$ to the basis $(b_1, \dots, b_N)$ is positive. An *orientation* in $\mathbb{R}^N$ is an equivalence class of ordered bases.

The *standard orientation* in $\mathbb{R}^N$ is the orientation containing the basis

$$(1, 0, \dots, 0), \quad (0, 1, \dots, 0), \quad (0, 0, \dots, 1) .$$

The orientation of a zero-dimensional manifold is denoted by $+1$ or -1.

A smooth manifold $\mathfrak{M}$ is said to be *oriented* if all of its tangent spaces $T\mathfrak{M}_x$ are consistently oriented. Consistency of orientations is defined as follows: if $\dim \mathfrak{M} = m > 0$, then, for every point $x \in \mathfrak{M}$, there exist a neighborhood U in $\mathfrak{M}$ and a diffeomorphism $h : U \to \mathbb{R}^m$ which preserves the orientation, i.e., for every $y \in U$ the isomorphism h'_y changes the chosen orientation of the space $T\mathfrak{M}_y$ into the standard orientation of the space $\mathbb{R}^m$.

In other words, if $a_1, \dots, a_m$ is a basis in $T\mathfrak{M}_y$ belonging to the chosen orientation of $T\mathfrak{M}_y$, then we require that the basis $h'_y a_1, \dots, h'_y a_m$ belongs to the standard orientation of the space $\mathbb{R}^m$.

Let $\mathfrak{M}$ and $\mathfrak{N}$ be smooth manifolds of dimensions m and n, respectively, with $n \leqslant m$. Consider a smooth mapping $f : \mathfrak{M} \to \mathfrak{N}$ and let C be the set of $x \in \mathfrak{M}$ such that the the derivative

$$f'_x : T\mathfrak{M}_x \to T\mathfrak{N}_x$$

has rank less than n. The points of C are *critical points*, the points of $f(C)$ are *critical values*, and the points of $\mathfrak{N} \setminus f(C)$ are *regular values* of f.

Lemma 2.4.1 (Sard). *The set of all critical values of a smooth mapping* $f : \mathfrak{M} \to \mathfrak{N}$ *has Lebesgue measure zero in* $\mathbb{R}^n$.

Corollary 1 (Brown). *The set of all regular values of a smooth mapping* $f : \mathfrak{M} \to \mathfrak{N}$ *is everywhere dense in* $\mathfrak{N}$.

Let $\mathfrak{M}$ and $\mathfrak{N}$ be N-dimensional oriented manifolds, $\mathfrak{M}$ being compact and $\mathfrak{N}$ connected. Consider a smooth mapping $f : \mathfrak{M} \to \mathfrak{N}$, and suppose that $x \in \mathfrak{M}$ is a regular point of f, i.e.,

$$f'_x : T\mathfrak{M}_x \to T\mathfrak{N}_{f(x)}$$

is a vector-space isomorphism of the oriented vector spaces $T\mathfrak{M}_x$ and $T\mathfrak{N}_{f(x)}$. We set

$$\operatorname{sign} f'_x = \begin{cases} 1 & \text{if the orientations of } T\mathfrak{M}_x \text{ and } T\mathfrak{N}_{f(x)} \text{ are the same },\\ -1 & \text{if the orientations of } T\mathfrak{M}_x \text{ and } T\mathfrak{N}_{f(x)} \text{ are opposite }. \end{cases}$$

The degree $\deg(f;y)$ of $f:\mathfrak{M}\to\mathfrak{N}$ at a regular value y is defined by

$$\deg(f;y) = \sum_{x\in f^{-1}(y)} \operatorname{sign} f'_x .$$

The classical implicit function theorem (see Sect. 1.6.2) implies that the complete preimage $f^{-1}(y)$ of a regular value y of f consists of isolated points. Moreover since $\mathfrak{M}$ is compact, the number of points in $f^{-1}(y)$ is finite: let

$$f^{-1}(y) = \{x_1,\dots,x_k\} .$$

Therefore

$$\deg(f;y) = \sum_{i=1}^{k} \operatorname{sign} f'_{x_i} .$$

The definition shows that $\deg(f;y)$ is an integer and is a locally constant function of y defined on an open subset of the manifold $\mathfrak{N}$ of full measure (by Sard's lemma (2.4.1)).

It turns out that $\deg(f;y)$ is independent of the choice of the regular value y, i.e., it is constant on the set of regular values. The value $\deg(f;y)$ is called the *degree* of $f:\mathfrak{M}\to\mathfrak{N}$ and it is denoted by $\deg f$.

Here are some properties of the degree of a mapping.

Two mappings $f_0:\mathfrak{M}\to\mathfrak{N}$ and $f_1:\mathfrak{M}\to\mathfrak{N}$ are *smoothly homotopic* ($f_0\sim f_1$) if there exists a smooth mapping $f:\mathfrak{M}\times[0,1]\to\mathfrak{N}$ such that

$$f(x;0)=f_0(x),\quad f(x;1)=f_1(x)\quad (x\in\mathfrak{M}) .$$

Homotopy invariance is a fundamental property of the degree of a mapping.

Theorem 2.4.1. *If* $f_0:\mathfrak{M}\to\mathfrak{N}$ *and* $f_1:\mathfrak{M}\to\mathfrak{N}$ *are smoothly homotopic mappings then* $\deg f_0=\deg f_1$.

The following result is a criterion for zero degree.

Theorem 2.4.2. *Suppose that the manifold* $\mathfrak{M}$ *is the boundary of a compact oriented manifold* X *and that the orientation of* $\mathfrak{M}$ *coincides with that of* ∂X. *If* $f:\mathfrak{M}\to\mathfrak{N}$ *is the restriction of the smooth mapping* $F:X\to\mathfrak{N}$ *to* $\mathfrak{M}$, *then* $\deg f=0$.

Let $\mathfrak{M}$, $\mathfrak{N}$, $\mathfrak{U}$ be smooth N-dimensional manifolds, $\mathfrak{M}$ and $\mathfrak{N}$ being compact and $\mathfrak{N}$ and $\mathfrak{U}$ connected. Consider smooth mappings

$$f:\mathfrak{M}\to\mathfrak{N},\quad g:\mathfrak{N}\to\mathfrak{U} .$$

Thus the composite

$$g \circ f : \mathfrak{M} \to \mathfrak{U}$$

is defined.

Theorem 2.4.3. $\deg(g \circ f) = \deg g \cdot \deg f$.

Let f be a diffeomorphism from a manifold $\mathfrak{M}$ onto a manifold $\mathfrak{N}$. Then all values of f are regular and every regular value has just one preimage. Therefore, if f does not change orientation (i.e., $T\mathfrak{M}_x$ and $T\mathfrak{N}_y$ have the same orientation), then $\deg f = 1$. If f reverses orientation, then $\deg f = -1$.

Next we define the degree of a continuous mapping.

Suppose that $\mathfrak{M}$ is a compact smooth N-dimensional manifold and, as usual, write S^N for the N-dimensional unit sphere. Let $f : \mathfrak{M} \to S^N$ be a continuous mapping and $f_\varepsilon : \mathfrak{M} \to S^N$ a smooth ε-approximation of f, i.e., a smooth mapping such that

$$|f(x) - f_\varepsilon(x)| \leqslant \varepsilon \quad (x \in \mathfrak{M}) .$$

If $\varepsilon > 0$ is sufficiently small then $\deg f_\varepsilon$ is independent of ε. Indeed, if f_ε^0 and f_ε^1 are smooth ε-approximations of f, then, for small ε, the mapping

$$f(x;\lambda) = \frac{(1-\lambda)f_\varepsilon^0(x) + \lambda f_\varepsilon^1(x)}{|(1-\lambda)f_\varepsilon^0(x) + \lambda f_\varepsilon^1(x)|} \quad (x \in \mathfrak{M},\ 0 \leqslant \lambda \leqslant 1)$$

is a smooth homotopy connecting f_0 and f_1, and so, by Theorem 2.4.1, the mappings f_ε^0 and f_ε^1 have the same degree.

The common degree of the smooth ε-approximations of a mapping f is called the *degree* of f and is denoted by $\deg f$.

We shall now define the notion of the rotation of a vector field and describe its main properties.

Let M be a subset of $\mathbb{R}^N$. A *vector field* on M is a mapping $\Phi : M \to \mathbb{R}^N$. It is convenient to think of a vector field as associating with each point $x \in M$ a vector $\Phi(x)$ emanating from x. As in Sect. 2.3.3, we may introduce the notions of a *continuous* vector field and a *smooth* vector field. In this section we consider only continuous vector fields.

A point x_0 at which $\Phi(x_0) = 0$ is called a *zero* of Φ. The zeros of Φ are often called *singular points* of Φ, although a singularity at x_0 is in fact a singularity not of Φ but of the related mapping $|\Phi|^{-1}\Phi : M \to S^{N-1}$.

A field Φ is *nondegenerate* on M if $\Phi(x) \neq 0$ for all $x \in M$.

Let Ω be a bounded domain in $\mathbb{R}^N$ with smooth boundary $\partial\Omega$. Then $\partial\Omega$ is a smooth $(N-1)$-dimensional manifold which is smoothly embedded into $\mathbb{R}^N$. In every tangent space $T\partial\Omega_x$ we choose an orientation such that every basis $e_1(x), \ldots, e_{N-1}(x)$ defining the orientation of $T\partial\Omega_x$, together with the outer normal vector $n(x)$ to $\partial\Omega$ at x, defines the standard orientation in $\mathbb{R}^N$, i.e., the determinant of the transition matrix from the basis $e_1(x), \ldots, e_{N-1}(x), n(x)$ to the standard basis $(1, 0, \ldots, 0), \ldots, (0, \ldots, 0, 1)$ is positive.

We orient the unit sphere S^{N-1} in $\mathbb{R}^N$ in the same manner.

Suppose that Φ is a nondegenerate continuous vector field defined on $\partial\Omega$. The *rotation* $\gamma(\Phi;\partial\Omega)$ of Φ on the boundary $\partial\Omega$ of Ω is the degree of the mapping

$$|\Phi|^{-1}\Phi : \partial\Omega \to S^{N-1} .$$

A continuous mapping

$$\Phi : M \times [0,1] \to \mathbb{R}^N$$

is called a *continuous deformation* of a field Φ_0 into a field Φ_1 if

$$\Phi(\cdot;0) = \Phi_0 \quad \text{and} \quad \Phi(\cdot;1) = \Phi_1 .$$

If in addition

$$\Phi(x;\lambda) \neq 0 \quad \text{for } x \in M,\ 0 \leqslant \lambda \leqslant 1 ,$$

then Φ is called a *nondegenerate deformation* or *homotopy* of Φ_0 into Φ_1. A nondegenerate deformation is sometimes called a *homotopy bridge* connecting Φ_0 and Φ_1 on the set M.

The most important property of rotations is their homotopy invariance.

Theorem 2.4.4. *If Φ_0, Φ_1 are fields that are homotopic on $\partial\Omega$, then their rotations are equal; that is,*

$$\gamma(\Phi_0;\partial\Omega) = \gamma(\Phi_1;\partial\Omega) .$$

The following result gives sufficient conditions for a vector field to have at least one zero.

Theorem 2.4.5 (principle of nonzero rotation). *Let Φ be a field defined on the closure $\overline{\Omega}$ of a domain Ω and nondegenerate on its boundary $\partial\Omega$, and suppose that $\gamma(\Phi;\partial\Omega) \neq \varnothing$. Then Φ has at least one zero in Ω.*

Let x_0 be an isolated zero of the field $\Phi : \Omega \to \mathbb{R}^N$. Consider some domain Ω_0 for which $\overline{\Omega}_0 \subset \Omega$, $x_0 \in \Omega_0$ and $\Phi(x) \neq 0$ for $x \in \overline{\Omega}_0 \setminus \{x_0\}$. Then the rotation $\gamma(\Phi;\partial\Omega_0)$ of the field Φ on $\partial\Omega_0$ does not depend on the choice of the domain Ω_0. This common rotation is called the *topological index* $\operatorname{ind}(x_0;\Phi)$ of the zero x_0 of Φ.

Theorem 2.4.6 (Kronecker). *Suppose that Φ is a field defined on the closure $\overline{\Omega}$ of a domain Ω, and that Φ is nondegenerate on $\partial\Omega$ and has finitely many zeros $x_1,\ldots,x_k$ in Ω. Then*

$$\gamma(\Phi;\partial\Omega) = \sum_{i=1}^{k} \operatorname{ind}(x_i;\Phi) .$$

A *linear* vector field is simply a linear mapping $A : \mathbb{R}^N \to \mathbb{R}^N$. We shall often identify a linear field A with its matrix with respect to the standard basis and also denote this matrix by A.

Theorem 2.4.7. *Let A be a linear field. Then the point 0 is an isolated zero of A if and only if $\det A \neq 0$, and then the topological index of the isolated zero of A at 0 is given by*

$$\operatorname{ind}(0; A) = \operatorname{sign}\det A\ .$$

Since

$$\operatorname{sign}\det A = (-1)^{\beta}\ , \tag{2.4.1}$$

where β is the sum of the multiplicities of the negative eigenvalues of the matrix A, Theorem 2.4.7 implies

Theorem 2.4.8. *If A is a linear field with $\det A \neq 0$ then the topological index of the isolated zero at 0 of A is given by*

$$\operatorname{ind}(0; A) = (-1)^{\beta}\ ,$$

where β is the sum of the multiplicities of the negative eigenvalues of the matrix A.

Suppose that the space $\mathbb{R}^N$ is the direct sum of vector subspaces R_0 and R_1, and let $P_0 : \mathbb{R}^N \to R_0$, $P_1 : \mathbb{R}^N \to R_1$ be the corresponding projection operators. Let Ω_0 and Ω_1 be bounded domains in R_0 and R_1. The *product*

$$\Omega = \Omega_0 \times \Omega_1$$

of Ω_0 and Ω_1 is the domain in $\mathbb{R}^N$ consisting of all sums $x = x_0 + x_1$ with $x_0 \in \Omega_0$ and $x_1 \in \Omega_1$.

Let Φ_0 and Φ_1 be continuous fields in R_0 and R_1 defined respectively on $\overline{\Omega}_0$ and $\overline{\Omega}_1$. The *direct sum* $\Phi = \Phi_0 \oplus \Phi_1$ of Φ_0 and Φ_1 is the vector field

$$\Phi(x) = \Phi_0(P_0 x) + \Phi_1(P_1 x)$$

defined on $\overline{\Omega}$.

Theorem 2.4.9. *Let Φ_0 and Φ_1 be nondegenerate fields on $\partial\Omega_0$ and $\partial\Omega_1$, respectively. Then the field $\Phi_0 \oplus \Phi_1$ is nondegenerate on $\partial\Omega$ and*

$$\gamma(\Phi_0 \oplus \Phi_1; \partial\Omega) = \gamma(\Phi_0; \partial\Omega_0) \cdot \gamma(\Phi_1; \partial\Omega_1)\ .$$

2.4.3 Hopf's Theorem

A fundamental property of the degree of a mapping is its homotopy invariance: homotopic mappings have the same degree. Although the converse statement is in general not true, there is a converse assertion for mappings between connected manifolds: if two mappings from a compact connected N-dimensional manifold $\mathfrak{M}$ to a connected N-dimensional manifold $\mathfrak{N}$ have the same degree, then they are homotopic. This remarkable theorem, often called

Hopf's classification theorem, plays a fundamental role in topology. There are several different proofs of this theorem; the one below is, apart from minor changes, virtually identical to Hopf's original proof.

We restrict consideration to mappings from a sphere to itself. The general case differs from this case only in technical details.

Theorem 2.4.10 (Hopf; see [3]). *Let $f_0 : S^N \to S^N$ and $f_1 : S^N \to S^N$ be continuous mappings of the same degree and let $N \geqslant 1$. Then f_0 and f_1 are homotopic on S^N.*

For the proof we require some auxiliary results.

Let M be a set in $\mathbb{R}^{N+1}$ and let $\varphi : M \to \mathbb{R}^{N+1}$ be a continuous mapping. A point $x_0 \in M$ is called a *zero* of φ if $\varphi(x_0) = 0$, and φ is *nondegenerate* on M if it has no zeros on M. We write B^{N+1} for the unit ball in $\mathbb{R}^{N+1}$.

Lemma 2.4.2. *Let $\varphi : S^N \to \mathbb{R}^{N+1}$ be a continuous nondegenerate mapping. Then φ can be extended to a continuous mapping $\Phi : B^{N+1} \to \mathbb{R}^{N+1}$ which has a unique zero.*

Proof. A suitable mapping Φ can be defined by

$$\Phi(x) = |x|\, f(x/|x|) \quad (x \in B^{N+1}) .$$

□

Lemma 2.4.2 implies

Corollary 1. *Let Ω be a domain in $\mathbb{R}^{N+1}$ whose closure $\overline{\Omega}$ is diffeomorphic to B^{N+1}. Then each continuous nondegenerate mapping $\varphi : \partial\Omega \to \mathbb{R}^{N+1}$ can be extended to a continuous mapping $\Phi : \overline{\Omega} \to \mathbb{R}^{N+1}$ which has a unique zero.*

Lemma 2.4.3. *Let T be the spherical layer*

$$T = \left\{x \in \mathbb{R}^{N+1} : \tfrac{1}{2} \leqslant |x| \leqslant 1\right\} .$$

Then each continuous nondegenerate mapping $\varphi : \partial T \to \mathbb{R}^{N+1}$ can be extended to a continuous mapping $\Phi : T \to \mathbb{R}^{N+1}$ which has a unique zero.

Proof. Let $\tilde{\Phi}$ be an arbitrary continuous extension of φ to T. Since φ is nondegenerate on ∂T, for some $\varepsilon \in \left(0, \frac{1}{2}\right)$ we have

$$|\widetilde{\Phi}(x)| > 2\varepsilon \quad (x \in T \setminus T_\varepsilon) ,$$

where

$$T_\varepsilon = \left\{x \in \mathbb{R}^{N+1} : \tfrac{1}{2} + \varepsilon < |x| < 1 - \varepsilon\right\} .$$

Let $\varphi_\varepsilon : T \to \mathbb{R}^{N+1}$ be a smooth ε-approximation for $\widetilde{\Phi}$. By Lemma 2.4.1, there exists an element $h \in \mathbb{R}^{N+1}$ with $|h| < \varepsilon$ such that 0 is a regular value of the mapping $\varphi_\varepsilon + h : T \to \mathbb{R}^{N+1}$. Therefore $\varphi_\varepsilon + h$ has only finitely many zeros $x_1, \dots, x_k$. Since

$$|\varphi_\varepsilon(x) + h| \geqslant |\widetilde{\Phi}(x)| - |\widetilde{\Phi}(x) - \varphi_\varepsilon(x)| - |h| > 2\varepsilon - \varepsilon - \varepsilon = 0$$

for $x \in T \setminus T_\varepsilon$, it follows that $x_i \in T_\varepsilon$ $(i = 1, \dots, k)$.

Since the spherical layer T_ε is connected, there exists an open set $\Omega \subset T_\varepsilon$ containing $\{x_1, \dots, x_k\}$ whose closure is diffeomorphic to the ball B^{N+1}. Consider the restriction g of the mapping $\varphi_\varepsilon + h$ to $\partial\Omega$. By Corollary 1, it can be extended to a continuous mapping $G : \overline{\Omega} \to \mathbb{R}^{N+1}$ which has a unique zero. Then a mapping $\Phi : T \to \mathbb{R}^{N+1}$ with the required properties can be defined by

$$\Phi(x) = \begin{cases} \left(1 + \dfrac{1}{2\varepsilon} - \dfrac{1}{\varepsilon}|x|\right) \varphi(x) + \left(\dfrac{1}{\varepsilon}|x| - \dfrac{1}{2\varepsilon}\right)(\varphi_\varepsilon(x) + h) \\ \qquad\qquad \text{for } \tfrac{1}{2} \leqslant |x| \leqslant \tfrac{1}{2} + \varepsilon\,, \\ \left(\dfrac{1}{\varepsilon} - \dfrac{1}{\varepsilon}|x|\right)(\varphi_\varepsilon(x) + h) + \left(1 - \dfrac{1}{\varepsilon} + \dfrac{1}{\varepsilon}|x|\right)\varphi(x) \\ \qquad\qquad \text{for } 1 - \varepsilon \leqslant |x| \leqslant 1\,, \\ \varphi_\varepsilon(x) + h \quad \text{for } x \in T_\varepsilon \setminus \overline{\Omega}\,, \\ G(x) \qquad\quad \text{for } x \in \overline{\Omega}\,. \end{cases}$$

□

Remark. This construction allows us to prove a more general result, namely, the assertion that a continuous nondegenerate mapping φ defined on the boundary $\partial\Omega$ of a connected domain Ω can be extended to a continuous mapping Φ which is defined on $\overline{\Omega}$ and has a unique zero.

We choose the standard orientation in $\mathbb{R}^{N+1}$ and consider the sphere S^N to be oriented as the boundary of the unit ball B^{N+1}.

In order to prove Hopf's theorem, it suffices to show that all mappings of degree zero are homotopic. Indeed, suppose that any two mappings of degree zero from S^N to itself are homotopic. Consider arbitrary mappings $f_0 : S^N \to S^N$ and $f_1 : S^N \to S^N$ for which $\deg f_0 = \deg f_1$. Define a mapping $\varphi : \partial T \to S^N$ by

$$\varphi(x) = \begin{cases} f_0(x) & \text{for } |x| = 1\,, \\ f_1(2x) & \text{for } |x| = \tfrac{1}{2}\,. \end{cases}$$

By Lemma 2.4.3, the mapping φ can be extended to a continuous mapping $\Phi : T \to \mathbb{R}^{N+1}$ which has a unique zero x_0.

We choose $\rho_0 > 0$ such that $B(\rho_0, x_0) \subset \operatorname{int} T$ and set

$$M = T \setminus B(\rho_0, x_0)\,,$$

$$H(x) = \frac{\Phi(x)}{|\Phi(x)|} \quad (x \in \overline{M}) .$$

Let

$$h_0 : S(\rho_0, x_0) \to S^N$$

be the restriction of H to $S(\rho_0, x_0)$ and

$$h_1 : \partial M \to S^N$$

be the restriction of this mapping to ∂M. Then

$$\deg h_1 = \deg \varphi - \deg h_0 .$$

Since $\deg \varphi = 0$ by construction, it follows that

$$\deg h_1 = -\deg h_0 ,$$

and since $\deg h_1 = 0$ by Theorem 2.4.2, we also have

$$\deg h_0 = 0 .$$

Therefore, by assumption, h_0 is homotopic on $S(\rho_0, x_0)$ to a constant mapping from $S(\rho_0, x_0)$ to a point $y_0 \in S^N$. Let h be an appropriate homotopy (so that $h(\cdot, 0) = h_0$, $h(\cdot, 1) \equiv y_0$), and set

$$F(x) = \begin{cases} H(x) & \text{for } x \in M , \\ h\left(x_0 + \rho_0 \dfrac{x - x_0}{|x - x_0|},\ 1 - \dfrac{|x - x_0|}{\rho}\right) & \text{for } x \in B(\rho_0, x_0) \setminus \{x_0\} , \\ y_0 & \text{for } x = x_0 . \end{cases}$$

We can finally define a homotopy

$$f : S^N \times [0, 1] \to S^N$$

of f_0 into f_1 by setting

$$f(x; \lambda) = F\left(x - \tfrac{1}{2}\lambda x\right) \quad (x \in S^N;\ 0 \leqslant \lambda \leqslant 1) .$$

Now we prove Hopf's theorem, by induction on the dimension N. First suppose that $N = 1$ and that

$$f_0 : S^1 \to S^1, \quad f_1 : S^1 \to S^1$$

are mappings of the same degree. Consider the parametrization

$$S^1 = \{(\cos t, \sin t) : 0 \leqslant t < 2\pi\}$$

of the circle S^1. Suppose that $(\varphi_0(t), \psi_0(t))$ and $(\varphi_1(t), \psi_1(t))$ are the components of f_0 and f_1 respectively at the point $(\cos t, \sin t)$. We can assume without loss of generality that

$$(\varphi_0(0), \psi_0(0)) = (\varphi_1(0), \psi_1(0)) = (1, 0) .$$

We set

$$\Theta_0(t) = \tan^{-1}\frac{\psi_0(t)}{\varphi_0(t)}, \quad \Theta_1(t) = \tan^{-1}\frac{\psi_1(t)}{\varphi_1(t)}$$

and denote by $\theta_0(t), \theta_1(t)$ continuous branches of the multivalued functions Θ_0 and Θ_1 for which

$$\theta_0(0) = \theta_1(0) = 0 .$$

Then we can define a homotopy

$$f(\cdot; \lambda) \quad (0 \leqslant \lambda \leqslant 1)$$

of f_0 into f_1 by

$$f(\cos t, \sin t; \lambda) = (\cos((1-\lambda)\theta_0(t) + \lambda\theta_1(t)), \sin((1-\lambda)\theta_0(t) + \lambda\theta_1(t)))$$

for $0 \leqslant \lambda \leqslant 1$.

Now assume that Hopf's theorem holds in dimension k. We shall prove that it holds in dimension $k+1$.

Consider a mapping $f : S^{k+1} \to S^{k+1}$ for which $\deg f = 0$. To prove Hopf's theorem, it suffices to show that f is homotopic to some mapping $g : S^{k+1} \to S^{k+1}$ such that the image of the sphere S^{k+1} is a proper subset of S^{k+1}. Indeed, each of these mappings is homotopic to some constant mapping $f_0 : S^{k+1} \to x_0 \in S^{k+1}$ and any two constant mappings are homotopic to each other. Furthermore, since f is homotopic to any of its smooth ε-approximations $f_\varepsilon : S^{k+1} \to S^{k+1}$ for small $\varepsilon > 0$, we can assume, without loss of generality, that$f$ is smooth.

Consider a regular value y_* of f and let

$$f^{-1}(y_*) = \{x_1^*, \ldots, x_m^*\} .$$

Since $\deg f = 0$, it follows that m is even and that for exactly half of the points x_i of the preimage $f^{-1}(y_*)$ we have $\operatorname{sign} f'_{x_i^*} = 1$. Write $m = 2l$ and suppose, for definiteness, that

$$\operatorname{sign} f'_{x_1^*} = \cdots = \operatorname{sign} f'_{x_l^*} = 1, \tag{2.4.2}$$

$$\operatorname{sign} f'_{x_{l+1}^*} = \cdots = \operatorname{sign} f'_{x_{2l}^*} = -1. \tag{2.4.3}$$

Let U be an open set containing $f^{-1}(y_*)$ such that $\overline{U}$ is homeomorphic to the ball B^{k+1} and

$$f(\overline{U}) \neq S^{k+1} .$$

Consider points $x_0, x_1 \in S^{k+1}$ such that $x_0 \notin \overline{U}$, $x_1 \notin f(\overline{U})$ and denote by x^0, x^1 the points diametrically opposite to x_0, x_1 on S^{k+1}. Let Π_0, Π_1 be the $(k+1)$-dimensional tangent planes to S^{k+1} at x^0, x^1 and p_0, p_1 be the stereographic projection operators from S^{k+1} to Π_0 and Π_1 respectively, determined by x_0, x_1. We choose coordinate systems $\{u_0^1, \ldots, u_0^{k+1}\}$ and $\{u_1^1, \ldots, u_1^{k+1}\}$ in Π_0 and Π_1 such that the bases $\{n_0; u_0^1, \ldots, u_0^{k+1}\}$ and $\{n_1; u_1^1, \ldots, u_1^{k+1}\}$, where n_0 and n_1 are outer normals to S^{k+1} at x_0 and x_1, belong to the standard orientation of $\mathbb{R}^{k+1}$, and such that the origin in Π_1 coincides with $p_1 y_*$. Let $R : \Pi_0 \to \Pi_1$ be an orientation-preserving isomorphism.

Consider the domain

$$W = V \setminus \bigcup_{i=1}^{2l} B^{k+1}(\rho, y_i^*)$$

in Π_0, where $V = p_0 U$ and $B^{k+1}(\rho, y_i^*)$ are balls in Π_0 of small radius ρ with centers at the points $y_i^* = p_0 x_i^*$. We define a mapping $\varphi : \overline{W} \to S^k$ by

$$\varphi(v) = \frac{R^{-1} \circ p_1 \circ f \circ p_0^{-1}(v)}{\left|R^{-1} \circ p_1 \circ f \circ p_0^{-1}(v)\right|} .$$

By Theorem 2.4.2, the degree of $\varphi : \partial W \to S^k$ is zero:

$$\deg \varphi = 0 . \tag{2.4.4}$$

On the other hand, by the additivity of degree we have

$$\deg \varphi = \deg \varphi_0 - \sum_{i=1}^{2l} \deg \varphi_i , \tag{2.4.5}$$

where φ_0 is the restriction of φ to ∂V and φ_i is the restriction of φ to $S^k(\rho, y_i^*)$. However, for small ρ the mappings

$$\varphi_i : S^k(\rho, y_i^*) \to S^k$$

are diffeomorphisms, and, by (2.4.2) and (2.4.3), these diffeomorphisms preserve orientation for $i = 1, \ldots, l$ and reverse it for $i = l+1, \ldots, 2l$. Therefore

$$\deg \varphi_i = 1 \quad (i = 1, \ldots, l) , \tag{2.4.6}$$

$$\deg \varphi_i = -1 \quad (i = l+1, \ldots, 2l) . \tag{2.4.7}$$

It follows from (2.4.5)–(2.4.7) that

$$\deg \varphi_0 = 0 .$$

Since the closure $\overline{V}$ of V is diffeomorphic to B^{k+1}, we can extend the mapping $\varphi_0 : \partial V \to S^k$ to a continuous mapping $\Phi_0 : \overline{V} \to S^k$. Indeed, let

$$r : \overline{V} \to B^{k+1}$$

be a diffeomorphism. Consider the mapping $\varphi_0 \circ r^{-1} : S^k \to S^k$ identifying S^k with the boundary of B^{k+1}. By Theorem 2.4.3 we have

$$\deg \varphi_0 \cdot r^{-1} = \deg \varphi_0 \cdot \deg r^{-1} = 0 \,.$$

Therefore by the induction hypothesis, $\varphi_0 \cdot r^{-1}$ is homotopic to a constant mapping from S^k to a point $z_0 \in S^k$. Let

$$h : S^k \times [0,1] \to S^k$$

be a homotopy, with

$$h(\cdot, 0) = \varphi_0 \cdot r^{-1}, \quad h(\cdot, 1) \equiv z_0 \,.$$

Then we can define $\Phi_0 : \overline{V} \to S^k$ by

$$\Phi_0(x) = \begin{cases} h\left(\dfrac{r(x)}{|r(x)|}, 1 - |r(x)|\right) & \text{for } r(x) \neq 0 \,, \\ z_0 & \text{for } r(x) = 0 \,. \end{cases}$$

Now we define $g : S^{k+1} \to S^{k+1}$ by

$$g(x) = \begin{cases} f(x) & \text{for } x \in S^{k+1} \setminus U \,, \\ p_1^{-1} \circ R \circ \Phi_0 \circ p_0(x) & \text{for } x \in U \,. \end{cases}$$

Since $y_* \notin g(S^{k+1})$ we have $g(S^{k+1}) \neq S^{k+1}$.

In order to complete the proof of Hopf's theorem, it remains to show that the mappings f and g are homotopic on S^{k+1}. A suitable homotopy can be defined by

$$f(x;\lambda) = \begin{cases} f(x) & \text{for } x \in S^{k+1} \setminus U, \ 0 \leqslant \lambda \leqslant 1 \,, \\ p_1^{-1}((1-\lambda)p_1 f(x) + \lambda p_1 g(x)) & \text{for } x \in U, \ 0 \leqslant \lambda \leqslant 1 \,. \end{cases}$$

□

2.4.4 Parusinski's Theorem

Two gradient vector fields ∇f_0 and ∇f_1 on B^N that are nondegenerate on S^{N-1} are said to be *gradient homotopic* if there exists a C^1-function $f(x;\lambda)$ on $B^N \times [0,1]$ satisfying the conditions

(1) $f(x;0) = f_0(x)$, $f(x;1) = f_1(x)$ $(x \in B^N)$,
(2) $\nabla f(x;\lambda) \neq 0$ $(x \in S^{N-1},\ 0 \leqslant \lambda \leqslant 1)$.

Lemma 2.4.4. *Let $f \in C^1(B^N)$. If the vector field ∇f is nondegenerate on B^N, then this vector field is gradient homotopic a constant vector field.*

Proof. Without loss of generality, we can assume that f is smooth. By Hadamard's lemma [31], we have

$$f(x) = f(0) + \sum_{i=1}^{N} x_i g_i(x) ,$$

where g_i are smooth functions, with

$$\nabla f(0) = (g_1(0), \dots, g_N(0)) .$$

Then the function

$$f(x,t) = f(0) + \sum_{i=1}^{N} x_i g_i(tx)$$

is a homotopy of the vector fields ∇f and $\nabla f(0)$ and the result follows. □

The following statement is the main result of this section.

Theorem 2.4.11 (Parusinski). *If two gradient vector fields on B^N are non-degenerate on S^{N-1} and are homotopic in the class of continuous vector fields, then they are gradient homotopic.*

2.4.5 Preparatory Lemmas

We shall consider the topological spaces

$$\begin{aligned}
\Gamma(N) &= \{v = (v_1, \dots, v_N) \in (C^0(B^N))^N : v(x) \neq 0 \ \text{ for } \ x \in S^{N-1}\} ,\\
\widetilde{\Gamma}(N) &= \{(v,h) \in C^0(S^{N-1}; TS^{N-1}) \times C^0(S^{N-1}) : |v|^2 + |h|^2 \neq 0 \\
&\qquad\qquad\qquad\qquad\qquad\qquad\qquad\qquad \text{for } x \in S^{N-1}\} ,\\
A(N) &= \{f \in C^1(B^N) : \nabla f \in \Gamma(N)\} ,\\
\widetilde{A}(N) &= \{(g,h) \in C^1(S^{N-1}) \times C^0(S^{N-1}) : (\nabla g, h) \in \widetilde{\Gamma}(N)\}
\end{aligned}$$

with topologies induced respectively from the spaces

$$(C^0(B^N))^N,\ C^0(S^{N-1};\ TS^{N-1}) \times C^0(S^{N-1}),$$

$$C^1(B^N),\ C^1(S^{N-1}) \times C^0(S^{N-1}) .$$

(We write $C^0(S^{N-1}; TS^{N-1})$ for the space of continuous sections of the tangent bundle TS^{N-1}, which can be regarded as the set of continuous vector fields defined on S^{N-1} and tangent to S^{N-1} at each point.)

Theorem 2.4.11 is clearly equivalent to the statement that the mapping $\nabla : A(N) \to \Gamma(N)$ induces a bijective mapping from the homotopy group $\pi_0(A(N))$ to $\pi_0(\Gamma(N))$. Write $\mathbf{n}(x)$ for the outer normal vector to S^{N-1} at x.

Lemma 2.4.5. *The mapping* $\varphi : A(N) \to \widetilde{A}(N)$ *defined by*

$$\varphi(f) = \left(f, \frac{\partial f}{\partial n}\right)$$

is a homotopy equivalence.

Proof. Choose a smooth mapping $\rho(t)$ from the interval $[0, 1]$ to itself which is equal to 0 in a neighborhood of zero and to 1 in a neighborhood of unity, and set

$$(\psi(g, h))(x) = \rho(|x|)\left(g\left(\frac{x}{|x|}\right) + (|x| - 1)h\left(\frac{x}{|x|}\right)\right) .$$

One can verify that $\varphi \circ \psi = \mathrm{id}_{\widetilde{A}(N)}$ and that $\psi \circ \varphi$ is linearly homotopic to the identity mapping $\mathrm{id}_{A(N)}$. □

Lemma 2.4.6. *The mapping* $\widetilde{\varphi} : \Gamma(N) \to \widetilde{\Gamma}(N)$ *defined by*

$$\widetilde{\varphi}(v) = (\widetilde{v}, h) ,$$

where

$$v|_{S^{N-1}} = \widetilde{v} + h\mathbf{n} ,$$

is a homotopy equivalence.

Lemma 2.4.6 can be proved using the same method as for Lemma 2.4.5.

2.4.6 Proof of Theorem 2.4.11

Consider the diagram

$$\begin{array}{ccc} A(N) & \xrightarrow{\varphi} & \widetilde{A}(N) \\ \nabla \downarrow & & \downarrow \widetilde{\nabla} \\ \Gamma(N) & \xrightarrow{\widetilde{\varphi}} & \widetilde{\Gamma}(N), \end{array}$$

where the mapping $\widetilde{\nabla}$ is defined by

$$\widetilde{\nabla}((g, h)) = (\nabla g, h) .$$

It is easy to check that that this diagram is commutative. Therefore, to prove Theorem 2.4.2, it suffices to show that $\widetilde{\nabla}$ induces a bijective mapping from $\pi_0(\widetilde{A}(N))$ to $\pi_0(\widetilde{\Gamma}(N))$. We shall establish injectivity by induction on N.

Consider the case $N = 2$. We represent the circle S^1 in the parametric form

$$\{(\cos s, \sin s) : s \in [0, 2\pi]\} .$$

It is easy to verify that the image of $\widetilde{\nabla}$ is the set

$$A' = \left\{(v, h) \in \widetilde{\Gamma}(2) : \int v_t ds = 0\right\},$$

where v_t is the tangent component of the field v on S^1, and that the mapping $\widetilde{\nabla} : \widetilde{A}(2) \to A'$ is a homotopy equivalence. Indeed, the mapping

$$(g, h) \mapsto ((\operatorname{grad} g, h), g(0))$$

is a homeomorphism from $\widetilde{A}(2)$ to $A' \times \mathbb{R}$, with inverse defined by

$$((v, h), c) \mapsto \left(c + \int_0^s v_t \, d\tau, h \right).$$

Thus it suffices to show that the embedding $i : A' \to \widetilde{\Gamma}(2)$ induces a bijective mapping from $\pi_0(A')$ to $\pi_0(\widetilde{\Gamma}(2))$. By Hopf's theorem, the connected components $\widetilde{\Gamma}_i(2)$ of $\widetilde{\Gamma}(2)$ are characterized by the degree of the mapping $v + h\mathbf{n}$:

$$\widetilde{\Gamma}_i(2) = \{(v, h) \in \widetilde{\Gamma}(2) : \deg(v + h\mathbf{n}) = i\} .$$

We begin by proving that for $i \neq 1$ the embedding $A' \cap \widetilde{\Gamma}_i(2) \to \widetilde{\Gamma}_i(2)$ is a homotopy equivalence. It suffices to show that $A' \cap \widetilde{\Gamma}_i(2)$ is a strict deformation retract of $\widetilde{\Gamma}_i(2)$. Let $\{v, h\} \in \widetilde{\Gamma}_i(2)$. We set

$$v_+(s) = \max\{0, v(s)\}, \quad v_-(s) = \min\{0, v(s)\} ,$$

$$c_+(v) = \int_{S^1} v_+(s), \quad c_-(v) = \int_{S^1} v_-(s) \, ds .$$

Since $i \neq 1$ we have $c_+(v) > 0$ and $c_-(v) < 0$. In addition, $c_+(v)$ and $c_-(v)$ depend continuously on v. A retraction with the required properties may be defined by

$$\Phi((v, h), \lambda) = \lambda(v, h) + (1 - \lambda) \left(Lv_+ + L^{-1}v_-, h \right) ,$$

where

$$L = L(v) = \sqrt{\frac{-c_-(v)}{c_+(v)}} .$$

For the case $i = 1$ we shall show that all elements $(v, h) \in A' \cap \widetilde{\Gamma}_1(2)$ are homotopic in A' to $(\sin s, 1)$.

Let $(v_0, h_0) \in A' \cap \widetilde{\Gamma}_1(2)$. Since

$$\int_{S^1} v_0(s) \, ds = 0 ,$$

the function v_0 vanishes on S^1. Without loss of generality, we can assume that $v_0(0) = 0$. Then, by the definition of $\widetilde{\Gamma}(2)$, we have $h_0(0) \neq 0$. Perturbing the element (v_0, h_0) in $A' \cap \widetilde{\Gamma}_1(2)$, we can assume that $(v_0, h_0) = (\sin s, 1)$ in a sufficiently small neighborhood of zero.

Consider the vector fields

$$F_0(s) = \sin s + i\ ,$$
$$F_1(s) = v_0(s) + ih_0(s)$$

on the unit circle of the complex plane. Since $i = 1$, the rotation of these fields is zero, and therefore the fields

$$\Phi_0(s) = \ln F_0(s)\ ,$$
$$\Phi_1(s) = \ln F_1(s)$$

are defined; in both cases we choose the branch of the logarithm such that

$$\Phi_0(0) = \Phi_1(0) = \frac{\pi i}{2}\ .$$

Consider the mapping

$$\Psi(s, \lambda) = \exp\{\lambda\Phi_0(s) + (1 - \lambda)\Phi_1(s)\} = \alpha(s, \lambda) + i\beta(s, \lambda)\ .$$

The mapping

$$\widetilde{\Psi}(s, \lambda) = \alpha(s, \lambda)\boldsymbol{\tau} + \beta\mathbf{n}$$

is a homotopy of the elements (v_0, h_0) and $(\sin s, 1)$ which lies in the set

$$\big\{(v, h) \in \widetilde{\Gamma}_1(2) : \min_{S^1} v < 0,\ \max_{S^1} v > 0\big\}\ .$$

To construct a homotopy lying in A', it suffices to repeat the arguments given in the case $i \neq 1$.

Now we establish the induction step. Let $(g, h) \in \widetilde{A}(N)$. We can assume, without loss of generality, that g is a smooth function and that all critical points of g are nondegenerate. Denote by P the set of critical points of g at which h is positive and by Q the set of critical points of g at which h is negative.

We shall calculate the degree of the mapping $\nabla g + h\mathbf{n}$ on S^{N-1}. Consider the spherical layer

$$U = \{x \in \mathbb{R}^N : 1 \leqslant |x| \leqslant 2\}\ ,$$

and the vector field

$$w(x) = v\left(\frac{x}{|x|}\right) - 2c(|x| - 1)n\left(\frac{x}{|x|}\right),$$

on U, where

$$v = \nabla g + h\mathbf{n}, \quad c = \max_{x \in S^{N-1}} h(x)\ .$$

If $c \leqslant 0$, then $\deg v = (-1)^N$ and $P = \varnothing$. If $c > 0$, then $w(x)$ has finitely many zeros in U which satisfy the conditions

$$\frac{x}{|x|} \in P, \quad |x| = \frac{h(x/|x|))}{2c} + 1 \, .$$

The index of each of these zeros is equal to $(-1)^{i(x/|x|)+1}$, where $i(x/|x|)$ is the Morse index of the critical point $x/|x|$ of g. Since the rotation of w on $S^{N-1}(2)$ is equal to $(-1)^n$, it follows that

$$\deg(\nabla g + h\mathbf{n}) + (-1)^n + \sum_{p \in P} (-1)^{i(p)} \, . \tag{2.4.8}$$

Suppose that (g_1, h_1), $(g_2, h_2) \in \widetilde{A}(N)$ are homotopic on $\widetilde{\Gamma}(N)$. Then

$$\deg(\nabla g_1 + h_1 n) = \deg(\nabla g_2 + h_2 n) \, . \tag{2.4.9}$$

Now we reduce (g_1, h_1) and (g_2, h_2) to a simpler form. Let $\varkappa$ be an isotopy of S^{N-1} that maps P into the set

$$\{x \in \mathbb{R}^N : x_N \geqslant \tfrac{1}{2}\}$$

and Q into the set

$$\{x \in \mathbb{R}^N : x_N \leqslant -\tfrac{1}{2}\} \, .$$

Such an isotopy exists when $N \geqslant 3$. Then (g_1, h_1) is homotopic to

$$(g_1(\varkappa_1^{-1}(x, 1)), \, h_1(\varkappa_1^{-1}(x, 1)))$$

in $\widetilde{A}(N)$, and (g_2, h_2) is homotopic to a similar element

$$(g_2(\varkappa_2^{-1}(x, 1)), \, h_2(\varkappa_2^{-1}(x, 1))) \, .$$

After this transformation, the linear homotopy

$$\lambda h(x) + (1 - \lambda) x_N \quad (0 \leqslant \lambda \leqslant 1)$$

becomes nondegenerate, and therefore we can assume that

$$h_1(x) = h_2(x) = x_N \, .$$

Let $P(1)$ and $P(2)$ be the sets of critical points of g_1 and g_2 at which $x_N > 0$. By (2.4.8) and (2.4.9), we have

$$\sum_{p \in P(1)} (-1)^{i(p)} = \sum_{p \in P(2)} (-1)^{i(p)} \, . \tag{2.4.10}$$

Let γ be an arbitrary diffeomorphism of B^{N-1} and the southern hemisphere S_-^{N-1} of S^{N-1}. It follows from (2.4.10) that the vector fields $\nabla(g_2 \circ \gamma)$ and

$\nabla(g_1 \circ \gamma)$ are homotopic in $\Gamma(N-1)$ and, by induction, are gradient homotopic. Suppose that the function

$$G : B^{N-1} \times [0,1] \to \mathbb{R}$$

realizes this homotopy. We consider a function on the set

$$(S_-^{N-1} \times [0,1]) \cup (S^{N-1} \times \{0,1\})$$

which agrees with $G \circ \gamma^{-1}$ on $S_-^{N-1} \times [0,1]$, with g_1 on $S^{N-1} \times \{0\}$, and with g_2 on $S^{N-1} \times \{1\}$. This function can be smoothly extended to a function $\widetilde{G}$ defined on $S^{N-1} \times [0,1]$, and it is easy to verify that $(\widetilde{G}; x_n)$ is a homotopy of (g_1, h_1) and (g_2, h_2) in $\widetilde{A}(N)$.

We have now proved that the mapping

$$\pi_0(\widetilde{\nabla}) : \pi_0(\widetilde{A}(N)) \to \pi_0(\widetilde{\Gamma}(N)) \tag{2.4.11}$$

is injective; to complete the proof of Theorem 2.4.2, we shall show that it is surjective. By Lemmas 2.4.5 and 2.4.6, it suffices to show that for each integer d and for $N \geqslant 2$ there exists a function $f \in A(N)$ satisfying the condition

$$\deg(\nabla f|_{S^{N-1}}) = d\,. \tag{2.4.12}$$

For $N = 2,\ d = 1$, the function

$$f_1(x_1, x_2) = x_1^2 + x_2^2$$

satisfies condition (2.4.12). For $N = 2$, $d \neq 1$, the function

$$f_d(x_1, x_2) = \rho\left(\sqrt{x_1^2 + x_2^2}\right) \Re((x_1 + ix_2)^{1-d})\,,$$

where $\rho : [0,1] \to [0,1]$ is a smooth mapping equal to 0 in a neighborhood of 0 and to 1 in a neighborhood of 1, satisfies condition (2.4.11). For $N > 2$ we set

$$f(x_1, \ldots, x_N) = f_d(x_1, x_2) + x_3^2 + \cdots + x_N^2\,.$$

This completes the proof of Theorem 2.4.11. $\square$

3 Infinite-Dimensional Problems

This chapter is devoted to generalizations of the results of Chapter 2 to infinite-dimensional problems of various types.

3.1 Deformations of Functionals on Hilbert Spaces

In this section, we establish a deformation principle for minimizers of functionals on Hilbert spaces.

3.1.1 H-Regular Functionals

Let H be a real separable Hilbert space with inner product $(\cdot,\cdot)$. We say that a continuously Fréchet differentiable functional f on H is *H-regular* if its gradient $\nabla f : H \to H$ is locally Lipschitzian and possesses the following *property* $(S)_+$: if a sequence (x_n) in H converges weakly to a point x_0 and

$$\overline{\lim_{n\to\infty}}\,(\nabla f(x_n), x_n - x_0) \leqslant 0\,, \tag{3.1.1}$$

then

$$\lim_{n\to\infty} \|x_n - x_0\| = 0\,. \tag{3.1.2}$$

3.1.2 The Deformation Principle for Minimizers

Recall that $x_* \in H$ is a *critical point* of a functional f if

$$\nabla f(x_*) = 0\,.$$

Let f_0, f_1 be H-regular functionals, and suppose that x_0 is a critical point of f_0 and x_1 is a critical point of f_1. We say that a one-parameter family of H-regular functionals $f(\cdot;\lambda)$ $(0 \leqslant \lambda \leqslant 1)$ is a *nondegenerate deformation* of $f_0 = f(\cdot;0)$ into $f_1 = f(\cdot;1)$ if

(1) both f and ∇f are continuous, the gradient $\nabla f = \nabla f(x;\lambda)$ being uniformly continuous in λ with respect to x in every ball

$$B(\rho) = \{x \in H : \|x\| \leqslant \rho\}\,,$$

(2) for every $\lambda \in [0,1]$ the functional $f(\cdot;\lambda)$ has a unique critical point $x(\lambda)$ which depends continuously on λ.

Theorem 3.1.1. *Suppose that f_0, f_1 are H-regular functionals, with critical points x_0, x_1 and that there exists a nondegenerate deformation $f(\cdot;\lambda)$ of f_0 to f_1. If x_0 is a local minimizer for f_0 then x_1 is a local minimizer for f_1.*

3.1.3 Preparatory Lemmas

Lemma 3.1.1. *If f is an H-regular functional then f is weakly lower semicontinuous.*

Proof. If we assume the contrary, then there exist a point x_0 and a sequence (x_n) which weakly converges to x_0 such that

$$\lim_{n\to\infty} f(x_n) < f(x_0)\ . \tag{3.1.3}$$

We apply the finite increments formula (see Sect. 1.5.2)

$$f(x_n) - f(x_0) = (\nabla f(x_0 + \theta_n(x_n - x_0)), x_n - x_0) \quad (0 < \theta_n < 1) \tag{3.1.4}$$

to the difference $f(x_n) - f(x_0)$. From (3.1.3) and (3.1.4) we have

$$\lim_{n\to\infty} (\nabla f(x_0 + \theta_n(x_n - x_0)), x_n - x_0) < 0\ .$$

This inequality and the $(S)_+$-property of the gradient ∇f imply that

$$\lim_{n\to\infty} \theta_n \|x_n - x_0\| = 0\ .$$

Therefore, by (3.1.4),

$$\lim_{n\to\infty} f(x_n) = f(x_0)\ ,$$

and this contradicts (3.1.3). □

Since Weierstrass's theorem holds for weakly lower semicontinuous functionals on weakly compact sets, Lemma 3.1.1 implies

Lemma 3.1.2. *Each H-regular functional f attains its greatest lower bound on every ball $B(\rho)$.*

Lemma 3.1.3. *Suppose that 0 is an absolute strict minimizer for an H-regular functional f on some ball $B(\rho)$. Then*

$$\inf_{x\in\partial B(\rho)} f(x) > f(0)\ . \tag{3.1.5}$$

Proof. If the contrary holds, then there exists a sequence (x_n) of points in $\partial B(\rho)$ for which

$$\lim_{n\to\infty} f(x_n) = f(0)\ . \tag{3.1.6}$$

Since $B(\rho)$ is weakly compact, we can assume without loss of generality that (x_n) weakly converges to a point $x_0 \in B(\rho)$. Therefore, since f is weakly lower semicontinuous, we have

$$\lim_{n\to\infty} f(x_n) \geqslant f(x_0) \,. \tag{3.1.7}$$

From (3.1.6) and (3.1.7) we have $f(x_0) \leqslant f(0)$. However, 0 is a strict minimizer for f on $B(\rho)$, and so $x_0 = 0$.

Applying the finite increments formula (1.5.1) to the difference $f(x_n) - f(\frac{1}{2}x_n)$, we obtain

$$f(x_n) - f(\tfrac{1}{2}x_n) = \tfrac{1}{2}(\nabla f(\theta_n x_n), x_n) \,, \tag{3.1.8}$$

where

$$\tfrac{1}{2} < \theta_n < 1 \,. \tag{3.1.9}$$

Since

$$\overline{\lim_{n\to\infty}} \, (f(x_n) - f(\tfrac{1}{2}x_n)) \leqslant 0 \,, \tag{3.1.10}$$

(3.1.8) and (3.1.10) imply that

$$\overline{\lim_{n\to\infty}} \, (\nabla f(v_n), v_n) \leqslant 0 \,,$$

where $v_n = \theta_n x_n$. Since the sequence (v_n) is weakly convergent to 0, we have

$$\lim_{n\to\infty} \|v_n\| = 0$$

by the $(S)_+$-property of the gradient ∇f. Therefore

$$\lim_{n\to\infty} \|x_n\| = 0$$

by (3.1.9), and this is a contradiction because the points x_n lie in $\partial B(\rho)$. □

3.1.4 Proof of the Main Theorem

We can assume without loss of generality that

$$f(0; \lambda) = 0 \quad (0 \leqslant \lambda \leqslant 1) \,,$$

$$x(\lambda) = 0 \quad (0 \leqslant \lambda \leqslant 1) \,.$$

Let Λ be the set of $\lambda \in [0, 1]$ for which 0 is a local minimizer for $f(\cdot; \lambda)$. The set Λ is nonempty; we show next that it is open.

Let $\lambda_0 \in \Lambda$. Then 0 is an absolute strict minimizer for $f(\cdot; \lambda_0)$ on some ball $B(\rho_0)$. By Lemma 3.1.3,

$$\inf_{x\in\partial B(\rho_0)} f(x; \lambda_0) > 0 \,.$$

Since $f(\cdot;\lambda)$ is uniformly continuous in λ with respect to $x \in B(\rho_0)$, we have

$$\inf_{x\in\partial B(\rho_0)} f(x;\lambda) > 0 \tag{3.1.11}$$

for $\lambda \in (\lambda_0 - \varepsilon, \lambda_0 + \varepsilon)$ provided that $\varepsilon > 0$ is sufficiently small. Let $x_*(\lambda)$ be an absolute minimizer for $f(\cdot;\lambda)$ on $B(\rho_0)$. Since

$$f(x_*(\lambda);\lambda) \leqslant 0 ,$$

we have

$$\|x_*(\lambda)\| < \rho_0$$

by (3.1.11). Thus

$$\nabla f(x_*(\lambda);\lambda) = 0$$

and so $x_*(\lambda) = 0$. Therefore Λ is open.

Now we show that Λ is closed. Write

$$M = \sup_{x\in B, 0\leqslant\lambda\leqslant 1} \|\nabla f(x,\lambda)\| \tag{3.1.12}$$

and

$$m = \inf_{1/2\leqslant\|x\|\leqslant 1, 0\leqslant\lambda\leqslant 1} \|\nabla f(x,\lambda)\| . \tag{3.1.13}$$

Since the gradient $\nabla f(\cdot;\lambda)$ satisfies a Lipschitz condition on B for every $\lambda \in [0,1]$ and is uniformly continuous in λ with respect to $x \in B$, it follows that $M < \infty$.

We claim that $m > 0$. If this is not true, then there exist a sequence (λ_n) in $[0,1]$ and a sequence (x_n) of points with $\frac{1}{2} \leqslant \|x_n\| \leqslant 1$ for which

$$\lim_{n\to\infty} \|\nabla f(x_n;\lambda_n)\| = 0 . \tag{3.1.14}$$

We can assume without loss of generality that (x_n) weakly converges to a point $x_0 \in B$ and that (λ_n) converges to a number $\lambda_0 \in [0,1]$. By (3.1.14) we have

$$\lim_{n\to\infty} (\nabla f(x_n;\lambda_0), x_n - x_0) = 0 .$$

Therefore, by the $(S)_+$-property of the gradient $\nabla f(\cdot;\lambda_0)$, the sequence (x_n) strongly converges to x_0. Hence

$$\nabla f(x_0;\lambda_0) = \lim_{n\to\infty} \nabla f(x_n;\lambda_n) = 0 ,$$

and so $x_0 = 0$. On the other hand,

$$\tfrac{1}{2} \leqslant \|x_0\| \leqslant 1 .$$

This is a contradiction, and we conclude that $m > 0$.

Now set

$$\rho = \frac{m}{4M} \,. \tag{3.1.15}$$

We show that for each $\lambda_0 \in \Lambda$ the point 0 is an absolute minimizer for $f(\cdot;\lambda_0)$ on $B(\rho)$.

Since 0 is a strict local minimizer for $f(\cdot;\lambda_0)$, this point is an asymptotically stable state of equilibrium of the differential equation

$$\frac{dx}{dt} = -\nabla f(x;\lambda_0)\,. \tag{3.1.16}$$

This follows from Theorem 1.6.4 and the inequality

$$\inf_{r\leqslant x\leqslant 1} \|\nabla f(x;\lambda_0)\| > 0\,, \tag{3.1.17}$$

which holds for every $r \in (0,1]$; in turn, (3.1.17) follows from the $(S)_+$-property of $\nabla f(\cdot;\lambda_0)$.

We consider the differential equation (3.1.16) on the open unit ball $\mathring{B}$. Let Ω be the *attraction set* of the zero state of equilibrium, i.e., the set of x for which

$$\lim_{t\to\infty} \|p(t,x)\| = 0\,,$$

and let $\mathring{B}(r)$ be the maximal open ball lying in Ω. To prove that 0 is an absolute minimizer for $f(x;\lambda_0)$ on $B(\rho)$, it suffices to prove that

$$r \geqslant \rho\,. \tag{3.1.18}$$

This inequality is clear if $r \geqslant \frac{1}{2}$. Suppose that $r < \frac{1}{2}$. Consider a sequence (x_n) of points in $\mathring{B}(r)$ and a sequence (y_n) of points not in Ω such that

$$\lim_{n\to\infty} \|x_n - y_n\| = 0\,.$$

From the definition of Ω, there is a sequence (t_n) satisfying

$$\tfrac{13}{16} \leqslant \|p(t_n,y_n)\| \leqslant \tfrac{15}{16} \quad (n = 1,2,\ldots)\,.$$

Since Ω contains a neighborhood of 0, we can assume without loss of generality that

$$t_n = t_0 \quad (n = 1,2,\ldots)\,.$$

Then, as $p(t_0,\cdot)$ is uniformly continuous, we have

$$\tfrac{3}{4} \leqslant \|p(t_0,x_n)\| \leqslant 1$$

for all sufficiently large n. Let x be a term of the sequence (x_n) for which this inequality holds.

We have $x \in \Omega$ by construction. Therefore

$$\lim_{t\to\infty} \|p(t,x)\| = 0 .$$

Consequently, there exist t_1 and t_2 such that $t_0 < t_1 < t_2$,

$$\|p(t_1,x)\| = \tfrac{3}{4} , \quad \|p(t_2,x)\| = \tfrac{1}{2} ,$$

$$\tfrac{1}{2} \leqslant \|p(t,x)\| \leqslant \tfrac{3}{4} \quad (t_1 \leqslant t \leqslant t_2) .$$

It is clear that

$$f(x;\lambda_0) - f(0;\lambda_0) \leqslant M\|x\| < Mr . \tag{3.1.19}$$

On the other hand,

$$\begin{aligned} f(x;\lambda_0) - f(0;\lambda_0) &> f(p(t_1,x);\lambda_0) - f(p(t_2,x);\lambda_0) \\ &= \int_{t_1}^{t_2} \|\nabla f(p(\tau,x);\lambda_0)\|^2 \, d\tau . \end{aligned} \tag{3.1.20}$$

We change the variable of integration to the length s of the curve $p(t,x)$ $(t_1 \leqslant t \leqslant t_2)$. Thus

$$s(t) = \int_{t_1}^{t} \|p'_\tau(\tau,x)\| \, d\tau ,$$

and

$$\int_{t_1}^{t_2} \|\nabla f(p(\tau,x);\lambda_0)\|^2 \, d\tau = \int_0^l \|\nabla f(p(\tau^{-1}(s),x);\lambda_0)\| \, ds \geqslant ml ,$$

where

$$l = s(t_2) > \|p(t_1,x) - p(t_2,x)\| .$$

Consequently,

$$f(x;\lambda_0) - f(0;\lambda_0) \geqslant m\|p(t_1,x) - p(t_2,x)\| \geqslant \tfrac{1}{4}m . \tag{3.1.21}$$

The inequalities (3.1.19) and (3.1.21) now give

$$r > \frac{m}{4M} .$$

We have thus shown that for all $\lambda \in \Lambda$ the point 0 is an absolute minimizer for $f(\cdot;\lambda)$ on $B(\rho)$, where ρ is defined by (3.1.15).

Now we prove that Λ is closed. Let (λ_n) be a sequence in Λ and suppose that $\lambda_n \to \lambda_0$. For each point $x_0 \in B(\rho)$ we have

$$f(x_0;\lambda_n) \geqslant f(0;\lambda_n) , \tag{3.1.22}$$

and passing to the limit as $n \to \infty$ we obtain

$$f(x_0; \lambda_0) \geqslant f(0; \lambda_0) .$$

Therefore 0 is a local minimizer for $f(\cdot; \lambda_0)$. Consequently, $\lambda_0 \in \Lambda$. It follows that Λ is closed; and as usual we conclude that $\Lambda = [0, 1]$. Therefore $1 \in \Lambda$ and the theorem is proved. □

Corollary 1. *Suppose that f_0, f_1 are H-regular functionals and that for all $x \neq 0$ the gradients $\nabla f_0(x)$ and $\nabla f_1(x)$ of f_0 and f_1 are not oppositely directed. If 0 is a local minimizer for f_0 then 0 is a local minimizer for f_1.*

To prove this corollary, it suffices to set

$$f(x; \lambda) = (1 - \lambda) f_0(x) + \lambda f_1(x)$$

and use Theorem 3.1.1.

Condition $(S)_+$ always holds if H is a finite-dimensional space. Therefore Theorem 3.1.1 generalizes Theorem 2.1.1.

Remark. Theorem 3.1.1 has an analog for H-regular functionals defined on a proper subset Ω of a Hilbert space H. The corresponding definition of a nondegenerate deformation and the formulation of the result are obvious. In particular, for the invariance of a local minimizer under deformations of H-regular functionals, we can relax the requirement that the critical point $x(\lambda)$ of each functional $f(\cdot, \lambda)$ be unique: it is sufficient that $x(\lambda)$ is isolated in H uniformly with respect to $\lambda \in [0, 1]$. It should also be pointed out that under the conditions of Theorem 3.1.1 the critical point $x(\lambda)$ is a local minimizer of each functional $f(\cdot; \lambda)$ $(0 \leqslant \lambda \leqslant 1)$.

3.1.5 The Property of H-Regularity

The assumption that the functionals under consideration are H-regular is essential for Theorem 3.1.1 to hold in infinite-dimensional spaces. For instance, this theorem breaks down even for deformations of quadratic functionals if the functionals are not H-regular. Here is an example.

Let $W_2^1(\mathbb{R})$ be the Hilbert space of all absolutely continuous functions $x : \mathbb{R} \to \mathbb{R}$ satisfying

$$\int_{-\infty}^{\infty} x^2(s)\, ds + \int_{-\infty}^{\infty} (x'(s))^2\, ds < \infty ,$$

with inner product defined by

$$(x, y) = \int_{-\infty}^{\infty} x(s) y(s)\, ds + \int_{-\infty}^{\infty} x'(s) y'(s)\, ds .$$

We define a one-parameter family of functionals in $W_2^1(\mathbb{R})$ by

$$f(x;\lambda) = \frac{1}{2}\int_{-\infty}^{\infty} ((x'(s))^2 - \lambda x^2(s))\,ds \quad (0 \leqslant \lambda \leqslant 1)\,.$$

The zero function is clearly a minimizer for $f(x;0)$ but not for $f(x;1)$. We claim that, for each $\lambda \in [0,1]$, this function is the only critical point of $f(x;\lambda)$. Indeed, the Euler equation for $f(x;\lambda)$ has the form

$$x'' + \lambda x = 0\,.$$

The solutions of this equation are the functions of the family

$$C_1 \sin\sqrt{\lambda}(t + C_2)$$

if $\lambda > 0$ and of the family

$$C_1 + C_2 t$$

if $\lambda = 0$. In each of these families the zero function is the only function belonging to $W_2^1(\mathbb{R})$.

3.2 Deformations of Functionals on Banach Spaces

In this section we establish that local minimizers are invariant under nondegenerate deformations of smooth functionals on reflexive Banach spaces.

3.2.1 E-Regular Functionals

Let E be a real separable reflexive Banach space. We shall say that a continuously Fréchet differentiable functional f on E is *E-regular* if its gradient $\nabla f : E \to E^*$ is bounded on bounded sets and satisfies the following condition: if the sequence (x_n) in E weakly converges to x_0 and

$$\overline{\lim_{n\to\infty}}\, \langle \nabla f(x_n), x_n - x_0\rangle \leqslant 0\,, \tag{3.2.1}$$

then

$$\lim_{n\to\infty} \|x_n - x_0\| = 0\,. \tag{3.2.2}$$

A one-parameter family of E-regular functionals $f(\cdot;\lambda)$ $(0 \leqslant \lambda \leqslant 1)$ is called a *nondegenerate deformation* of $f_0 = f(\cdot;0)$ into $f_1 = f(\cdot;1)$ if

(1) both $f(\cdot;\cdot)$ and the gradient $\nabla f(\cdot;\cdot)$ are continuous on $E \times [0,1]$, and $\nabla f(\cdot;\cdot)$ is uniformly continuous in λ with respect to x from every bounded set,

(2) for every $\lambda \in [0,1]$, the functional $f(\cdot;\lambda)$ has a unique critical point $x(\lambda)$ which depends continuously on λ.

Theorem 3.2.1. *Suppose that f_0, f_1 are E-regular functionals and that there exists a nondegenerate deformation $f(\cdot;\lambda)$ of f_0 into f_1. If $x_0 = x(0)$ is a local minimizer for f_0 then $x_1 = x(1)$ is a local minimizer for f_1.*

3.2.2 Preparatory Lemmas

Lemma 3.2.1. *If f is an E-regular functional then f is weakly lower semi-continuous on E.*

Lemma 3.2.2. *Let $x = 0$ be an absolute strict minimizer of an E-regular functional f on some ball $B(\rho)$. Then*

$$\inf_{\|x\|=\rho} f(x) > f(0) . \tag{3.2.3}$$

The proofs of Lemmas 3.2.1 and 3.2.2 are similar to the proofs of the corresponding results for functionals on Hilbert spaces (Lemmas 3.1.1 and 3.1.3).

Let

$$E_1 \subset E_2 \subset \cdots \subset E_n \subset \cdots$$

be an an increasing sequence of finite-dimensional subspaces of E such that

$$\overline{\bigcup_{n\geqslant 1} E_n} = E . \tag{3.2.4}$$

Let $f : E \times [0,1] \to \mathbb{R}$ be a nondegenerate deformation of an E-regular functional f_0 into an E-regular functional f_1 and let

$$x(\lambda) = 0 \quad (0 \leqslant \lambda \leqslant 1) .$$

Write f_n for the restriction of f to $E_n \times [0,1]$ for each n.

Lemma 3.2.3. *Let*

$$T_n(\rho) = T(\rho, 1) \cap E_n .$$

Then there exists $n(\rho) \in \mathbb{N}$ for which

$$\inf_{x \in T_n(\rho), n \geqslant n(\rho), 0 \leqslant \lambda \leqslant 1} \|\nabla f_n(x; \lambda)\| > 0 . \tag{3.2.5}$$

Proof. Suppose that the assertion is false. Then, for some $\rho \in (0,1]$, there exist a sequence (x_k) in $T_{n_k}(\rho)$ and a sequence (λ_k) in $[0,1]$ such that

$$\lim_{k\to\infty} \|\nabla_x f_{n_k}(x_k, \lambda_k)\| = 0 . \tag{3.2.6}$$

Since $\|x_k\| \leqslant 1$, we can assume without loss of generality that (x_k) weakly converges to some point x_0 with $\|x_0\| \leqslant 1$; we can also assume that $(\lambda_k) \to \lambda_0$.

Consider points $v_k \in E_{n_k}$ for which

$$\lim_{k\to\infty} \|v_k - x_0\| = 0 . \tag{3.2.7}$$

Then

$$\begin{aligned}
&\langle \nabla f(x_k;\lambda_0), x_k - x_0\rangle \\
&\quad = \langle \nabla f(x_k;\lambda_0) - \nabla f(x_k;\lambda_k), x_k - x_0\rangle + \langle \nabla f(x_k;\lambda_k), x_k - x_0\rangle \\
&\quad = \langle \nabla f(x_k;\lambda_0) - \nabla f(x_k;\lambda_k), x_k - x_0\rangle + \langle \nabla f_{n_k}(x_k;\lambda_k), x_k - v_k\rangle \\
&\qquad +\langle \nabla f(x_k;\lambda_k), v_k - x_0\rangle \;.
\end{aligned} \tag{3.2.8}$$

Since

$$\begin{aligned}
&\lim_{k\to\infty} |\langle \nabla f(x_k;\lambda_0) - \nabla f(x_k;\lambda_k), x_k - x_0\rangle| \\
&\quad \leqslant \lim_{k\to\infty} \|\nabla f(x_k;\lambda_0) - \nabla f(x_k;\lambda_k)\| \cdot \|x_k - x_0\| \\
&\quad \leqslant 2 \lim_{k\to\infty} \|\nabla f(x_k;\lambda_0) - \nabla f(x_k;\lambda_k)\| = 0 \;,
\end{aligned} \tag{3.2.9}$$

$$\begin{aligned}
&\lim_{k\to\infty} |\langle \nabla f_{n_k}(x_k;\lambda_k), x_k - v_k\rangle| \\
&\quad \leqslant \lim_{k\to\infty} \|\nabla f_{n_k}(x_k;\lambda_k)\| \cdot \|x_k - v_k\| \\
&\quad \leqslant 2 \lim_{k\to\infty} \|\nabla f_{n_k}(x_k;\lambda_k)\| = 0 \;,
\end{aligned} \tag{3.2.10}$$

and

$$\begin{aligned}
&\lim_{k\to\infty} |\langle \nabla f(x_k;\lambda_k), v_k - x_0\rangle| \\
&\quad \leqslant \lim_{k\to\infty} \|\nabla f(x_k;\lambda_k)\| \cdot \|v_k - x_0\| = 0 \;,
\end{aligned} \tag{3.2.11}$$

we conclude from (3.2.8) that

$$\lim_{k\to\infty} \langle \nabla f(x_k;\lambda_0), x_k - x_0\rangle = 0 \;. \tag{3.2.12}$$

By (3.2.12) and the $(S)_+$-property of the gradient $\nabla f(x;\lambda_0)$, the sequence x_k strongly converges to x_0.

Let $h \in E$ and let (h_k) be a sequence with $h_k \in E_{n_k}$ that strongly converges to h. Then

$$\begin{aligned}
&|\langle \nabla f(x_0;\lambda_0), h\rangle| \leqslant \lim_{k\to\infty} |\langle \nabla f(x_0;\lambda_0) - \nabla f(x_k;\lambda_k), h\rangle| \\
&\quad + \lim_{k\to\infty} |\langle \nabla f(x_k;\lambda_k), h - h_k\rangle| + \lim_{k\to\infty} |\langle \nabla f(x_k;\lambda_k), h_k\rangle| \;.
\end{aligned} \tag{3.2.13}$$

Since

$$\begin{aligned}
&\lim_{k\to\infty} |\langle \nabla f(x_0;\lambda_0) - \nabla f(x_k;\lambda_k), h\rangle| \\
&\quad \leqslant \lim_{k\to\infty} \|\nabla f(x_0;\lambda_0) - \nabla f(x_k;\lambda_0)\| \cdot \|h\| \\
&\qquad + \lim_{k\to\infty} \|\nabla f(x_k;\lambda_0) - \nabla f(x_k;\lambda_k)\| \cdot \|h\| = 0 \;,
\end{aligned} \tag{3.2.14}$$

$$\begin{aligned}
&\lim_{k\to\infty} |\langle \nabla f(x_k;\lambda_k), h - h_k\rangle| \\
&\quad \leqslant \lim_{k\to\infty} \|\nabla f(x_k;\lambda_k)\| \cdot \lim_{k\to\infty} \|h - h_k\| = 0 \;,
\end{aligned} \tag{3.2.15}$$

and

$$\begin{aligned}\lim_{k\to\infty} |\langle \nabla f(x_k;\lambda_k), h_k\rangle| &= \lim_{k\to\infty} |\langle \nabla f_{n_k}(x_k;\lambda_k)\,, h_k\rangle| \\ &\leqslant \lim_{k\to\infty} \|\nabla f_{n_k}(x_k;\lambda_k)\| \cdot \|h_k\| = 0\,,\end{aligned} \tag{3.2.16}$$

we conclude from (3.2.13) that

$$\langle \nabla f(x_0;\lambda_0), h\rangle = 0\,.$$

Therefore

$$\nabla f(x_0;\lambda_0) = 0\,,$$

and $x_0 = 0$. On the other hand, $x_0 \in T(\rho, 1)$. This gives a contradiction and the lemma is proved. □

3.2.3 Proof of the Deformation Theorem

We can assume without loss of generality that

$$f(0;\lambda) = 0, \quad x(\lambda) = 0 \quad (0 \leqslant \lambda \leqslant 1)$$

and that 0 is an absolute minimizer of the functional $f_0 = f(\cdot;0)$ on some unit ball B.

We set

$$M(s) = 2 \sup_{\substack{x\in B(s)\\ 0\leqslant\lambda\leqslant 1}} |f(x;\lambda)| \quad (0 \leqslant s \leqslant 1)\,.$$

Clearly for each n we have

$$\max_{0\leqslant\lambda\leqslant 1} |f_n(x;\lambda)| < M(\|x\|) \quad (x \in E_n,\ 0 < \|x\| \leqslant 1)\,, \tag{3.2.17}$$

where E_n is any increasing sequence of finite-dimensional subspaces E satisfying condition (3.2.4). By Lemma 3.2.3, there exists a positive monotonic increasing function m on $(0, 1]$ with the following property: for every $r \in (0, 1]$ there exists some $n(r) \in \mathbb{N}$ such that

$$\min_{0\leqslant\lambda\leqslant 1} \|\nabla f_n(x;\lambda)\| > m(\|x\|) \quad (x \in E_n, r \leqslant \|x\| \leqslant 1) \tag{3.2.18}$$

for all $n \geqslant n(r)$. Furthermore, by Lemma 3.2.2, there exists a positive monotonic increasing function k on $[0, 1]$ such that

$$f_n(x;0) > k(\|x\|) \quad (x \in E_n, 0 < \|x\| < 1) \tag{3.2.19}$$

for all n. We choose $r > 0$ such that

$$\int_r^{\rho} m(s)\, ds - 2M(r) > 0 \tag{3.2.20}$$

and

$$k(\rho) > M(r) , \tag{3.2.21}$$

where

$$\rho = M^{-1}\left(\int_r^1 m(s)\,ds - M(r)\right) . \tag{3.2.22}$$

Then, for this r, we can find some $n_0 \in \mathbb{N}$ such that (3.2.18) holds for $n \geqslant n_0$. Then, for $n \geqslant n_0$, every one-parameter family of functions $f_n : E_n \times [0,1] \to \mathbb{R}$ satisfies the hypothesis of Lemma 2.2.2. Hence

$$\min_{\|x\|=\rho} f_n(x;1) > M(r) \quad (x \in E_n)$$

for $n \geqslant n_0$, and so

$$\inf_{\|x\|=\rho} f_1(x) \geqslant M(r) \quad (x \in E) . \tag{3.2.23}$$

Let x_1 be an absolute minimizer of f_1 on $B(\rho)$. Since

$$f_1(0) = 0 ,$$

it follows from (3.2.23) that $x_1 \in \mathring{B}(\rho)$. Therefore

$$\nabla f_1(x_1) = 0 .$$

We conclude that $x_1 = 0$, and the theorem is proved. □

3.3 Global Deformations of Functionals

In this section we apply the homotopy method to the study of global extrema.

In the theorems proved so far in this chapter on the invariance of minimizers under nondegenerate deformations, we have established invariance only for local minimizers. Direct analogs for the invariance of *global* minimizers break down even for functions in finitely many variables, as pointed out in Sect. 2.1.2.

Below we study a class of H-regular functionals having a unique critical point which is a global minimizer.

3.3.1 Growing Functionals

Suppose that H is a Hilbert space. Recall that a functional $f : H \to \mathbb{R}$ is said to be growing if

$$\lim_{\|x\|\to\infty} f(x) = \infty .$$

In applications, we sometimes have to establish that a functional f is growing just from information about the behavior of its gradient ∇f. Below we give a useful criterion for f to have this property, formulated in terms of the rate of increase of its gradient at infinity.

Let $f : H \to \mathbb{R}$ be a differentiable functional whose gradient is locally Lipschitzian and which has a unique critical point. We may assume, for simplicity, that this point is located at the point 0 of the space H.

Theorem 3.3.1. *If 0 is a local minimizer for a functional $f : H \to \mathbb{R}$ and if there is a positive continuous function $\alpha : [0, \infty) \to \mathbb{R}$ such that*

$$\|\nabla f(x)\| \geqslant \alpha(\|x\|) \tag{3.3.1}$$

and

$$\int_0^\infty \alpha(s)\, ds = \infty \,, \tag{3.3.2}$$

then f is growing.

For the proof we require an auxiliary result.

Let $\mathcal{D}$ be a domain in H. We consider the ordinary differential equation

$$\frac{dx}{dt} = -\nabla f(x) \tag{3.3.3}$$

on $\mathcal{D}$, and denote by $p(t, x)$ the solution of this equation with the initial condition $p(0, x) = x$ (this solution exists and is unique since the right-hand side of the equation satisfies a local Lipschitz condition).

Lemma 3.3.1. *Suppose that*

$$\|\nabla f(x)\| \geqslant \alpha > 0 \quad (x \in \mathcal{D}) \tag{3.3.4}$$

and

$$p(t, x) \in \mathcal{D} \quad (0 \leqslant t \leqslant t_0)\,. \tag{3.3.5}$$

Then

$$f(x) - f\left(p(t_0, x)\right) \geqslant \alpha \|x - p(t_0, x)\|\,. \tag{3.3.6}$$

The proof of this lemma is similar to the proof of its finite-dimensional analog, namely, Lemma 2.1.1.

Now we prove Theorem 3.3.1. For simplicity, we assume that $f(0) = 0$.

Since 0 is a local minimizer for f, it is an absolute strict minimizer for f on some ball $B(r)$ with $r > 0$. We shall show that

$$\inf_{\|x\|=\rho} f(x) \geqslant \mu(\rho) > 0 \tag{3.3.7}$$

for every $\rho \in (0, r)$.

Fix an arbitrary point $x \in S(\rho)$ and let $p(t,x)$ be the solution of Eq. (3.3.3) with the initial condition $p(0,x) = x$. We denote by T the spherical layer

$$T = \left\{x \in H : \tfrac{1}{2}\rho < \|x\| < r\right\} .$$

We show that

$$p(t_0, x) \in \partial T \tag{3.3.8}$$

for some $t_0 > 0$.

If this is false, then

$$p(t,x) \in T \quad (0 \leqslant t < \infty) . \tag{3.3.9}$$

Therefore, by (3.3.1), we have

$$\lim_{t\to\infty} f(p(t,x)) = f(x) - \lim_{t\to\infty} \int_0^t \|\nabla f(p(t,x))\|^2 \, dt = -\infty . \tag{3.3.10}$$

On the other hand, since ∇f is locally Lipschitzian, f is bounded on T, and by (3.3.9) we have

$$\inf_{0\leqslant t<\infty} f(p(t,x)) > -\infty . \tag{3.3.11}$$

The assertions (3.3.10) and (3.3.11) contradict each other. Therefore (3.3.8) holds for some t_0. It now follows by Lemma 3.3.1 that

$$f(x) - f(p(t_0,x)) \geqslant \alpha\|x - p(t_0,x)\| , \tag{3.3.12}$$

where

$$\alpha = \min_{\rho/2\leqslant s\leqslant r} \alpha(s) > 0 .$$

From (3.3.8) and (3.3.12) we have

$$\begin{aligned} f(x) &\geqslant f(p(t_0,x)) + \alpha\|x - p(t_0,x)\| \\ &\geqslant \alpha\|x - p(t_0,x)\| \geqslant \alpha \min\left(\tfrac{1}{2}\rho, r-\rho\right) . \end{aligned} \tag{3.3.13}$$

It follows that (3.3.7) holds for

$$\mu(\rho) = \alpha \min\left(\tfrac{1}{2}\rho, r-\rho\right) \quad (0 < \rho < r) . \tag{3.3.14}$$

Thus f is a Lyapunov functional for Eq. (3.3.3), defined in a neighborhood of the zero state of equilibrium and satisfying (1.6.6) and (1.6.7), with μ and ν defined by (3.3.14) and $\nu(\rho) = \alpha^2(\rho)$ $(0 < \rho < r)$. Therefore the zero state of equilibrium of Eq. (3.3.3) is asymptotically stable. Let $\mathcal{H}$ be the attraction set of the zero state of equilibrium. Clearly $\mathcal{H}$ is nonempty and open, and using the same method as in Theorem 2.1.4 we can prove that it is closed. Therefore $\mathcal{H} = H$.

We have thus shown that the zero state of equilibrium of Eq. (3.3.3) is stable in the large. Hence, fixing a point $x \in H$ and considering the trajectory $p(t,x)$, we conclude that

$$\lim_{t\to\infty} \|p(t,x)\| = 0\,. \tag{3.3.15}$$

Therefore

$$f(x) = f(0) - \int_0^\infty \frac{d}{ds} f\left(p(s,x)\right)\, ds = -\int_0^\infty \left\|\nabla f\left(p(s,x)\right)\right\|^2\, ds\,. \tag{3.3.16}$$

Estimating the last integral using the same method as for the integral $J(t)$ in Theorem 2.1.4, we obtain

$$\int_0^\infty \left\|\nabla f\left(p(s,x)\right)\right\|^2\, ds \geqslant \int_0^{\|x\|} \alpha(s)\, ds\,. \tag{3.3.17}$$

Now (3.3.2), (3.3.16), and (3.3.17) imply

$$\lim_{\|x\|\to\infty} f(x) = \infty\,,$$

and the theorem is proved. □

3.3.2 Global Deformations

Theorems 3.1.1 and 3.3.1 imply

Theorem 3.3.2. *Suppose that f_0, f_1 are H-regular functionals and that there exists a nondegenerate deformation $f : H \times [0,1] \to \mathbb{R}$ of $f_0 = f(0)$ into $f_1 = f(1)$. Let x_0 be a local minimizer for f_0 and*

$$\|\nabla f_1(x)\| \geqslant \alpha(\|x - x_0\|)\,, \tag{3.3.18}$$

where $\alpha(s)$ $(0 < s < \infty)$ is a continuous positive function for which

$$\int_0^\infty \alpha(s)\, ds = \infty\,. \tag{3.3.19}$$

Then the critical point x_1 of f_1 is a global minimizer, and

$$\lim_{\|x\|\to\infty} f_1(x) = \infty\,.$$

If in the situation of the Theorem 3.3.2 conditions (3.3.18) and (3.3.19) are not satisfied, then x_1 need not be a global minimizer for f_1, even if f_0 is growing. However, if the deformation $f(\cdot;\lambda)$ is linear, i.e., has the form

$$f(x;\lambda) = (1-\lambda)f_0(x) + \lambda f_1(x) \quad (x \in H, 0 \leqslant \lambda \leqslant 1)\,, \tag{3.3.20}$$

then a global minimizer is preserved.

Theorem 3.3.3. *Suppose that f_0, f_1 are H-regular functionals whose unique critical point is 0, and suppose that for $x \neq 0$ the vectors $\nabla f_0(x)$ and $\nabla f_1(x)$ are not oppositely directed. If*

$$\lim_{\|x\|\to\infty} f_0(x) = \infty\,, \tag{3.3.21}$$

then 0 is a global minimizer for f_1.

Proof. We may suppose for simplicity that $f_0(0) = f_1(0) = 0$. We fix a point $x_0 \in H$ and show that

$$f_1(0) \leqslant f_1(x_0)\,.$$

Indeed, there exists a number $c > 0$ such that

$$x_0 \in \Omega\,, \tag{3.3.22}$$

where

$$\Omega = \{x \in H : f_0(x) < c\}\,.$$

By (3.3.21), the set $\overline{\Omega}$ is weakly compact. Therefore, by Lemma 3.1.2, the function f_1 has a minimizer x_1 in $\overline{\Omega}$. We shall show that $x_1 \in \Omega$. Indeed, if $x_1 \in \partial\Omega$, then, by Theorem 1.7.7, we have

$$\nabla f_1(x_1) = -\lambda_0 \nabla f_0(x_1) \quad (\lambda_0 > 0)\,.$$

Since $x_1 \neq 0$, it follows from this that the vectors $\nabla f_0(x_1)$ and $\nabla f_1(x_1)$ are oppositely directed, contradicting the hypothesis of the theorem. Thus $x_1 \in \Omega$. However, then

$$\nabla f_1(x_1) = 0$$

and, consequently, $x_1 = 0$. Therefore by (3.3.22) we have

$$f_1(0) = \min_{x\in\Omega} f_1(x) \leqslant f_1(x_0)\,,$$

as required. $\square$

3.3.3 Generalizations

The following theorem holds.

Theorem 3.3.4. *Let f_0, f_1 be functionals that are Gâteaux differentiable and weakly lower semicontinuous. Suppose that f_0 is growing, i.e.*

$$\lim_{\|x\|\to\infty} f_0(x) = \infty ,$$

and that the set $\mathfrak{M}$ of critical points of f_1 is bounded. If for all points x outside of some ball the vectors $\nabla f_0(x)$ and $\nabla f_1(x)$ are not oppositely directed, then some point in $\mathfrak{M}$ is a global minimizer of f_1 .

This theorem can be proved using in the same way as Theorem 3.3.3. Theorem 3.3.4 implies

Corollary 1. *Let f be a functional that is Gâteaux differentiable and weakly lower semicontinuous and suppose that the set $\mathfrak{M}$ of critical points of f is contained in some ball $\overset{\circ}{B}(r)$. If for all x in the sphere $\partial B(r)$ the vectors $\nabla f(x)$ are not opposite in direction to the vectors x, then some point in $\mathfrak{M}$ is an absolute minimizer for f on $B(r)$.*

3.4 Deformations of Lipschitzian Functionals

In this section we prove a deformation principle for minimizers of certain Lipschitzian functionals.

3.4.1 (P, S)-Regular Functionals

Suppose that E is a real Banach space, with dual E^*, and that $f : E \to \mathbb{R}$ is a locally Lipschitzian functional. As usual, for $x \in E$ we write $f^0(x; v)$ for the generalized derivative of f at x in the direction of v, and $\partial f(x) \subset E^*$ for the generalized gradient of f at x. Recall that a point x_* is a critical point of f if

$$0 \in \partial f(x_*) .$$

We say that a locally Lipschitzian functional f satisfies the *Palais–Smale condition* or is a (P, S)*-regular functional* if

$$\inf_{y\in\partial f(x), x\in M} \|y\|_{E^*} > 0 \tag{3.4.1}$$

for every bounded closed subset M of E that contains no critical points of f. The class of functionals satisfying the Palais–Smale condition is very extensive. It contains, for instance, all functionals on reflexive Banach spaces whose generalized gradients satisfy the following generalized condition $(S)_+$: if the sequence (x_n) weakly converges to x_0 and

$$\overline{\lim_{n\to\infty}} \inf_{y\in\partial f(x_n)} \langle y, x_n - x_0\rangle \leqslant 0 ,$$

then

$$\lim_{n\to\infty} \|x_n - x_0\| = 0 .$$

In particular, the E-regular functionals of Sect. 3.2.1, satisfy the Palais–Smale condition.

3.4.2 The Deformation Theorem

Let f_0 and f_1 be (P,S)-regular functionals having unique critical points x_0 and x_1 respectively. We say that a one-parameter family $f(\cdot;\lambda) : E \to \mathbb{R}$ $(0 \leqslant \lambda \leqslant 1)$ of (P,S)-regular functionals is a *nondegenerate deformation* of $f_0 = f(\cdot;0)$ into $f_1 = f(\cdot;1)$ if

(1) the functional $f(x;\lambda)$ is uniformly continuous in λ with respect to x on every ball $B(r)$ in E,

(2) the multivalued mapping

$$\partial_x f : E \times [0,1] \to E^*$$

is upper semicontinuous in λ uniformly with respect to x on every ball $B(r)$ in E,

(3) for every λ the functional $f(\cdot;\lambda)$ has a unique critical point $x(\lambda)$ which continuously depends on $\lambda \in [0,1]$.

Theorem 3.4.1. *Suppose that f_0, f_1 are (P,S)-regular functionals and that there exists a nondegenerate deformation $f(\cdot;\lambda)$ $(0 \leqslant \lambda \leqslant 1)$ of f_0 into f_1. If the point $x_0 = x(0)$ is a local minimizer for f_0 then $x_1 = x(1)$ is a local minimizer for f_1.*

The proof of Theorem 3.4.1 will be given in Sect. 3.4.4. In the proof we may assume for simplicity that $x(\lambda) = 0$ and $f(0;\lambda) = 0$.

3.4.3 Preparatory Lemmas

We fix $r > 0$, $\lambda \in [0,1]$, and denote by $\mathfrak{M}(\lambda)$ the set of continuous *piecewise-linear splines* $x = x(t)$ $(0 \leqslant t \leqslant t(x))$ satisfying the following conditions:

(1) $x(t) \in B(r)$ $(0 \leqslant t \leqslant t(x)$;

(2) the function

$$\varphi(t) = f(x(t);\lambda) \quad (0 \leqslant t \leqslant t(x))$$

is monotone decreasing;

(3) every link of the spline is a segment of a straight line: for some $v_1, v_2, \dots$ we have

$$x(t) = x(t_{n-1}) + (t - t_n)v_{n-1} \quad (t_{n-1} \leqslant t \leqslant t_n;\ n = 1, 2, \dots) .$$

The numbers $t_0, t_1, \dots$ in the definition of x will be called the *nodes* of x. The technique for constructing splines $x \in \mathfrak{M}(\lambda)$ with properties (1)–(3) will be described below.

Define $m : [0, 1] \to \mathbb{R}$ by

$$m(s) = \inf_{y \in \partial_x f(x;\lambda), s \leqslant \|x\| \leqslant r, 0 \leqslant \lambda \leqslant 1} \|y\|_{E^*} . \tag{3.4.2}$$

By condition (2) in the definition of a nondegenerate deformation, we have

$$m(s) > 0 \quad (0 < s \leqslant r) . \tag{3.4.3}$$

It follows from the Separation Theorem 1.3.2 that for every nonzero point $x \in B(r)$ there exists a vector $v(x) \in E$ of unit norm for which

$$f^0(x; v(x); \lambda) < -\tfrac{1}{2} m(\|x\|) .$$

Furthermore, since the generalized derivative $f^0(x; v; \lambda)$ is upper semicontinuous for every nonzero $x \in B(r)$, there is a positive number $r(x)$ such that

$$f^0(u; v(x); \lambda) \leqslant -\tfrac{1}{2} m(\|x\|) \tag{3.4.4}$$

for all $u \in B(r(x), x)$.

Finally, for every nonzero $x \in B(r)$ let $S(x)$ be the set of elements v of unit norm such that for some $t_0 > 0$ and $t \in [0, t_0)$ we have

$$x + tv \in B(r) \tag{3.4.5}$$

and

$$f^0(x + tv; v; \lambda) \leqslant -\tfrac{1}{2} m\left(\tfrac{3}{4}\|x + tv\|\right) . \tag{3.4.6}$$

Let $t(v)$ be the largest value of t for which (3.4.5) and (3.4.6) are satisfied.

Now we fix a nonzero point $x_0 \in B(r)$ and we construct a spline $x(t)$ in $\mathfrak{M}(\lambda)$ emanating from x_0 according to the following rule.

We take a vector $v_0 \in S(x_0)$ for which

$$t(v_0) \geqslant \tfrac{1}{2} \sup\nolimits_{v \in S(x_0)} t(v) \tag{3.4.7}$$

as the direction which defines the first link of the spline and make the length of the first link of the spline equal to $t(v_0)$. We then take the endpoint of the first link of the spline as the starting point for constructing the second link of the spline according to the same rule, etc. If at some time τ the spline $x(t)$ meets the sphere $S(r)$, then the spline terminates at this point.

Lemma 3.4.1. *If the spline $x \in \mathfrak{M}(\lambda)$ lies in the ball $\mathring{B}(r)$ on the interval $[0, \tau]$, then the set of nodes of x has no condensation points in $[0, \tau]$.*

Proof. If the contrary holds, then there exists a sequence (t_n) of nodes converging to a point $t_* \in [0, \tau]$. Then, by (3.4.4), there exist a number

$$r_* \in (0, \tfrac{1}{3}\|x\|)$$

and a direction

$$v_* = v(x_*)$$

such that

$$f^0(u; v_*; \lambda) \leqslant -\tfrac{1}{2}m(\|x_*\|)$$

for all points $u \in B(r_*, x_*)$. Therefore the points $x_n = x(t_n)$ lie in $B(\frac{1}{8}r_*, x_*)$ for sufficiently large n, and this contradicts the construction of the spline since inequality (3.4.7) does not hold for these values of n. □

Lemma 3.4.2. *Suppose that for some $\lambda \in [0, 1]$ the point 0 is an absolute strict minimizer for the functional $f(x; \lambda)$ on some ball $B(\rho_0)$. Then*

$$\inf_{\|x\|=\rho} f(x; \lambda) > 0 , \tag{3.4.8}$$

for all $\rho \in (0, \rho_0)$.

Proof. If the contrary holds, then there exists a point x_0 on some sphere $S(\rho)$ with $0 < \rho < \rho_0$ such that

$$f(x_0; \lambda) < \tfrac{1}{4}\delta m\left(\tfrac{3}{4}(\rho - \delta)\right) , \tag{3.4.9}$$

where

$$0 < \delta < \rho < \rho + \delta < \rho_0 .$$

Let x be a spline in $\mathfrak{M}$ with starting point

$$x(0) = x_0 ,$$

constructed as described above. We set

$$\varphi(t) = f(x(t); \lambda) .$$

The function φ is differentiable almost everywhere and satisfies

$$\frac{d\varphi}{dt} \in \langle \partial f(x_{n-1} + tv_{n-1}; \lambda), v_{n-1} \rangle \tag{3.4.10}$$

on the link (t_{n-1}, t_n) of the spline x. By the construction of x we have

$$\sup_{y \in \partial f(x_{n-1} + tv_{n-1}, \lambda)} \langle y, v_{n-1} \rangle \leqslant -\tfrac{1}{2}m\left(\tfrac{3}{4}\|x_{n-1} + tv_{n-1}\|\right) . \tag{3.4.11}$$

Consequently, at some time $t = t_*$, the spline x reaches the boundary ∂T of the spherical layer

$$T = \{x \in E : \rho - \delta \leqslant \|x\| \leqslant \rho + \delta\} .$$

From (3.4.9)–(3.4.11) we have

$$f(x(t_*); \lambda) = \varphi(t_*) = f(x_0; \lambda) + \int\limits_0^{t_*} \varphi'(\tau)\, d\tau < -\tfrac{1}{4}\delta m\left(\tfrac{3}{4}(\rho - \delta)\right),$$

and this contradicts the assumption that the point 0 is an absolute strict minimizer of $f(\cdot; \lambda)$ on $B(\rho_0)$. □

3.4.4 Proof of the Deformation Theorem

Let Λ be the set of $\lambda \in [0, 1]$ for which 0 is a local minimizer of $f(\cdot; \lambda)$. By the hypothesis of the theorem, Λ is nonempty. We shall show that it is open.

Let $\lambda_0 \in \Lambda$. Since 0 is the only critical point of $f(\cdot; \lambda_0)$, it is an absolute strict minimizer for $f(\cdot; \lambda_0)$ on $B(\rho_0)$. Consequently, by Lemma 3.4.2,

$$\inf_{\|x\|=\rho_0/2} f(x; \lambda) = a > 0 .$$

Then, for some $\varepsilon > 0$, we have

$$\inf_{\|x\|=\rho_0/2} f(x; \lambda) > \tfrac{1}{2}a \tag{3.4.12}$$

for all $\lambda \in [\lambda_0 - \varepsilon, \lambda_0 + \varepsilon] \cap [0, 1]$.

We fix

$$\lambda_1 \in [\lambda_0 - \varepsilon, \lambda_0 + \varepsilon] \cap [0, 1]$$

and choose $\rho_1 > 0$ such that

$$\sup_{\|x\|\leqslant\rho_1} f(x; \lambda_1) < \tfrac{1}{2}a . \tag{3.4.13}$$

Let $x_0 \in B(\rho_1)$. By (3.4.12) and (3.4.13), the spline $x \in \mathfrak{M}(\lambda_1)$ which begins at x_0 does not leave the ball $B(\frac{1}{2}\rho_0)$. Therefore, by (3.4.6), for the function

$$\varphi(t) = f(x(t); \lambda_1)$$

there exists a sequence $t_k \to \infty$ such that

$$\lim_{k\to\infty} \varphi'(t_k) = 0 .$$

It follows that

$$\lim_{k\to\infty} \|x(t_k)\| = 0$$

and so

$$f(x_0;\lambda_1) \geqslant \lim_{k\to\infty} f(x(t_k);\lambda_1) = 0 \; .$$

Thus 0 is a minimizer for $f(\cdot;\lambda_1)$ on $B(\rho_1)$, so that $\lambda_1 \in \Lambda$. Consequently Λ is open.

Now we show that Λ is closed.

Suppose that $\lambda \in \Lambda$ and $x = x(t) \in \mathfrak{M}(\lambda)$ $(0 \leqslant t \leqslant t(x))$ is a spline. From this spline, a number $\varepsilon > 0$, and an interval $[0,\tau] \subset [0,t(x)]$, we construct a set of continuous piecewise-linear splines y according to the following rules:

(1) the nodes $t_n \in [0,\tau]$ of the splines x and y and the directions of the links coincide;

(2) the inequality

$$\max_{0\leqslant t\leqslant \tau} |x(t) - y(t)| < \varepsilon \tag{3.4.14}$$

is satisfied;

(3) the inequality

$$f^0(y(t);v_{n-1};\lambda) \leqslant -\tfrac{1}{4}m\left(\tfrac{1}{2}\,\|y(t)\|\right) \quad (t_{n-1} \leqslant t \leqslant t_n) \tag{3.4.15}$$

holds for every pair t_{n-1}, t_n of consecutive nodes;

(4) if $t > \tau$, then y can be extended by using the same construction as for x.

Denote the set of splines constructed in this way by $\widehat{\mathfrak{M}}(\lambda)$. By construction, $\mathfrak{M}(\lambda) \subset \widehat{\mathfrak{M}}(\lambda)$. Let $\widetilde{\mathfrak{M}}(\lambda)$ be the set of all continuous splines in $\widehat{\mathfrak{M}}(\lambda)$ having finitely many links.

We set

$$M = \sup_{y\in\partial f_x(x;\lambda),x\in B(r),0\leqslant\lambda\leqslant 1} \|y\|_{E^*} \; .$$

By condition (2) in the definition of a nondegenerate deformation, we have $M < \infty$.

Let

$$r_0 = \frac{rm(r/4)}{16M} \; .$$

We shall show that for any $\lambda \in \Lambda$ the point 0 is an absolute minimizer for $f(\cdot;\lambda)$ on $B(r_0)$, and prove in this way that the set Λ is closed.

Fix $\lambda \in \Lambda$. Since 0 is a strict local minimizer for $f(\cdot;\lambda)$, there exists a maximal open set $\Omega(\lambda)$ containing 0 such that the splines $y \in \widetilde{\mathfrak{M}}(\lambda)$ with the initial conditions from $\Omega(\lambda)$ converge to 0.

Let $\mathring{B}(r_1)$ be the maximal open ball contained in Ω. In order to prove the inequality

$$f(x;\lambda) \geqslant 0 \quad (x \in \mathring{B}(r_0)) \; ,$$

it suffices to prove that $r_0 \leqslant r_1$. Since this is clear if $r_1 \geqslant \frac{1}{2}r$, suppose that

$$r_1 < \tfrac{1}{2}r \; .$$

Consider a point $y_0 \notin \Omega$ on the sphere $S(r_1)$. For some $\tau_0 > 0$, the point $y(\tau_0)$ of the spline

$$y \in \widetilde{\mathfrak{M}}(\lambda)\,, \quad y(0) = y_0$$

satisfies

$$\tfrac{3}{4}r < \|y(\tau_0)\| < r\,.$$

Therefore there exists a spline

$$x \in \widetilde{\mathfrak{M}}(\lambda)$$

with the initial condition

$$x(0) \in \mathring{B}(r_1)$$

for which

$$\tfrac{3}{4}r < \|x(\tau_0)\| < r\,.$$

Since $x(0) \in \Omega$, there exist τ_1 and τ_2 with $\tau_1 < \tau_2$ such that

$$\|x(\tau_1)\| = \tfrac{3}{4}r, \quad \|x(\tau_2)\| = \tfrac{1}{2}r \tag{3.4.16}$$

and

$$\tfrac{1}{2}r \leqslant \|x(t)\| \leqslant \tfrac{3}{4}r \quad (\tau_1 \leqslant t \leqslant \tau_2)\,. \tag{3.4.17}$$

Clearly we have

$$f(x(0);\lambda) - f(0;\lambda) \leqslant M\,\|x(0)\| < Mr_1\,. \tag{3.4.18}$$

On the other hand,

$$\begin{aligned} f(x(0);\lambda) - f(0;\lambda) &> f(x(\tau_1);\lambda) - f(x(\tau_2);\lambda) \\ &= -\int_{\tau_1}^{\tau_2} \varphi'(\tau)\,d\tau \geqslant \frac{r}{16} m\,(r/4))\,. \end{aligned} \tag{3.4.19}$$

From (3.4.18) and (3.4.19) we obtain $r_0 < r_1$ and the theorem follows. □

3.5 Deformations of Nonsmooth Problems with Constraints

3.5.1 Statement of the Problem

Let E be a real Banach space and $f,\, g : E \to \mathbb{R}$ be locally Lipschitzian functionals. Consider the problem of minimization of f on the set

$$M = \{x \in E : g(x) \leqslant 0\}\,.$$

We shall write this problem in the form

$$\begin{cases} f(x) \to \min\,, \\ g(x) \leqslant 0\,. \end{cases} \tag{3.5.1}$$

Recall that a point x_* is a *local minimizer* of problem (3.5.1) if for some $\rho > 0$ the inequality

$$f(x) \geqslant f(x_*)$$

holds for all $x \in M \cap B(\rho)$. If this inequality is strict for $x \in M \cap B(\rho)$ and $x \neq x_*$, then x_* is a *strict local minimizer* for problem (3.5.1).

A point x_* is an *extremum* of problem (3.5.1) if

$$0 \in \partial f(x_*) + \mu \partial g(x_*) \tag{3.5.2}$$

for some $\mu \geqslant 0$ and the condition of *complementary slackness*

$$\mu g(x_*) = 0$$

holds.

The local minimizers of problem (3.5.1) are critical points.

In this section, we apply the deformation method to the investigation of problems of the form (3.5.1). More general problems of the form

$$\begin{cases} f(x) \to \min\,, \\ g_i(x) \leqslant 0 \quad (i = 1, \ldots, n) \end{cases} \tag{3.5.3}$$

can be reduced to this form, since if we define

$$g(x) = \max_{1 \leqslant i \leqslant n} g_i(x)\,,$$

then problem (3.5.3) takes the form (3.5.1).

Now we consider two problems,

$$\begin{cases} f_0(x) \to \min\,, \\ g_0(x) \leqslant 0 \end{cases} \tag{3.5.4}$$

and

$$\begin{cases} f_1(x) \to \min\,, \\ g_1(x) \leqslant 0\,, \end{cases} \tag{3.5.5}$$

where f_0, f_1, g_0, $g_1 : E \to \mathbb{R}$ are locally Lipschitzian functionals.

A one-parameter family of problems

$$\begin{cases} f(x; \lambda) \to \min\,, \\ g(x; \lambda) \leqslant 0\,, \end{cases} \quad (0 \leqslant \lambda \leqslant 1) \tag{3.5.6}$$

is called a *nondegenerate deformation* of problem (3.5.4) into problem (3.5.5) if

(1) the functionals $f(\cdot;\lambda)$ and $g(\cdot;\lambda)$ are locally Lipschitzian with respect to x for every $\lambda \in [0,1]$ and continuous in λ uniformly with respect to x on every ball of the space E,

(2) the multivalued mappings

$$\partial f : E \times [0,1] \to E^*, \quad \partial g : E \times [0,1] \to E^*$$

are upper semicontinuous in λ uniformly with respect to x on every ball of E,

(3) for every $\lambda \in [0,1]$ and every point $x \in E$ for which

$$g(x;\lambda) = 0\ ,$$

the functional $g(\cdot;\lambda)$ is regular, i.e., its generalized Clarke derivative $g^0(x;\lambda;v)$ in the direction of v (defined in Section 1.5.3) coincides with the classical derivative $g'(x;\lambda;v)$ in the direction of v,

(4) for every $\lambda \in [0,1]$ and every $\mu > 0$ the functional

$$h(x;\lambda) = f(x;\lambda) + \mu g_+(x;\lambda)$$

satisfies the Palais–Smale (P,S)-condition, where

$$g_+(x;\lambda) = \max\{0, g(x;\lambda)\}\ ,$$

(5) for every $\lambda \in [0,1]$, problem (3.5.6) has a unique critical point $x(\lambda)$ $(0 \leqslant \lambda \leqslant 1)$ which depends continuously on λ.

3.5.2 The Deformation Theorem

We can now state and prove the main result of this section.

Theorem 3.5.1. *Suppose that there exists a nondegenerate deformation* (3.5.6) *of problem* (3.5.4) *into problem* (3.5.5). *Suppose that the condition*

$$0 \notin \partial g(x(\lambda);\lambda) \tag{3.5.7}$$

holds for every $\lambda \in [0,1]$. *If the point* $x_0 = x(0)$ *is a local minimizer of problem* (3.5.4) *then* $x_1 = x(1)$ *is a local minimizer of problem* (3.5.5).

Proof. We can assume without loss of generality that

$$x(\lambda) = 0, \quad f(0;\lambda) = 0 \quad (0 \leqslant \lambda \leqslant 1)\ .$$

We set

$$h(x;\lambda;k) = f(x;\lambda) + k g_+(x;\lambda)\ , \tag{3.5.8}$$

$$\mathcal{D}(\lambda) = \{x \in E : g_+(x;\lambda) = 0\}\ .$$

In order to prove the theorem, it suffices to show that for large $k > 0$ the family (3.5.8) satisfies the hypothesis of Theorem 3.4.1.

By condition (2) we have

$$\sup_{p \in \partial f(x;\lambda),\ x \in B,\ 0 \leqslant \lambda \leqslant 1} \|p\|_{E^*} = M < \infty . \tag{3.5.9}$$

Moreover it follows from the regularity condition (3) that for all x and λ such that $g(x;\lambda) = 0$ we have

$$\partial g_+(x;\lambda) = \mathrm{co}\{0, \partial g(x;\lambda)\} ,$$

where

$$g_+(x;\lambda) = \max\{0, g(x;\lambda)\} .$$

We conclude from this and condition (2) in the definition of a nondegenerate deformation that the multivalued mapping

$$\partial g_+ : E \times [0,1] \to E^*$$

is upper semicontinuous in λ uniformly with respect to x on every ball of E.

Condition (2) and (3.5.7) imply that for some $\rho \in (0, \frac{1}{2})$ we have

$$\inf_{p \in \partial g_+(x;\lambda),\ x \in B(\rho) \setminus \mathcal{D}(\lambda),\ 0 \leqslant \lambda \leqslant 1} \|p\|_{E^*} \geqslant b > 0 . \tag{3.5.10}$$

Now we show that the family (3.5.8) satisfies the conditions of Theorem 3.4.1 with

$$k > \frac{3M}{b} .$$

Suppose that 0 is a minimizer for f_0 on the set $\mathcal{D}_0 \cap B(r)$ (where $\mathcal{D}_0 = \mathcal{D}(0)$), and let

$$2r < \rho .$$

We can assume without loss of generality that 0 belongs to the boundary of $\mathcal{D}_0$. Let $\mathfrak{M}$ be the set of splines constructed for the functional

$$g_+(\cdot) = g_+(\cdot;0)$$

and the set $B(2r) \setminus \mathcal{D}_0$ according to the procedure used for constructing the splines in the set $\mathfrak{M}$ in the proof of Theorem 3.4.1. Thus on the nth link of the continuous spline $x \in \mathfrak{M}$ we have

$$\dot{x}(t) = v_{n-1} \quad (t_{n-1} < t < t_n;\ \|v_{n-1}\| = 1)$$

and

$$g_+^0(x(t); v_{n-1}) \leqslant -\tfrac{1}{2}\nu(\tfrac{3}{4}\|x(t)\|) , \tag{3.5.11}$$

where

$$\nu(s) = \inf_{p \in \partial g_+(x),\ s \leqslant \|x\| \leqslant 1,\ x \notin \mathcal{D}_0} \|p\|_{E^*} .$$

For this link of the spline, (3.5.9) and (3.5.11) yield

$$h^0(x;0;k;v_{n-1}) \leqslant f_0^0(x;v_{n-1}) + kg_+^0(x;v_{n-1}) < 0 .$$

Therefore for almost all t we have

$$\dot{\gamma}(t) < 0 , \tag{3.5.12}$$

$$\dot{\eta}(t) \leqslant -\tfrac{1}{2}\nu(\tfrac{3}{4}\|x(t)\|) , \tag{3.5.13}$$

where

$$\gamma(t) = h(x(t);0;k), \quad \eta(t) = g_+(x(t)) .$$

If the spline x begins at a point

$$x_0 \in B(r) \setminus \mathcal{D}_0 ,$$

then two cases arise: either

$$x(T_0) \in B(2r) \cap \mathcal{D}_0$$

for some T_0 (finite or infinite) or

$$\|x(T_1)\| = 2r$$

for some T_1. In the first case, (3.5.12) yields that

$$h(x_0;0;k) > f_0(0) . \tag{3.5.14}$$

In the second case, (3.5.13) yields that

$$kg_+(x_0) > k_1 r , \tag{3.5.15}$$

where

$$k_1 > M .$$

It follows from (3.5.15) that

$$kg_+(x_0) > k_1 d_{\mathcal{D}_0}(x_0) , \tag{3.5.16}$$

where $d_{\mathcal{D}_0}$ is the distance functional for the subset $\mathcal{D}_0$. It is shown in [65] that the point 0 is a minimizer of the functional

$$f_0(x) + k_1 d_{\mathcal{D}_0}(x)$$

on $B(r)$. It now follows from (3.5.14) and (3.5.16) that in both cases

$$h(x_0;0;k) > f_0(0) ,$$

i.e., the point 0 is a minimizer of the functional $h(\cdot;0;k)$ on $B(r)$.

Let $\lambda \in [0,1]$. By (3.5.9), we have

$$\inf_{p\in\partial h(x;\lambda;k)} \|p\|_{E^*} \geqslant \inf_{p\in\partial f(x;\lambda)+k\partial g_+(x;\lambda)} \|p\|_{E^*} \geqslant kb - M > 0 \qquad (3.5.17)$$

for $x \in B(\rho) \setminus \mathcal{D}(\lambda)$. If

$$x \in B(\rho) \cap \mathcal{D}(\lambda) \setminus \{0\}\,,$$

then

$$0 \notin \partial f(x;\lambda) + k\partial g_+(x;\lambda)$$

since the extremum is unique. This, together with (3.5.7) and conditions (2), (4) in the definition of a nondegenerate deformation, implies that

$$\inf_{p\in\partial f(x;\lambda)+k\partial g_+(x;\lambda),\ 0<s\leqslant\|x\|\leqslant\rho,\ 0\leqslant\lambda\leqslant1} \|p\|_{E^*} > 0\,.$$

Thus the family of functionals $h(\cdot;\lambda;k)$ satisfies the hypothesis of Theorem 3.4.1 and the theorem follows. □

3.5.3 Linear Deformations and Conditions for Global Minimizers

Suppose that E is a reflexive Banach space and f_0, f_1, $g : E \to \mathbb{R}$ are locally Lipschitzian and weakly lower semicontinuous functionals. Consider the problems

$$\begin{cases} f_0(x) \to \min\,, \\ g(x) \leqslant 0 \end{cases} \qquad (3.5.18)$$

and

$$\begin{cases} f_1(x) \to \min\,, \\ g(x) \leqslant 0\,. \end{cases} \qquad (3.5.19)$$

Suppose that the point 0 is an extremum of these problems.

Theorem 3.5.2. *Suppose that for every $\lambda \in [0,1]$ the problem*

$$\begin{cases} f(x;\lambda) = (1-\lambda)f_0(x) + \lambda f_1(x) \to \min\,, \\ g(x) \leqslant 0 \end{cases} \qquad (3.5.20)$$

has no nonzero extrema. Let f_0 be a growing functional defined on the admissible set

$$\mathcal{D} = \{x \in E : g(x) \leqslant 0\}\,,$$

i.e.,

$$\lim_{\|x\|\to\infty,\ x\in\mathcal{D}} f_0(x) = \infty\,. \qquad (3.5.21)$$

Suppose, finally, that

$$0 \notin \partial g(x) \quad (x \neq 0)\,. \qquad (3.5.22)$$

Then the zero extremum is a global minimizer of problem (3.5.19).

Proof. We fix some $C > 0$ and consider the problem

$$\begin{cases} f_1(x) \to \min , \\ g(x) \leqslant 0 , \\ f_0(x) - C \leqslant 0 . \end{cases} \tag{3.5.23}$$

The set

$$G = \{x \in E : g(x) \leqslant 0,\ f_0(x) \leqslant C\}$$

is weakly compact, and therefore (3.5.23) has a solution x_*. We shall show that $x_* = 0$.

Suppose that $x_* \neq 0$. Since x_* is an extremum of the problem (3.5.23), we have

$$0 \in \alpha \partial f_1(x_*) + \beta \partial f_0(x_*) + \gamma \partial g(x_*) , \tag{3.5.24}$$

$$\gamma g(x_*) = 0 , \tag{3.5.25}$$

$$\beta(f_0(x_*) - C) = 0 , \tag{3.5.26}$$

where $\alpha, \beta, \gamma \geqslant 0$, and $\alpha + \beta + \gamma > 0$. It follows from condition (3.5.22) that

$$\alpha + \beta > 0 ,$$

and we can rewrite (3.5.24) and (3.5.25) as

$$0 \in (1 - \lambda_*)\partial f_*(x_*) + \lambda_0 \partial f_1(x_*) + \gamma_0 \partial g(x_*) ,$$

$$\gamma_0 g(x_*) = 0 ,$$

where

$$\lambda_* = \frac{\alpha}{\alpha + \beta}, \quad \gamma_* = \frac{\gamma}{\alpha + \beta} .$$

Therefore x_* is an extremum of problem (3.5.20) for $\lambda = \lambda_*$, and so $x_* = 0$.

Since the constant C is arbitrary, it follows by condition (3.5.21) that the zero extremal is a global minimizer for problem (3.5.19), and the theorem is proved. □

4 Conley Index

4.1 Conley Index in Finite-Dimensional Problems

4.1.1 Flows in Finite-Dimensional Spaces

Let Ω be an open set in $\mathbb{R}^m$. A *flow* in Ω is a continuous mapping p from a subset of $\mathbb{R} \times \Omega$ to Ω, with domain of the form $\bigcup_{x\in\Omega}(\tau_1(x), \tau_2(x)) \times \{x\}$ for some functions τ_1, $\tau_2 : \Omega \to \mathbb{R} \cup \{\pm\infty\}$, satisfying the following conditions:

(1) for each $x \in \Omega$ we have $0 \in (\tau_1(x), \tau_2(x))$ and $p(0, x) = x$;

(2) if $x \in \Omega$ and we have $t \in (\tau_1(x), \tau_2(x))$ and $t' \in (\tau_1(p(t, x)), \tau_2(p(t, x)))$, then $t + t' \in (\tau_1(x), \tau_2(x))$ and $p(t + t', x) = p(t', p(t, x))$.

We extend the notation $p(t, x)$ by writing

$$\begin{aligned} p(T,x) &= \{p(t,x) : t \in T\} && (T \subset \mathbb{R}) , \\ p(t,X) &= \{p(t,x) : x \in X\} && (X \subset \Omega) , \\ p(T,X) &= \{p(t,x) : t \in T, x \in X\} && (T \subset \mathbb{R},\ X \subset \Omega) . \end{aligned}$$

This extended notation will only be used when the natural condition holds on the points and sets concerned; for example, if we write something like $p(T, X)$, where T is an arbitrary set of values of t and X is an arbitrary set of values of x, we always assume that $p(t, x)$ is defined for all $x \in X$ and all $t \in T$.

We recall that the flow p is assumed to be continuous. By this we mean that at every point (t_0, x_0) of the domain of definition of p and for any neighborhood U of $p(t_0, x_0)$ there are neighborhoods V_t of t_0 and V_x of x_0 such that $p(V_t, V_x)$ is a subset of U. Here it is implicitly assumed that $p(V_t, V_x)$ is defined.

A subset of Ω of the form $p(I, x)$ with I a subinterval of $(\tau_1(x), \tau_2(x))$ is called a *trajectory* of p. By entire trajectory we mean $p(\mathbb{R}, x)$.

We use the following result repeatedly in the sequel.

Lemma 4.1.1. *Suppose that the trajectory $p([t_1, t_2], x_1)$ lies in an open subset U of Ω. Then x_1 has a neighborhood V such that $p([t_1, t_2], V) \subset U$.*

Proof. We fix $\theta \in [t_1, t_2]$. The continuity of p implies that there are open neighborhoods W_θ^x of x_1 in U and W_θ^t of θ in $\mathbb{R}$ such that

$$p(W_\theta^t, W_\theta^x) \subset U .$$

We choose from the open cover

$$\{W_\theta^t : \theta \in [t_1, t_2]\}$$

of the compact set $[t_1, t_2]$ a finite subcover

$$\{W_{\theta_i}^t : i = 1, 2, \ldots, n\}$$

and set

$$V = \bigcap_{i=1}^{n} W_{\theta_i}^x .$$

By construction, $p(W_{\theta_i}^t, V) \subset U$ for all i. Since

$$[t_1, t_2] \subset \bigcup_{i=1}^{n} W_{\theta_i}^t ,$$

the set V satisfies the conditions of the lemma. □

A subset S of Ω is called an *invariant set* if $p(\mathbb{R}, x) \subset S$ for all $x \in S$. A closed bounded invariant set $S \subset \Omega$ is said to be *isolated* if there exists a compact subset $\mathfrak{U}$ of Ω with $S \subset \operatorname{int} \mathfrak{U}$ such that S contains all trajectories lying entirely in $\mathfrak{U}$. Such a compact set is called an *isolating neighborhood* of S.

We fix an isolated invariant set S and an isolating neighborhood $\mathfrak{U}$. Consider the sets

$$A^+ = \{x \in \mathfrak{U} : p([0, +\infty), x) \subset \mathfrak{U}\} ,$$

$$A^- = \{x \in \mathfrak{U} : p((-\infty, 0], x) \subset \mathfrak{U}\} .$$

It follows from the definition of an isolating neighborhood that

$$A^+ \cap A^- = S .$$

Lemma 4.1.2. *The subsets A^+ and A^- are closed.*

Proof. Suppose that (x_n) is a sequence in A^+ with limit $x_* \notin A^+$. Then $p(t_*, x_*) \notin \mathfrak{U}$ for some $t_* > 0$. Let W be a neighborhood of $p(x_*, t_*)$ which lies in Ω and is disjoint from $\mathfrak{U}$. By the continuity of p there exists a neighborhood U of x_* such that $p(t_*, U) \subset W$. Since $x_n \in A^+$ for each n, the neighborhood U contains no elements of the sequence (x_n) and so x_* cannot be its limit. This contradiction shows that A^+ is closed. A similar argument shows that A^- is closed, and the lemma follows. □

Lemma 4.1.3. *Let Z be a closed subset of $\mathfrak{U}$ such that that $Z \cap A^+ = \varnothing$. Then there exists a number $T > 0$ such that for each $x \in Z$ the trajectory $p([0,T],x)$ does not lie entirely in $\mathfrak{U}$.*

Proof. We fix $x \in Z$. Since $x \notin A^+$, we have

$$p(t(x),x) \notin \mathfrak{U}$$

for some $t(x) > 0$. (We recall that this implies that $t(x) < \tau_2$ where τ_2 is as in the definition of a flow.) Then x has a neighborhood V_x such that

$$p(t(x),V_x) \cap \mathfrak{U} = \varnothing\ .$$

From the open cover

$$\{V_x : x \in Z\}$$

of the compact set Z we choose a finite subcover

$$\{V_{x_i} : i = 1,2,\ldots,n\}$$

and we set

$$T = \max_{1 \leqslant i \leqslant N} t(x_i)\ .$$

This number T has the properties required by the lemma. □

Lemma 4.1.4. *Suppose that Z is a closed subset of $\mathfrak{U}$ and that $Z \cap A^- = \varnothing$. Then there exists a number $T > 0$ such that for all $x \in Z$ the trajectory $p([-T,0],x)$ does not lie entirely in $\mathfrak{U}$.*

The proof of this lemma is similar to the proof of Lemma 4.1.3.

A closed subset Z of $\mathfrak{U}$ is called *positive invariant* relative to the isolating neighborhood $\mathfrak{U}$ if each trajectory $p([0,t],x)$ with $x \in Z$ that lies in $\mathfrak{U}$ is contained in Z. The condition that Z is positive invariant can be expressed by writing $\mathfrak{P}(Z) = Z$, where $\mathfrak{P}$ is the operation on subsets defined by

$$\mathfrak{P}(Z) = \{p(t,x) : x \in Z,\ t \geqslant 0,\ p([0,t],x) \subset \mathfrak{U}\}\ .$$

Below we give conditions which ensure that sets $\mathfrak{P}(Z)$ are closed.

Lemma 4.1.5. *If Z is a closed subset of $\mathfrak{U}$ and $Z \cap A^+ = \varnothing$, then $\mathfrak{P}(Z)$ is closed.*

Proof. Since $Z \cap A^+ = \varnothing$ we have

$$p([0,+\infty),x) \not\subset \mathfrak{U}$$

for all $x \in Z$. By Lemma 4.1.3, there exists a number $T > 0$ such that

$$p([0,T],x) \not\subset \mathfrak{U}$$

for all $x \in Z$.

Suppose that $(p(t_n, x_n))$ is a sequence in $\mathfrak{P}(Z)$ that converges to a point y. We shall prove that $y \in \mathfrak{P}(Z)$. Since the sequences (x_n) and (t_n) are bounded, we can suppose without loss of generality that they are convergent, to x_* and t_*, say. Let us prove that $p([0, t_*], x_*) \subset \mathfrak{U}$. We assume that $p(\theta, x_*) \notin \mathfrak{U}$ for some $\theta \in (0, t_*)$. Without loss of generality, we can suppose that $p([0, \theta], x_*) \subset \Omega$. Then, by continuity of the flow, we have $p(\theta, x_n) \notin \mathfrak{U}$ for all sufficiently large n. Since $t_n \geqslant \theta$ for all sufficiently large n, this contradicts the condition that $p([0, t_n], x_n) \subset \mathfrak{U}$. Therefore $p([0, t_*], x_*) \subset \mathfrak{U}$. Since the flow is continuous, we have $p(t_*, x_*) = y$. Consequently $y \in \mathfrak{P}(Z)$ and the lemma follows. □

Lemma 4.1.6. *If Z is a closed subset of $\mathfrak{U}$ and $A^- \subset Z$, then $\mathfrak{P}(Z)$ is closed.*

Proof. Suppose that the sequence $(p(t_n, x_n))$ in $\mathfrak{P}(Z)$ converges to a point y. We shall prove that $y \in \mathfrak{P}(Z)$. Since $A^- \subset Z \subset \mathfrak{P}(Z)$, it suffices to consider the case $y \notin A^-$. In this case, there exists a number $T > 0$ such that $p(-T, y) \notin \mathfrak{U}$, and we can assume that $p([-T, 0], y) \subset \Omega$. Since the flow is continuous, there exists a neighborhood U of y such that $p(-T, U) \cap \mathfrak{U} = \varnothing$. We have $p(t_n, x_n) \in U$ for all sufficiently large n; therefore $t_n < T$ for all sufficiently large n. Thus the sequences (x_n) and (t_n) are bounded, and so we can assume that they are convergent, say to x_* and t_* respectively. Arguing as in the proof of Lemma 4.1.5, we can show that $p([0, t_*], x_*) \subset \mathfrak{U}$. We have $p(t_*, x_*) = y$ by continuity of the flow. Consequently $y \in \mathfrak{P}(Z)$, and the lemma is proved. □

We call a subset Z_1 of $\mathfrak{U}$ a *$\mathfrak{U}$-neighborhood* of the set $Z_2 \subset \mathfrak{U}$ if every point $x \in Z_2$ has an open neighborhood V_x such that $V_x \cap \mathfrak{U} \subset Z_1$.

Lemma 4.1.7. *Let Z be a closed subset of $\mathfrak{U}$ that is a $\mathfrak{U}$-neighborhood of the nonempty set A^-. There exists a set $Z_1 \subset \mathfrak{U}$ which satisfies $\mathfrak{P}(Z_1) \subset Z$ and which is a $\mathfrak{U}$-neighborhood of the set A^-.*

Proof. Let K be the closure of $\mathfrak{U} \setminus Z$. Since Z is a $\mathfrak{U}$-neighborhood of A^-, it follows that $K \cap A^- = \varnothing$. Then, by Lemma 4.1.4, there exists a number $T > 0$ such that $p([-T, 0], x) \not\subset \mathfrak{U}$ for all $x \in K$.

We shall prove that every point $x \in A^-$ has a neighborhood V_x such that

$$\mathfrak{P}(V_x \cap \mathfrak{U}) \subset Z .$$

We shall consider two cases.

Case 1. Suppose that the trajectory $p([0, T], x)$ lies in $\mathfrak{U}$. Then, by Lemma 4.1.1, there exists a neighborhood V_x of x such that

$$p([0, T], V_x) \cap K = \varnothing .$$

For every $\xi \in V_x$, the trajectory $p([0,T],\xi)$ does not meet K, and so

$$\{p(t,\xi) : t \in [0,T],\ p([0,t],\xi) \subset \mathfrak{U}\} \subset Z\ .$$

Now if $t > T$ and $p([0,t],\xi) \subset \mathfrak{U}$, then $p((T,t],\xi) \subset Z$ by the definition of T. Therefore $\mathfrak{P}(V_x \cap \mathfrak{U}) \subset Z$.

Case 2. Suppose that $p([0,T],x)$ does not lie in B. We set

$$t_* = \sup_{p([0,t],x)\subset\mathfrak{U}} t\ .$$

Suppose that $t' > t_*$ is so close to t_* that

$$p([0,t'],x) \cap K = \varnothing$$

and $p(t',x) \notin \mathfrak{U}$. Then, by Lemma 4.1.1, there exists a neighborhood V_x of x such that

$$p([0,t'],V_x) \cap K = \varnothing, \quad p(t',V_x) \cap \mathfrak{U} = \varnothing\ .$$

For this neighborhood we have

$$\mathfrak{P}(V_x \cap \mathfrak{U}) \subset Z\ .$$

The set

$$V = \bigcup_{x\in A^-} V_x$$

is open, contains A^-, and satisfies the condition

$$\mathfrak{P}(V \cap \mathfrak{U}) \subset Z\ .$$

We can now take as Z_1 any closed subset of V which is an $\mathfrak{U}$-neighborhood of A^-. □

4.1.2 Index Pairs for Invariant Sets

An *index pair* for an isolated invariant set S relative to an isolating neighborhood $\mathfrak{U}$ is a pair $\langle N_1, N_2\rangle$ of closed subsets $N_1,\ N_2 \subset \mathfrak{U}$ satisfying the following conditions:

(1) $\mathfrak{P}(N_1) = N_1$, $\mathfrak{P}(N_2) = N_2$;

(2) $S \cap N_2 = \varnothing$, $N_1 \setminus \mathfrak{U}$ is an $\mathfrak{U}$-neighborhood of the set A^-;

(3) if $x \in N_1 \setminus N_2$ and $p(t,x) \notin \mathfrak{U}$ for some $t > 0$, then for some $t_0 \in (0,t)$ we have

$$p([0,t_0),x) \subset N_1 \setminus N_2, \quad p(t_0,x) \in N_2\ .$$

In this section, we shall prove that every closed bounded invariant set has an index pair relative to each of its isolating neighborhoods. We shall need the following result.

Lemma 4.1.8. *If $\langle N_1, N_2\rangle$ is an index pair, then $A^+ \cap N_2 = \varnothing$.*

Proof. Since S and N_2 are closed and disjoint, S has an open neighborhood U_S which is disjoint from N_2. There are no trajectories lying entirely in $\widetilde{\mathfrak{U}} = \mathfrak{U} \setminus U_S$. We claim that $\widetilde{\mathfrak{U}}$ can be written as a union of closed sets $\widetilde{\mathfrak{U}}_1$ and $\widetilde{\mathfrak{U}}_2$ such that $p([0,+\infty), x) \not\subset \mathfrak{U}$ for $x \in \widetilde{\mathfrak{U}}_1$ and $p((-\infty;0], x) \not\subset \mathfrak{U}$ for $x \in \widetilde{\mathfrak{U}}_2$. Since $\widetilde{\mathfrak{U}} \cap A^+$ and $\widetilde{\mathfrak{U}} \cap A^-$ are disjoint, these sets have disjoint open neighborhoods W^+ and W^-. We set

$$\widetilde{\mathfrak{U}}_1 = \widetilde{\mathfrak{U}} \setminus W^+ , \quad \widetilde{\mathfrak{U}}_2 = \widetilde{\mathfrak{U}} \setminus W^- .$$

These two sets satisfy the required conditions since

$$\widetilde{\mathfrak{U}}_1 \cup \widetilde{\mathfrak{U}}_2 = (\widetilde{\mathfrak{U}} \setminus W^+) \cup (\widetilde{\mathfrak{U}} \setminus W^-) = \widetilde{\mathfrak{U}} \setminus (W^+ \cap W^-) = \widetilde{\mathfrak{U}} .$$

By Lemmas 4.1.3 and 4.1.4, there exists a number $T > 0$ such that

$$p([0,T], x) \not\subset \mathfrak{U} \quad (x \in \widetilde{\mathfrak{U}}_1) ,$$

$$p([-T,0], x) \not\subset \mathfrak{U} \quad (x \in \widetilde{\mathfrak{U}}_2) .$$

Thus

$$p([-T,T], x) \not\subset \mathfrak{U} \quad (x \in \widetilde{\mathfrak{U}}) .$$

Let

$$\xi \in A^+ \cap N_2 .$$

Then $\xi \in \widetilde{\mathfrak{U}}$. From what was proved above, the trajectory $p([0,3T], \xi)$ does not lie entirely in $\widetilde{\mathfrak{U}}$. Consequently some point on it is not in N_2, and this contradicts the positive invariance of N_2 (see condition (1) in the definition of an index pair). □

Now we prove the the existence of index pairs.

Theorem 4.1.1. *Suppose that S is a nonempty isolated invariant set and that W is an arbitrary open set containing S. For any isolating neighborhood $\mathfrak{U}$ of S there exists an index pair $\langle N_1, N_2\rangle$ such that $N_1 \setminus N_2 \subset W$.*

Proof. Since S and $\overline{\Omega \setminus \mathfrak{U}}$ are closed and disjoint, they have disjoint open neighborhoods W_S and W'_S. Since $\overline{W_S} \subset \operatorname{int} \mathfrak{U}$, the set S has an open neighborhood whose closure lies in $\operatorname{int} \mathfrak{U}$. Therefore, when proving the theorem, we can assume that $\overline{W} \subset \operatorname{int} \mathfrak{U}$.

Since $A^+ \setminus W$ and $A^- \setminus W$ are closed and disjoint, they have disjoint open neighborhoods, V^+ and V^-, say.

The sets A^+ and A^- have neighborhoods U^+ and U^- satisfying

$$U^+ \cap U^- \subset W ;$$

we can take, for instance,

$$U^+ = W \cup V^+ \,, \quad U^- = W \cup V^- \,.$$

By Lemma 4.1.5, the set

$$N_2 = \mathfrak{P}(\mathfrak{U} \setminus U^+)$$

is closed, and by construction,

$$N_2 \cap A^+ = \varnothing \,.$$

By Lemma 4.1.7, there exists a closed subset Z of $\mathfrak{U}$ which is a $\mathfrak{U}$-neighborhood of A^- and satisfies

$$\mathfrak{P}(Z) \subset U^- \,.$$

We set $N_1 = \mathfrak{P}(Z)$; from Lemma 4.1.6 this set is closed. We shall show that the pair $\langle N_1, N_2 \rangle$ satisfies all conditions of the theorem. Since

$$N_1 \subset U^- \,, \quad \mathfrak{U} \setminus U^+ \subset N_2 \,,$$

we have

$$N_1 \setminus N_2 = N_1 \cap (\mathfrak{U} \setminus N_2) \subset U^- \cap U^+ \subset W \,.$$

It remains to verify condition (3) in the definition of an index pair.

Let $x \in N_1 \setminus N_2$ and $p(t,x) \notin \mathfrak{U}$ for some $t > 0$. We set

$$t_* = \sup_{p([0,t],x) \subset N_1 \setminus N_2} t \,.$$

Since N_1 is closed, we have

$$p([0,t_*],x) \subset N_1 \,.$$

Since $\overline{W} \subset \operatorname{int} \mathfrak{U}$ it follows that

$$p(t_*,x) \in \operatorname{int} \mathfrak{U} \,.$$

Consequently the trajectory $p([0,t_*],x)$ can be extended to a trajectory $p([0,t_*+\varepsilon],x)$ lying in $\mathfrak{U}$. Since N_1 is positive invariant, this trajectory lies in N_1. By construction, there exists a monotone nonincreasing sequence (t_n) in $[t_*, t_*+\varepsilon]$ which tends to t_* and satisfies

$$p(t_n,x) \notin N_1 \setminus N_2 \,.$$

Since, in addition, $p(t_n,x) \in N_1$ for all n, it follows that $p(t_n,x) \in N_2$. Therefore we have

$$p(t_*,x) = \lim_{n\to\infty} p(t_n,x) \in N_2$$

because N_2 is closed. Thus condition (3) in the definition of an index pair holds for the pair $\langle N_1, N_2 \rangle$, and the theorem is proved. □

Remark. If $S = \varnothing$, then the pair of sets

$$\langle N_1, N_2 \rangle = \langle \mathfrak{U}, \mathfrak{U} \rangle$$

is an index pair for S.

4.1.3 Conley Index

The *Conley index* of an isolated invariant set S is the homotopy type of the quotient pointed space N_1/N_2, where $\langle N_1, N_2\rangle$ is an arbitrary index pair for S.

In this section, we show that the Conley index of S is well defined, i.e., that the homotopy type of N_1/N_2 is independent of the choice of isolating neighborhood and index pair.

We fix an arbitrary index pair $\langle N_1, N_2\rangle$ and define

$$N_1^t = \{p(t,x) : x \in N_1, p([0,t],x) \subset \mathfrak{U}\} ,$$

$$N_2^{-t} = \{p((-\tau),x) : x \in N_2, \tau \in [0,t], p([-\tau,0],x) \subset \mathfrak{U}\} .$$

Lemma 4.1.9. *The sets N_1^t and N_2^{-t} are closed for all $t \geqslant 0$.*

Proof. Suppose that the sequence $(p(t,x_n))$ in N_1^t converges to a point y. We can assume without loss of generality that $x_n \to x_* \in N_1$. Let us prove that the trajectory $p([0,t],x_*)$ lies in $\mathfrak{U}$. Suppose instead that $p(\theta,x_*) \notin \mathfrak{U}$ (and $p([0,\theta],x_*) \subset \Omega$). Since $x_n \to x_*$, it follows that $p(\theta,x_n) \notin \mathfrak{U}$ for all sufficiently large n. This contradicts the assumption that $p(t,x_n) \in N_1^t$ for each $n \in \mathbb{N}$. Since

$$y = \lim_{n\to\infty} p(t,x_n) = p(t,x_*) ,$$

we have $y \in N_1^t$ by definition. Therefore N_1^t is closed.

Now suppose that the sequence $(p(-\tau_n,x_n))$ in N_2^{-t} converges to a point $y \in \mathfrak{U}$. We shall prove that $y \in N_2^{-t}$. We can assume without loss of generality that (x_n) and (τ_n) converge to $x_* \in N_2$ and $\tau_* \in [0,t]$ respectively.

We claim that the trajectory $p([-\tau_*,0],x_*)$ lies in $\mathfrak{U}$. Suppose that $p(\theta,x_*) \notin \mathfrak{U}$ for some $\theta \in (-\tau_*,0)$. Since $x_n \to x_*$, it follows that $p(\theta,x_n) \notin \mathfrak{U}$ for all sufficiently large n. However,

$$p([-\tau_n,0],x_n) \subset \mathfrak{U}$$

by construction. These two assertions contradict each other because $\tau_n > -\theta$ for all sufficiently large n. Thus our claim above follows.

Since

$$p(-\tau_*,x_*) = \lim_{n\to\infty} p(-\tau_n,x_n) = y ,$$

we find that $y \in N_2^{-t}$, and we conclude that N_2^t is closed. □

Lemma 4.1.10. *For all $t \geqslant 0$ the embedding i: $N_1^t/N_2 \to N_1/N_2$ is a homotopy equivalence, with homotopy inverse h: $N_1/N_2 \to N_1^t/N_2$ defined by*

$$h([x]) = \begin{cases} [p(t,x)] & \text{if } p([0,t],x) \subset N_1 \setminus N_2 , \\ [N_2] & \text{if } p([0,t],x) \not\subset N_1 \setminus N_2 . \end{cases}$$

Proof. Consider the mapping

$$\mathcal{H} : N_1/N_2 \times [0,1] \to N_1/N_2$$

defined by

$$\mathcal{H}([x],\theta) = \begin{cases} [p(\theta t, x)] & \text{if } p([0,\theta t], x) \subset N_1 \setminus N_2 \,, \\ [N_2] & \text{if } p([0,\theta t], x) \not\subset N_1 \setminus N_2 \,. \end{cases}$$

We shall prove that $\mathcal{H}$ is continuous by showing that the complete preimage $\mathcal{H}^{-1}(U)$ of an arbitrary open set $U \subset N_1/N_2$ is open in $N_1/N_2 \times [0,1]$.

Let $\widetilde{U}$ be the union of all classes belonging to U. From the definition of the quotient topology, this set is open in N_2, i.e., of form

$$\approx{U} \cap N_1$$

with $\approx{U}$ open in Ω.

We fix an element $([x],\theta) \in \mathcal{H}^{-1}(U)$. First consider the case $[x] = [N_2]$. In this case $[N_2] \in U$, and so

$$N_1 \cap N_2 \subset \approx{U} \,.$$

We shall construct an open neighborhood $\approx{V}$ of $N_1 \cap N_2$ satisfying

$$p([0,t], \approx{V} \cap N_1) \cap \mathfrak{U} \subset \approx{U} \,.$$

We fix $\xi \in N_1 \cap N_2$. If

$$p([0,t],\xi) \subset \mathfrak{U} \,,$$

then

$$p([0,t],\xi) \subset N_1 \cap N_2$$

by the definition of an index pair, and ξ has a neighborhood $\approx{V}_\xi$ satisfying

$$p([0,t], \approx{V}_\xi) \subset \approx{U} \,,$$

by Lemma 4.1.1. If instead

$$p([0,t],\xi) \not\subset \mathfrak{U} \,,$$

then we can find $\tau \in [0,t]$ such that

$$p([0,\tau],\xi) \subset \approx{U} \quad \text{and} \quad p(\tau,\xi) \notin \mathfrak{U} \,.$$

Now ξ has a neighborhood $\approx{V}_\xi$ such that

$$p([0,\tau], \approx{V}_\xi) \subset \approx{U} \quad \text{and} \quad p(\tau, \approx{V}_\xi) \cap \mathfrak{U} = \varnothing \,,$$

and writing

$$\widetilde{\widetilde{V}} = \bigcup_{\xi \in N_1 \cap N_2} \widetilde{\widetilde{V}}_\xi ,$$

we have by construction

$$p([0,t], \widetilde{\widetilde{V}}) \cap \mathfrak{U} \subset \widetilde{\widetilde{U}} .$$

We set

$$V = \bigcup_{\xi \in N_1 \cap \widetilde{\widetilde{V}}} \{[\xi]\} .$$

This set is open in N_1/N_2 and

$$V \times [0,1] \subset \mathcal{H}^{-1}(U) .$$

Now we pass to the case $[x] \neq [N_2]$, i.e., $\{x\} \neq N_2$. Suppose that

$$p([0, \theta t], x) \not\subset \mathfrak{U} .$$

Then $\theta > 0$ and

$$\mathcal{H}([x], \theta) = [N_2] .$$

For sufficiently small $\varepsilon > 0$ we have

$$p([0, (\theta - \varepsilon)t], x) \not\subset \mathfrak{U} .$$

There exists a neighborhood $\widetilde{\widetilde{V}}_x$ of x disjoint from N_2 such that

$$p([0, (\theta - \varepsilon)t], \xi) \not\subset \mathfrak{U} \quad (\xi \in \widetilde{\widetilde{V}}_x) .$$

We set

$$V = \bigcup_{\xi \in N_1 \cap \widetilde{\widetilde{V}}_x} \{[\xi]\} .$$

By definition, for all $[\xi] \in V$ and $\theta' \in (\theta - \varepsilon, \theta + \varepsilon) \cap [0,1]$ we have

$$\mathcal{H}([\xi], \theta') = [N_2] .$$

Thus

$$V \times ((\theta - \varepsilon, \theta + \varepsilon) \cap [0,1]) \subset \mathcal{H}^{-1}(U) .$$

Suppose now that $x \notin N_2$ but

$$p([0, \theta t], x) \subset \mathfrak{U} \quad \text{and} \quad p([0, \theta t], x) \not\subset N_1 \setminus N_2 .$$

In this case $\theta > 0$. Since $\mathcal{H}([x], \theta) = [N_2]$, it follows that $N_1 \cap N_2 \subset \widetilde{\widetilde{U}}$. Therefore

$$p(\theta t, x) \in \widetilde{\widetilde{U}} .$$

For some neighborhood $\tilde{\tilde{V}}_x$ of x and some $\varepsilon > 0$ we have

$$p([(\theta-\varepsilon)t, (\theta+\varepsilon)t], \tilde{\tilde{V}}_x) \subset \tilde{\tilde{U}} \ .$$

We set

$$V = \bigcup_{\xi \in N_1 \cap \tilde{\tilde{V}}_x} \{[\xi]\} \ .$$

By definition,

$$\mathcal{H}([\xi], \theta') \subset U$$

for all $[\xi] \in V$ and

$$\theta' \in (\theta-\varepsilon, \theta+\varepsilon) \cap [0,1] \ .$$

Finally, suppose that

$$p([0, \theta t], x) \subset N_1 \setminus N_2 \ ,$$

where

$$\mathcal{H}([x], \theta) = [p(\theta t, x)] \neq [N_2] \ .$$

Since

$$\mathcal{H}([x], \theta) \in U \ ,$$

it follows that

$$p(\theta t, x) \in \tilde{\tilde{U}} \ .$$

For some neighborhood $\tilde{\tilde{V}}_x$ of x and $\varepsilon > 0$ we have

$$p([0, (\theta+\varepsilon)t], \tilde{\tilde{V}}_x) \cap N_2 = \varnothing, \quad p([(\theta-\varepsilon)t, (\theta+\varepsilon)t], \tilde{\tilde{V}}_x) \subset \tilde{\tilde{U}} \ .$$

We set

$$V = \bigcup_{\xi \in N_1 \cap \tilde{\tilde{V}}_x} \{[\xi]\} \ .$$

By definition,

$$\mathcal{H}([\xi], \theta') \subset U$$

for all $[\xi] \in V$, $\theta' \in (\theta-\varepsilon, \theta+\varepsilon) \cap [0,1]$.

This completes the proof that $\mathcal{H}$ is continuous.

Now consider the mapping

$$\mathcal{H}^t : N_1^t/N_2 \times [0,1] \mapsto N_1^t/N_2$$

defined by

$$\mathcal{H}^t([x],\theta) = \begin{cases} [p(\theta t, x)] & \text{if } \; p([0,\theta t],x) \subset N_1^t \setminus N_2 \,, \\ [N_2] & \text{if } \; p([0,\theta t],x) \not\subset N_1^t \setminus N_2 \,. \end{cases}$$

An argument similar to the one establishing the continuity of $\mathcal{H}$ shows that $\mathcal{H}^t$ is continuous. By definition we have

$$i \circ h = \mathcal{H}(\cdot,1), \quad h \circ i = \mathcal{H}^t(\cdot,1) \,.$$

Since the mappings $\mathcal{H}(\cdot,1)$ and $\mathcal{H}^t(\cdot,1)$ are homotopic to identity mappings (by homotopies that were constructed earlier), it follows that i is a homotopy equivalence with homotopy inverse h, and the lemma is proved. □

Lemma 4.1.11. *Suppose that X_1, X_2, X_3, X_4 are topological spaces with base point and that φ_{12}, φ_{23}, φ_{34} are base-point preserving continuous mappings*

$$X_1 \xrightarrow{\varphi_{12}} X_2 \xrightarrow{\varphi_{23}} X_3 \xrightarrow{\varphi_{34}} X_4 \,.$$

If $\varphi_{23} \circ \varphi_{12}$ and $\varphi_{34} \circ \varphi_{23}$ are homotopy equivalences, then so are φ_{12}, φ_{23}, φ_{34}.

Proof. Let φ_{31} and φ_{42} be homotopy inverses of $\varphi_{23} \circ \varphi_{12}$ and $\varphi_{34} \circ \varphi_{23}$. By definition,

$$\varphi_{23} \circ \varphi_{12} \circ \varphi_{31} \sim \mathrm{id}_{(X_3)} \,, \tag{4.1.1}$$

$$\varphi_{31} \circ \varphi_{23} \circ \varphi_{12} \sim \mathrm{id}_{(X_1)} \,, \tag{4.1.2}$$

$$\varphi_{34} \circ \varphi_{23} \circ \varphi_{42} \sim \mathrm{id}_{(X_4)} \,, \tag{4.1.3}$$

$$\varphi_{42} \circ \varphi_{34} \circ \varphi_{23} \sim \mathrm{id}_{(X_2)} \,, \tag{4.1.4}$$

where $\mathrm{id}_{(X_n)}$ denotes the identity mapping from X_n into X_n and the notation $f \sim g$ means that the mappings f and g are homotopic.

It follows from (4.1.2) that $\varphi_{31} \circ \varphi_{23}$ is a left homotopy inverse of φ_{12}. Consider the mapping $\varphi_{12} \circ \varphi_{31} \circ \varphi_{23}$:

$$\begin{aligned} \varphi_{12} \circ \varphi_{31} \circ \varphi_{23} &\sim \mathrm{id}_{(X_2)} \circ \varphi_{12} \circ \varphi_{31} \circ \varphi_{23} \\ &\sim \varphi_{42} \circ \varphi_{34} \circ \varphi_{23} \circ \varphi_{12} \circ \varphi_{31} \circ \varphi_{23} \\ &\sim \varphi_{42} \circ \varphi_{34} \circ \mathrm{id}_{(X_3)} \circ \varphi_{23} \sim \mathrm{id}_{(X_2)} \,. \end{aligned} \tag{4.1.5}$$

Thus φ_{12} is a homotopy equivalence with homotopy inverse $\varphi_{31} \circ \varphi_{23}$.

By (4.1.1) and (4.1.5), $\varphi_{12} \circ \varphi_{31}$ is a homotopy inverse of φ_{23}.

It follows from (4.1.3) that $\varphi_{23} \circ \varphi_{42}$ is a right homotopy inverse of φ_{34}. By (4.1.4), we have

$$\begin{aligned}\varphi_{23} \circ \varphi_{42} \circ \varphi_{34} &\sim \varphi_{23} \circ \varphi_{42} \circ \varphi_{34} \circ \varphi_{23} \circ \varphi_{12} \circ \varphi_{31} \\ &\sim \varphi_{23} \circ \mathrm{id}_{(X_2)} \circ \varphi_{12} \circ \varphi_{31} \sim \mathrm{id}_{(X_3)} .\end{aligned}$$

Thus $\varphi_{23} \circ \varphi_{42}$ is a homotopy inverse of φ_{34}. □

The following elementary result is well known; cf. [116].

Lemma 4.1.12 (factorization of a continuous mapping). *Let $f : X \to Y$ be a continuous mapping of topological spaces and suppose that $X_1 \subset X$, $Y_1 \subset Y$. If $f(X_1) \subset Y_1$, then the mapping from X/X_1 to Y/Y_1 induced by f is continuous.*

Lemma 4.1.13. *The mapping $g : N_1/N_2^{-t} \mapsto N_1^t/N_2$ defined by*

$$g([x]) = \begin{cases} [p(t,x)] & \textit{for } \ p([0,t],x) \subset N_1 \setminus N_2 , \\ [N_2] & \textit{for } \ p([0,t],x) \not\subset N_1 \setminus N_2 \end{cases}$$

is a homeomorphism.

Proof. Direct verification shows that g is a bijection. In Lemma 4.1.10 we proved the continuity of the mapping

$$h : N_1/N_2 \to N_1^t/N_2$$

defined by

$$h([x]) = \begin{cases} [p(t,x)] & \text{for } \ p([0,t],x) \subset N_1 \setminus N_2 , \\ [N_2] & \text{for } \ p([0,t],x) \not\subset N_1 \setminus N_2 . \end{cases}$$

Since h takes the set N_2^{-t}/N_2 to $\{[N_2]\}$, the continuity of g follows from Lemma 4.1.12.

Now consider the mapping

$$\widetilde{g} : N_1^t \to N_1, \quad \widetilde{g}(x) = p(-t,x) .$$

If $x \in N_2$, then $\widetilde{g}(x) \in N_2^{-t}$. Therefore, by Lemma 4.1.12, the mapping

$$\approx{g} : N_1^t/N_2 \to N_1/N_2^{-t}, \quad \approx{g}([x]) = [p(-t,x)]$$

is continuous. Since $\approx{g}$ is the inverse of g, the lemma follows. □

Lemma 4.1.14. *For all $t \geqslant 0$ the embedding*

$$j : N_1/N_2 \to N_1/N_2^{-t}$$

is a homotopy equivalence; moreover $i \circ g$ is a homotopy inverse of j.

Proof. Consider the chain of mappings

$$N_1/N_2 \xrightarrow{j} N_1/N_2^{-t} \xrightarrow{g} N_1^t/N_2 \xrightarrow{i} N_1/N_2 \,.$$

Being the composite of two homotopy equivalences, $i \circ g$ is a homotopy equivalence. The mapping $g \circ j$ coincides with the mapping h in the hypothesis of Lemma 4.1.10. Then it follows from Lemma 4.1.11 that j is a homotopy equivalence. By Lemma 4.1.10,

$$(i \circ g) \circ j = i \circ h \sim \mathrm{id}_{(N_1/N_2)} \,,$$

and

$$\begin{aligned} j \circ (i \circ g) &\sim (i \circ g)^{-1} \circ i \circ g \circ j \circ i \circ g \sim (i \circ g)^{-1} \circ i \circ h \circ i \circ g \\ &\sim (i \circ g)^{-1} \circ i \circ g \sim \mathrm{id}_{(N_1/N_2^{-t})} \,. \end{aligned}$$

The lemma follows. □

Lemma 4.1.15. *Let $\langle N_1, N_2 \rangle$ and $\langle \widetilde{N}_1, \widetilde{N}_2 \rangle$ be two index pairs. Then*

$$N_1^t \subset \widetilde{N}_1, \quad N_1^t \cap N_2 \subset \widetilde{N}_1 \cap \widetilde{N}_1^{-t}$$

for all sufficiently large $t \geqslant 0$.

Proof. Let K be the closure of $\mathfrak{U} \setminus \widetilde{N}_1$. By condition (2) in the definition of an index pair,

$$K \cap A^- = \varnothing \,.$$

By Lemma 4.1.4, there exists a number $T > 0$ such that

$$p([-T, 0], x) \not\subset \mathfrak{U}$$

for all $x \in K$. Consequently,

$$N_1^t \cap K = \varnothing$$

for all $t \geqslant T$. Therefore $N_1^t \subset \widetilde{N}_1$ for all $t \geqslant T$.

Now N_2 is closed and, by Lemma 4.1.8, disjoint from A^+. From Lemma 4.1.3, for some $T_1 \geqslant T$ we have $p([0, T_1], x) \not\subset \mathfrak{U}$ for all $x \in N_2$. Let $t \geqslant T_1$. If $x \in N_1^t \cap N_2$, then, by what was proved above, $x \in \widetilde{N}_1$, and since we have $p([0, t], x) \not\subset \mathfrak{U}$, it follows that $x \in \widetilde{N}_2^{-t}$. □

Let us fix two index pairs $\langle N_1, N_2 \rangle$ and $\langle \widetilde{N}_1, \widetilde{N}_2 \rangle$. By Lemma 4.1.15, for all sufficiently large t we have embeddings of pairs

$$\begin{aligned} (N_1^t, N_1^t \cap N_2) &\subset (\widetilde{N}_1, \widetilde{N}_2 \cap \widetilde{N}_2^{-t}) \,, \\ (\widetilde{N}_1^t, \widetilde{N}_1^t \cap \widetilde{N}_2) &\subset (N_1, N_1 \cap N_2^{-t}) \,, \end{aligned} \tag{4.1.6}$$

and they induce embeddings of quotient spaces

$$i_* : N_1^t/N_2 \mapsto \widetilde{N}_1/\widetilde{N}_2^{-t}, \quad \widetilde{i}_* : \widetilde{N}_1^t/\widetilde{N}_2 \mapsto N_1/N_2^{-t} . \tag{4.1.7}$$

Below we shall show that the embeddings (4.1.7) are homotopy equivalences. It will follow that the Conley index of an invariant set is independent of the choice of index pair for a fixed isolating neighborhood.

Lemma 4.1.16. *The embeddings* (4.1.7) *are homotopy equivalences.*

Proof. Consider the chain of mappings

$$\begin{aligned} N_1^t/N_2 \xrightarrow{i_*} \widetilde{N}_1/\widetilde{N}_2^{-t} \xrightarrow{g} \widetilde{N}_1^t/\widetilde{N}_2 \xrightarrow{\widetilde{i}_*} N_1/N_2^{-t} \\ \xrightarrow{g} N_1^t/N_2 \xrightarrow{i_*} \widetilde{N}_1/\widetilde{N}_2^{-t} . \end{aligned} \tag{4.1.8}$$

Direct verification shows that the composite $\widetilde{i}_* \circ g \circ i_*$ of the first three mappings in (4.1.8) is equal to the composite of the mappings in the chain

$$\begin{aligned} N_1^t/N_2 \xrightarrow{i} N_1/N_2 \xrightarrow{j} N_1/N_2^{-t} \xrightarrow{g} N_1^t/N_2 \\ \xrightarrow{i} N_1/N_2 \xrightarrow{j} N_1/N_2^{-t} . \end{aligned} \tag{4.1.9}$$

Since each mapping in (4.1.9) is a homotopy equivalence, $\widetilde{i}_* \circ g \circ i_*$ is a homotopy equivalence. Similarly, the mapping $i_* \circ g \circ \widetilde{i}_*$ in (4.1.8) is a homotopy equivalence. Therefore by Lemma 4.1.11 the mapping $\widetilde{i}_*$ in (4.1.8) is a homotopy equivalence. □

Next we show now that the Conley index of an invariant set is independent of the choice of isolating neighborhood. Let $\mathfrak{U}$ and $\widehat{\mathfrak{U}}$ be isolating neighborhoods of the invariant set S. We can assume without loss of generality that $\widehat{\mathfrak{U}} \subset \mathfrak{U}$.

Lemma 4.1.17. *Let $\langle N_1, N_2 \rangle$ be an index pair for S with respect to $\mathfrak{U}$. If the closure of $N_1 \setminus N_2$ consists of interior points of $\widehat{\mathfrak{U}}$, then $\langle N_1 \cap \widehat{\mathfrak{U}}, N_2 \cap \widehat{\mathfrak{U}} \rangle$ is an index pair for S with respect to $\widehat{\mathfrak{U}}$. The embedding*

$$\widehat{i} : (N_1 \cap \widehat{\mathfrak{U}})/N_2 \to N_1/N_2 \tag{4.1.10}$$

is a homotopy equivalence.

Proof. Since the sets N_1, N_2, $\widehat{\mathfrak{U}}$ are closed, so are the sets $N_1 \cap \widehat{\mathfrak{U}}$ and $N_2 \cap \widehat{\mathfrak{U}}$. We shall verify that $N_1 \cap \widehat{\mathfrak{U}}$ is positive invariant. Let

$$x \in N_1 \cap \widehat{\mathfrak{U}} \quad \text{and} \quad p([0,t], x) \subset \widehat{\mathfrak{U}} .$$

Since $\widehat{\mathfrak{U}} \subset \mathfrak{U}$, by the positive invariance of N_1 we have

$$p([0,t],x) \subset N_1 \, .$$

Thus

$$p([0,t],x) \subset N_1 \cap \widehat{\mathfrak{U}} \, .$$

The positive invariance of N_2 can be verified similarly.

We set

$$\widehat{A}^- = \{x \in \widehat{\mathfrak{U}} : p([-\infty,0],x) \subset \widehat{\mathfrak{U}}\} \, .$$

Clearly $\widehat{A}^- \subset A^-$. If $x \in \widehat{A}^-$ then $x \in A^-$, and x has a neighborhood V such that

$$V \cap \mathfrak{U} \subset N_1 \, ;$$

thus

$$V \cap \widehat{\mathfrak{U}} \subset N_1 \cap \widehat{\mathfrak{U}} \, ,$$

and so $N_1 \cap \widehat{\mathfrak{U}}$ is a $\mathfrak{U}$-neighborhood of $\widehat{A}^-$.

Now we verify condition (3) in the definition of an index pair. Suppose that

$$x \in (N_1 \cap \widehat{\mathfrak{U}}) \setminus N_2$$

and that $p(T,x) \notin \widehat{\mathfrak{U}}$ for some $T > 0$. We set

$$T_1 = \sup_{p([0,t],x) \subset (N_1 \cap \widehat{\mathfrak{U}}) \setminus N_2} t \, .$$

The trajectory $p([0,T_1],x)$ lies in $N_1 \cap \widehat{\mathfrak{U}}$. By definition,

$$p([0,T_1],x) \subset \overline{N_1 \setminus N_2} \subset \operatorname{int} \widehat{\mathfrak{U}} \, .$$

Consequently $p([0,T_1],x)$ can be extended to a trajectory $p([0,T_1+\nu],x)$ lying in $\widehat{\mathfrak{U}}$. Since N_1 is positive invariant, we have

$$p([0,T_1+\nu],x) \subset N_1 \cap \widehat{\mathfrak{U}} \, .$$

However, from the definition of N_1, there exists a decreasing sequence (t_n) in $[T_1, T_1+\nu]$ with $t_n \to T_1$ whose elements do not belong to $(N_1 \cap \widehat{\mathfrak{U}}) \setminus N_2$. Consequently $t_n \in N_2$ for each n, and so

$$p(T_1,x) = \lim_{n\to\infty} p(t_n,x) \in N_2 \cap \widehat{\mathfrak{U}} \, .$$

Thus $\langle N_1 \cap \widehat{\mathfrak{U}}, N_2 \cap \widehat{\mathfrak{U}} \rangle$ is an index pair for S with respect to $\widehat{\mathfrak{U}}$. Direct verification shows that $\widehat{i}$ is a bijection.

Next we prove that $\widehat{i}$ is continuous. Let V be an open subset of N_1/N_2, and $U = \widehat{i}^{-1}(V)$. We can write the set

$$\widehat{V} = \bigcup_{[x]\in V} \{x\}$$

in the form $\widehat{\widehat{V}} \cap N_1$, with $\widehat{\widehat{V}}$ open. We shall prove that U is open. It suffices to show that the set

$$\widehat{U} = \bigcup_{[x] \in U} \{x\}$$

is equal to $\widehat{\widehat{V}} \cap N_1 \cap \widehat{\mathfrak{U}}$. If $[N_2] \notin V$, then, by definition,

$$\widehat{V} = \widehat{U} \subset \widehat{\mathfrak{U}} ,$$

so that

$$\widehat{U} = \widehat{\widehat{V}} \cap N_1 = \widehat{\widehat{V}} \cap N_1 \cap \widehat{\mathfrak{U}} .$$

If $[N_2] \in V$, then

$$N_1 \cap N_2 \subset \widehat{V} ,$$

and in this case,

$$\widehat{U} = \widehat{V} \cap \widehat{\mathfrak{U}} = \widehat{\widehat{V}} \cap N_1 \cap \widehat{\mathfrak{U}} .$$

The continuity of $\widehat{i}$ follows: the continuity of the inverse of $\widehat{i}$ can be established similarly. □

We have now shown that the Conley index is well defined, i.e., that the homotopy type of the quotient space N_1/N_2 is independent of the choice of the isolating neighborhood and the index pair.

Next we establish several important properties of index pairs.

Theorem 4.1.2. *Let $\langle N_1, N_2 \rangle$, $\langle \widetilde{N}_1, \widetilde{N}_2 \rangle$ be two index pairs for the invariant set S with respect to the same isolating neighborhood $\mathfrak{U}$. If*

$$N_1 \subset \widetilde{N}_1 \quad \textit{and} \quad N_1 \cap N_2 \subset \widetilde{N}_1 \cap \widetilde{N}_2 ,$$

then the induced embedding

$$i_0 : N_1/N_2 \to \widetilde{N}_1/\widetilde{N}_2$$

is a homotopy equivalence.

Proof. It follows from Lemma 4.1.15 that

$$\widetilde{N}_1^t \subset N_1 \quad \text{and} \quad \widetilde{N}_1^t \cap \widetilde{N}_2 \subset N_1 \cap N_2^{-t}$$

for t large enough, and, by Lemma 4.1.16, the embedding

$$\widetilde{i}_* : \widetilde{N}_1^t/\widetilde{N}_2 \to N_1/N_2^{-t}$$

is a homotopy equivalence. Let φ be the composite of the mappings in the chain

$$\widetilde{N}_1/\widetilde{N}_2 \xrightarrow{j} \widetilde{N}_1/\widetilde{N}_2^{-t} \xrightarrow{\widetilde{g}} \widetilde{N}_1^t/\widetilde{N}_2 \xrightarrow{\widetilde{i}_*} N_1/N_2^{-t} \xrightarrow{g} N_1^t/N_2 \xrightarrow{i} N_1/N_2 .$$

Direct verification shows that $\varphi \circ i_0$ and $i_0 \circ \varphi$ are equal respectively to the composite of the mappings

$$N_1/N_2 \xrightarrow{h} N_1^{2t}/N_2 \xrightarrow{i} N_1/N_2$$

and the composite of the mappings

$$\widetilde{N}_1/\widetilde{N}_2 \xrightarrow{h} \widetilde{N}_1^{2t}/\widetilde{N}_2 \xrightarrow{i} \widetilde{N}_1/\widetilde{N}_2 \, .$$

By Lemma 4.1.10,

$$\varphi \circ i_0 \sim \mathrm{id}_{(N_1/N_2)}, \quad i_0 \circ \varphi \sim \mathrm{id}_{(\widetilde{N}_1/\widetilde{N}_2)} \, .$$

Thus i_0 and φ are inverse homotopy equivalences, as required. □

We shall now prove a theorem on the Conley index of a Cartesian product of invariant sets. It is often useful for calculating the Conley index of specific invariant sets.

Suppose that flows p_1 and p_2 are defined in open sets $\Omega_1 \subset \mathbb{R}^{m_1}$ and $\Omega_2 \subset \mathbb{R}^{m_2}$. Let S_1 and S_1 be isolated invariant sets in Ω_1 and Ω_2 associated with isolating neighborhoods $\mathfrak{U}_1$ and $\mathfrak{U}_2$. Suppose that $\langle N_1, N_2 \rangle$ is an index pair for S_1 with respect to $\mathfrak{U}_1$ and $\langle P_1, P_2 \rangle$ is an index pair for S_2 with respect to $\mathfrak{U}_2$. We define a flow p in $\Omega_1 \times \Omega_2$ by

$$p(t, (x, y)) = (p_1(t, x), p_2(t, y)) \, .$$

Recall that the product of two quotient spaces X/X_0 and Y/Y_0 is the space

$$(X/X_0) \wedge (Y/Y_0) \stackrel{\text{def}}{=} X \times Y/((X_0 \times Y) \cup (X \times Y_0)) \, .$$

Theorem 4.1.3. *The set $S_1 \times S_2$ is an isolated invariant set with isolating neighborhood $\mathfrak{U}_1 \times \mathfrak{U}_2$. The pair*

$$\langle N_1 \times P_1, (N_1 \times P_2) \cup (N_2 \times P_1) \rangle \tag{4.1.11}$$

is an index pair for $S_1 \times S_2$.

Proof. All required properties except condition (3) in the definition of an index pair follow directly from the corresponding definitions.

Let

$$(x, y) \in (N_1 \times P_1) \setminus ((N_1 \times P_2) \cup (N_2 \times P_1))$$

and let $p(T, (x, y)) \notin \mathfrak{U}_1 \times \mathfrak{U}_2$ for some $T > 0$. Then $x \in N_1 \setminus N_2$, $y \in P_1 \setminus P_2$, and we have either $p_1(T, x) \notin \mathfrak{U}_1$ or $p_2(T, y) \notin \mathfrak{U}_2$. Assume, for definiteness, that $p_1(T, x) \notin \mathfrak{U}_1$.

Since $\langle N_1, N_2 \rangle$ is an index pair, we can find $t_1 \in [0, T]$ such that

$$p_1([0, t_1), x) \subset N_1 \setminus N_2 \quad \text{and} \quad p(t_1, x) \in N_2 \, .$$

If

$$p_2([0,t_1),y) \subset P_1 \setminus P_2$$

then

$$p([0,t_1),(x,y)) \subset (N_1 \times P_1) \setminus (N_1 \times P_2 \cup N_2 \times P_1)\ ,$$

and

$$p(t_1,(x,y)) \in N_2 \times P_1\ .$$

Suppose instead that

$$p_2([0,t_1),y) \not\subset P_1 \setminus P_2\ .$$

Then, by condition (2) in the definition of an index pair,

$$p_2([0,+\infty),y) \not\subset \mathfrak{U}_2\ .$$

Consequently, there exists a number $t_2 \geqslant 0$ such that

$$p_2([0,t_2),y) \subset P_1 \setminus P_2, \quad p_2(t_2,y) \in P_2\ .$$

Clearly $t_2 < t_1$. Since

$$p_1([0,t_2),x) \subset N_1 \setminus N_2\ ,$$

we have

$$p([0,t_2),(x,y)) \subset (N_1 \times P_1) \setminus (N_1 \times P_2 \cup N_2 \times P_1)$$

and

$$p(t_2,(x,y)) \in N_1 \times P_2\ .$$

We conclude that condition (3) in the definition of an index pair holds for the pair of sets (4.1.11). □

To conclude this section, we show that Conley index is preserved under homeomorphisms of flows.

Along with the flow p defined on the open set Ω, we consider the flow q defined on an open set $\widehat{\Omega}$ by

$$q(t,y) = T(p(t,T^{-1}(y)))\ , \tag{4.1.12}$$

where $T : \Omega \to \widehat{\Omega}$ is a homeomorphism. It is easily verified that the mapping q satisfies the conditions for a flow.

Lemma 4.1.18. *If the set $S \subset \Omega$ is invariant with respect to p, then the set $\widehat{S} = T(S)$ is invariant with respect to q.*

Proof. Let $y \in T(S)$. Then $y = T(x)$ for some $x \in S$. Therefore

$$q(t,y) = T(p(t,x)) \in T(S)$$

for all $t \in \mathbb{R}$ since $p(t,x) \in S$ for all $t \in \mathbb{R}$. □

Let N be an isolating neighborhood of an invariant set S in Ω. It turns out that $\widehat{N} = T(N)$ is an isolating neighborhood of $\widehat{S} = T(S)$. Since T is a homeomorphism, $\widehat{N} = T(N)$ is closed and $T(\operatorname{int} N)$ is open and contains $\widehat{S}$. Therefore $\widehat{S} \subset \operatorname{int} \widehat{N}$. If some trajectory $q((-\infty, +\infty), y_0)$ lies in $\widehat{N}$, then by (4.1.12) the trajectory $p((-\infty, +\infty), T^{-1}(y_0))$ lies in N, and since N is an isolating neighborhood for S, we have

$$p((-\infty, +\infty), T^{-1}(y_0)) \subset S .$$

Consequently $T^{-1}(y_0) \in S$ and $y_0 \in T(S) = \widehat{S}$. It follows from Lemma 4.1.18 that

$$q((-\infty, +\infty), y_0) \subset \widehat{S} .$$

Thus $\widehat{N}$ is an isolating neighborhood for $\widehat{S}$.

Theorem 4.1.4. *The Conley index of an invariant set S relative to the flow p coincides with the Conley index of the invariant set $\widehat{S} = T(S)$ relative to the flow q.*

Proof. We set

$$A^- = \{x \in N : p((-\infty, 0), x) \subset N\} ,$$

$$A^+ = \{x \in N : p((0, +\infty), x) \subset N\} ,$$

$$\widehat{A}^- = \{y \in \widehat{N} : q((-\infty, 0), y) \subset \overline{N}\} ,$$

$$\widehat{A}^+ = \{y \in \widehat{N} : q((0, +\infty), y) \subset \overline{N}\} .$$

Direct verification shows that $\widehat{A}^+ = T(A^+)$, $\widehat{A}^- = T(A^-)$. Let $\langle N_1, N_2 \rangle$ be an index pair for S with respect to N. It follows from the definition of an index pair that the pair $\langle \widehat{N}_1, \widehat{N}_2 \rangle$, where $\widehat{N}_1 = T(N_1)$, $\widehat{N}_2 = T(N_2)$, is an index pair for $\widehat{S}$ with respect to $\widehat{N}$. Thus the mapping T induces a homeomorphism of the quotient spaces N_1/N_2 and $\widehat{N}_1/\widehat{N}_2$, and the theorem follows. □

4.1.4 Homotopy Invariance of the Conley Index

Let $p(\lambda)$, $\lambda \in [0, 1]$, be a family of flows that depend continuously on λ, and are defined in a domain Ω in $\mathbb{R}^m$. Write $p(t, x; \lambda)$ for the operation of shifting the point $x \in \Omega$ for the time t for a given λ. We shall use notation such as

$$p([t_1, t_2], x; \lambda) = \{p(\tau, x; \lambda) : \tau \in [t_1, t_2]\} ,$$

$$p([t_1, t_2], X; \lambda) = \{p(\tau, x; \lambda) : \tau \in [t_1, t_2],\ x \in X\}$$

when appropriate, similar to the notation introduced in Sect. 4.1.1.

Let $\mathfrak{U}$ be a compact subset of Ω and suppose that it is a common isolating neighborhood of a family of invariant sets $S(\lambda)$ ($\lambda \in [0, 1]$). In this section we shall prove that the invariant sets $S(\lambda)$ all have the same Conley index.

We set

$$A^+(\lambda) = \{x \in \mathfrak{U} : p([0, +\infty), x; \lambda) \subset \mathfrak{U}\} ,$$
$$A^-(\lambda) = \{x \in \mathfrak{U} : p((-\infty, 0], x; \lambda) \subset \mathfrak{U}\}$$

and prove that the sets $A^+(\lambda)$ and $A^-(\lambda)$ depend continuously on λ.

Lemma 4.1.19. *Let W be an open neighborhood of $A^+(\lambda_0)$ for some fixed λ_0. Then $A^+(\lambda) \subset W$ for all λ which are sufficiently close to λ_0.*

Proof. Set $K = \mathfrak{U} \setminus W$. Then K is compact and disjoint from $A^+(\lambda_0)$. Consequently,

$$p([0, +\infty), x; \lambda_0) \not\subset \mathfrak{U}$$

for all $x \in K$.

We fix $x \in K$. We can find some $t = t(x) > 0$ for which

$$p([0, t(x)], x; \lambda_0) \subset \Omega \quad \text{and} \quad p(t(x), x; \lambda_0) \notin \mathfrak{U} .$$

Since the flow is continuous, there exist a neighborhood V_x of x and a number $\varepsilon = \varepsilon(x) > 0$ such that

$$p([0, t(x)], y; \lambda) \subset \Omega, \quad p(t(x), y; \lambda) \notin \mathfrak{U}$$

$$(y \in V_x, \quad \lambda \in (\lambda_0 - \varepsilon, \lambda_0 + \varepsilon) \cap [0, 1]) .$$

The family of open sets $\{V_x : x \in K\}$ forms a cover of the compact set K. Let

$$\{V_{x_k} : k = 1, 2, \dots, n\}$$

be a finite subcover and set

$$\widehat{\varepsilon} = \min\{\varepsilon(x_1), \varepsilon(x_2), \dots, \varepsilon(x_n)\} .$$

For every $x \in K$ we have

$$p([0, +\infty), x; \lambda) \not\subset \mathfrak{U} \quad (\lambda \in (\lambda_0 - \widehat{\varepsilon}, \lambda_0 + \widehat{\varepsilon}) \cap [0, 1]) ,$$

and hence

$$A^+(\lambda) \cap K = \varnothing .$$

Thus $A^+(\lambda) \subset W$ for all $\lambda \in (\lambda_0 - \widehat{\varepsilon}, \lambda_0 + \widehat{\varepsilon}) \cap [0, 1]$. □

We fix $\lambda_0 \in [0, 1]$ and $\varepsilon_0 > 0$. Let $\mathfrak{U}$ be a subset of Ω and assume that for all $\lambda \in (\lambda_0 - \varepsilon_0, \lambda_0 + \varepsilon_0) \cap [0, 1]$ the set $\mathfrak{U}$ is an isolating neighborhood of some invariant set $S(\lambda)$.

Lemma 4.1.20. *The Conley index of the invariant sets $S(\lambda)$ is the same for all λ which are sufficiently close to λ_0.*

Proof. If $S(\lambda_0) = \varnothing$, then, by Lemmas 4.1.3 and 4.1.4, we have $S(\lambda) = \varnothing$ if λ is sufficiently close to λ_0, and the assertion of the lemma follows.

Suppose that $S(\lambda_0) \neq \varnothing$. By Theorem 4.1.1 there is an index pair $\langle N_1, \widehat{N}_1 \rangle$ for $S(\lambda_0)$ such that

$$\overline{N_1 \setminus \widehat{N}_1} \subset \operatorname{int} \mathfrak{U} \,.$$

The set N_1 is a $\mathfrak{U}$-neighborhood of $A^-(\lambda_0)$. Let $Z \subset \operatorname{int} N_1$ also be a $\mathfrak{U}$-neighborhood of $A^-(\lambda_0)$. By Lemma 4.1.7, there exists a compact set N_2, which is positive invariant relative to $\mathfrak{U}$, for which

$$N_2 \subset Z \subset \operatorname{int} N_1$$

and such that, in addition, N_2 is a $\mathfrak{U}$-neighborhood of $A^-(\lambda_0)$. Using this procedure again, we may construct a set N_3 having similar properties and satisfying

$$N_3 \subset \operatorname{int} N_2 \,.$$

The set $\widehat{N}_1$ is disjoint from $A^+(\lambda_0)$. Therefore there exists a closed set $\widehat{Z}$ satisfying the conditions

$$\widehat{N}_1 \subset \operatorname{int} \widehat{Z}, \quad \widehat{Z} \cap A^+(\lambda_0) = \varnothing \,.$$

By Lemma 4.1.5, the set $\widehat{N}_2 = \mathfrak{P}_{\lambda_0}(\widehat{Z})$ is closed, positive invariant with respect to $\mathfrak{U}$, and disjoint from $A^+(\lambda_0)$. Thus

$$\widehat{N}_1 \subset \operatorname{int} \widehat{N}_2 \,.$$

Using this procedure again, we may construct a set $\widehat{N}_3$ such that

$$\widehat{N}_2 \subset \operatorname{int} \widehat{N}_3 \,.$$

For all $i, j \in \{1, 2, 3\}$ the pair $\langle N_i, \widehat{N}_j \rangle$ is an index pair for $S(\lambda_0)$. Indeed, conditions (1), (2) in the definition of an index pair hold by construction, and condition (3) follows from the inclusions

$$\overline{N_i \setminus \widehat{N}_j} \subset \overline{N_1 \setminus \widehat{N}_1} \subset \operatorname{int} \mathfrak{U}$$

and standard arguments used in the proof of Theorem 4.1.1 (on the existence of an index pair).

We shall prove that for all values of λ that are sufficiently close to λ_0 there exists an index pair $\langle P_1(\lambda), \widehat{P}_1(\lambda) \rangle$ for $S(\lambda)$ in $\mathfrak{U}$ satisfying

$$N_2 \subset P_1(\lambda) \subset N_1 \,, \tag{4.1.13}$$

$$\widehat{N}_1 \subset \widehat{P}_1(\lambda) \subset \widehat{N}_2 \,. \tag{4.1.14}$$

Set $K = \overline{\mathfrak{U} \setminus N_1}$. Since $N_2 \subset \operatorname{int} N_1$ by construction we have

$$K \cap N_2 = \varnothing \,.$$

We fix $x \in K$. Since $x \notin A^-(\lambda_0)$, we have

$$p((-\infty, 0], x; \lambda_0) \not\subset \mathfrak{U} .$$

Set

$$t_x = \sup\{t : p([-t, 0], x; \lambda_0) \subset \mathfrak{U}\} .$$

Since the set N_2 is positive invariant with respect to $\mathfrak{U}$, the trajectory $p([-t_x, 0], x; \lambda_0)$ does not meet N_2. Let $T_x > t_x$ be so close to t_x that

$$p([-T_x, 0], x; \lambda_0) \subset \Omega$$

and, in addition,

$$p(-T_x, x; \lambda_0) \notin \mathfrak{U} \quad \text{and} \quad p([-T_x, 0], x; \lambda_0) \cap N_2 = \varnothing .$$

Since the flow is continuous, there exist a neighborhood V_x of x and a number $\varepsilon = \varepsilon(x) > 0$ such that whenever

$$\xi \in V_x \quad \text{and} \quad \lambda \in (\lambda_0 - \varepsilon(x), \lambda_0 + \varepsilon(x)) \cap [0, 1]$$

we have

$$p([-T_x, 0], \xi; \lambda) \cap N_2 = \varnothing, \quad p(-T_x, \xi; \lambda) \notin \mathfrak{U} .$$

We choose a finite subcover

$$\{V_x : k = 1, 2, \ldots, n\}$$

from the open cover $\{V_x : x \in K\}$ of the compact set K and define

$$\varepsilon = \min_{1 \leqslant k \leqslant n} \varepsilon(x_k) .$$

For every $x \in K$ we have $x \in V_{x_l}$ for some l, and, consequently,

$$\begin{gathered} p([-T_{x_l}, 0], x; \lambda) \cap N_2 = \varnothing , \\ p(-T_{x_l}, x; \lambda) \notin \mathfrak{U} \quad (\lambda \in (\lambda_0 - \varepsilon, \lambda_0 + \varepsilon) \cap [0, 1]) . \end{gathered}$$

Thus $x \notin \mathfrak{P}_\lambda(N_2)$ for all $\lambda \in (\lambda_0 - \varepsilon, \lambda_0 + \varepsilon) \cap [0, 1]$. Therefore (4.1.13) holds for $P_1(\lambda) = \mathfrak{P}_\lambda(N_2)$. Similarly we can show that (4.1.14) holds for $\widehat{P}_1(\lambda) = \mathfrak{P}_\lambda(\widehat{N}_1)$ if λ is sufficiently close to λ_0.

Let us show that $\langle P_1(\lambda), \widehat{P}_1(\lambda) \rangle$ is an index pair for $S(\lambda)$ for all λ which are sufficiently close to λ_0. Since $\langle N_2, \widehat{N}_2 \rangle$ is an index pair for $S(\lambda_0)$, we have

$$A^-(\lambda_0) \subset \operatorname{int} N_2 .$$

By Lemma 4.1.19,

$$A^-(\lambda) \subset \operatorname{int} N_2$$

for all λ which are sufficiently close to λ_0. Therefore, by Lemma 4.1.6, the set $P_1(\lambda) = \mathfrak{P}_\lambda(N_2)$ is closed. From Lemma 4.1.5 the set $\widehat{P}_1(\lambda)$ is closed. Since

$$\overline{P_1(\lambda) \setminus \widehat{P}_1(\lambda)} \subset \overline{N_1 \setminus \widehat{N}_1} \subset \operatorname{int} \mathfrak{U} \,,$$

we can establish condition (3) in the definition of an index pair using arguments like those used in the proof of Theorem 4.1.1.

We can use the method described above to construct index pairs $\langle P_2(\lambda), \widehat{P}_2(\lambda)\rangle$ satisfying the conditions

$$N_3 \subset P_2(\lambda) \subset N_2, \quad \widehat{N}_2 \subset \widehat{P}_2(\lambda) \subset \widehat{N}_3 \,.$$

Thus we have chains of embeddings

$$N_3 \subset P_2(\lambda) \subset N_2 \subset P_1(\lambda) \subset N_1 \,,$$
$$\widehat{N}_1 \subset \widehat{P}_1(\lambda) \subset \widehat{N}_2 \subset \widehat{P}_2(\lambda) \subset \widehat{N}_3 \,.$$

These embeddings induce a chain of embeddings of quotient spaces

$$N_3/\widehat{N}_1 \xrightarrow{i_1} P_2/\widehat{P}_1 \xrightarrow{i_2} N_2/\widehat{N}_2 \xrightarrow{i_3} P_1/\widehat{P}_2 \,.$$

By Theorem 4.1.2, the mappings $i_2 \circ i_1$ and $i_3 \circ i_2$ are homotopy equivalences. Therefore, by Lemma 4.1.11, the embeddings i_1, i_2, i_3 are homotopy equivalences.

Direct verification shows that $\langle P_2, \widehat{P}_1\rangle$ and $\langle P_1, \widehat{P}_2\rangle$ are index pairs for $S(\lambda)$. Thus the Conley index of $S(\lambda)$ coincides with the Conley index of $S(\lambda_0)$ if λ is sufficiently close to λ_0, as required. □

Now we can easily prove the main result of this section.

Theorem 4.1.5. *Suppose that $\mathfrak{U} \subset \Omega$ is an isolating neighborhood of the invariant set $S = S(\lambda)$ for all $\lambda \in [0,1]$. Then the sets $S(\lambda)$ $(\lambda \in [0,1])$ all have the same Conley index.*

Proof. Consider the function $\varphi : [0,1] \mapsto \mathbb{R}$ which assumes the value 0 at the point λ if the Conley indices of the sets $S(0)$ and $S(\lambda)$ are the same and the value 1 otherwise. By Lemma 4.1.20, this function is constant in some neighborhood of each point of the interval $[0,1]$. Consequently, $\varphi(\lambda) = 0$ for all $\lambda \in [0,1]$. □

4.1.5 The Conley Index of Nondegenerate Critical Points

In some cases, the Conley index of an isolated invariant set can be computed explicitly. We shall consider one of these cases in this section.

Suppose that we have a system of differential equations

$$\frac{dx}{dt} = g(x) \tag{4.1.15}$$

defined in an open domain Ω in $\mathbb{R}^N$ and that the mapping $g : \mathbb{R}^N \to \mathbb{R}^N$ satisfies a Lipschitz condition. Then we can define a flow $p(t, x)$ in Ω, as the solution of the Cauchy problem for Eq. (4.1.15) with the initial condition

$$p(0, x) = x .$$

The flow p is well defined, from classical existence and uniqueness theorems for the Cauchy problem.

A special case of system (4.1.15) is the gradient system

$$\frac{dx}{dt} = \nabla f(x) , \tag{4.1.16}$$

where $f : \mathbb{R}^N \mapsto \mathbb{R}$ is a function whose gradient satisfies a Lipschitz condition. In this case, the critical points of f are invariant sets relative to the flow defined by (4.1.16). Therefore we can speak about the Conley index of a critical point of a smooth function.

Let x_0 be a nondegenerate critical point of f. We denote by $\nabla^2 f$ the Hessian of f: thus $\nabla^2 f(x) = [h_{ij}(x)]$, where

$$h_{ij}(x) = \frac{\partial^2 f}{\partial x_i \partial x_j}(x) \quad (i, j = 1, \ldots, N) .$$

We claim that each nondegenerate point x_0 of f is isolated. Indeed,

$$\nabla f(x_0 + h) = \nabla^2 f(x_0)h + \omega(h) ,$$

where

$$\lim_{|h| \to 0} |h|^{-1}\omega(h) = 0$$

and

$$|\nabla^2 f(x_0)h| \geqslant a|h| \quad (h \in \mathbb{R}^N)$$

for some $a > 0$; thus

$$|\nabla f(x_0 + h)| \geqslant |\nabla^2 f(x_0)h| - |\omega(h)| \geqslant a|h| - |\omega(h)| > 0$$

for all sufficiently small $h \neq 0$, and our claim follows.

Lemma 4.1.21. *If $f \in C^1$ then each isolated critical point of f is an isolated invariant set of the system* (4.1.16).

Proof. Suppose that x_0 is the only critical point of f in some ball $B(\rho, x_0)$. We shall show that no trajectory of the system (4.1.16) except $\{x_0\}$ lies entirely in this ball.

Assume that

$$p((-\infty, +\infty), x_1) \subset B(\rho, x_0)$$

for some $x_1 \neq x_0$. Consider the function

$$\varphi(t) = f(p(t, x_1)) \ .$$

By definition,

$$\frac{d}{dt}\varphi(t) = |\nabla f(p(t, x_1))|^2 \ .$$

Therefore the function φ increases monotonically. Since f is bounded on $B(\rho, x_0)$, the integral

$$\int\limits_{-\infty}^{+\infty} \frac{d}{dt}\varphi(t)\, dt = \int\limits_{-\infty}^{+\infty} |\nabla f(p(t, x_1))|^2\, dt$$

converges. Since

$$-\infty < \lim_{t\to-\infty} \varphi(t) < \lim_{t\to+\infty} \varphi(t) < +\infty \ , \tag{4.1.17}$$

at least one of the limits in (4.1.17) is different from $f(x_0)$. Suppose, for instance, that

$$\lim_{t\to+\infty} \varphi(t) \neq f(x_0) \ . \tag{4.1.18}$$

Consider a sequence $(p(t_n, x_1))$ such that $t_n \to +\infty$ and

$$\lim_{n\to\infty} |\nabla f(p(t_n, x_n))| = 0 \ .$$

Choose a subsequence of the bounded sequence $(p(t_n, x_1))$ that converges to a point $x_2 \in B(\rho, x_0)$. By (4.1.18), we have $x_2 \neq x_0$, and x_2 is a state of equilibrium of the system (4.1.16). However x_0 is the only critical point of f in $B(\rho, x_0)$, and the resulting contradiction completes the proof of the lemma. □

Thus the Conley index is defined for nondegenerate critical points of a smooth function f. In order to calculate it, we define the function

$$\widetilde{f}(x) = \tfrac{1}{2}(\nabla^2 f(x_0)(x - x_0), x - x_0) \ .$$

The point x_0 is the only critical point of $\widetilde{f}$. Choose $\rho > 0$ so small that the ball $B(\rho, x_0)$ is an isolating neighborhood for the invariant set $\{x_0\}$ for both f and $\widetilde{f}$. Consider the family of functions

$$f(x; \lambda) = \lambda \widetilde{f}(x) + (1 - \lambda) f(x) \ .$$

Since

$$\begin{aligned}\nabla f(x;\lambda) &= \lambda \nabla^2 f(x_0)(x - x_0) + (1-\lambda)(\nabla^2 f(x_0)(x - x_0) + \omega(x - x_0)) \\ &= \nabla^2 f(x_0)(x - x_0) + (1-\lambda)\omega(x - x_0) \ ,\end{aligned}$$

it follows that for $\lambda \in [0,1]$ the ball $B(\rho, x_0)$ contains no critical points of $f(\cdot;\lambda)$ except x_0. Therefore $B(\rho, x_0)$ is an isolating neighborhood of the invariant set $\{x_0\}$ for all $\lambda \in [0,1]$. By Theorem 4.1.5, the Conley indices of the critical point x_0 of the functions f and $\widetilde{f}$ are the same. Thus we have reduced the calculation of the Conley index of a nondegenerate critical point to the calculation of the Conley index of the nondegenerate critical point 0 of a quadratic function of the form

$$\varphi(x) = (Hx, x) ,$$

where H is a nondegenerate symmetric matrix. We can write H in the form $\Gamma\mathcal{D}\Gamma^*$, where Γ is an orthogonal matrix and $\mathcal{D}$ is a diagonal matrix whose diagonal entries are the eigenvalues $\lambda_1, \ldots, \lambda_m$ of H. Since the Conley index of the critical point 0 of φ does not change under small perturbations of H, we can assume without loss of generality that all principal minors of Γ are nonzero. We can write Γ as a product LU of a lower triangular matrix L and an upper triangular matrix U; let

$$L = \begin{bmatrix} l_{11} & & 0 \\ \vdots & \ddots & \\ l_{N1} & \ldots & l_{NN} \end{bmatrix} \quad \text{and} \quad U = \begin{bmatrix} u_{11} & \ldots & u_{1N} \\ & \ddots & \vdots \\ 0 & & u_{NN} \end{bmatrix} .$$

Let L_1, U_1 be the diagonal matrices with the same diagonal entries as L, U and let $L_2 = L - L_1$, $U_2 = U - U_1$; thus L_2, U_2 are respectively lower triangular and upper triangular matrices with zero diagonal entries.

Since

$$\det \Gamma = u_{11} \ldots u_{NN} \cdot l_{11} \ldots l_{NN} \neq 0 ,$$

for all $\lambda \in [0,1]$ the function

$$\varphi_\lambda(x) = (\Phi x, x) ,$$

where

$$\Phi = (\lambda L_1 + L_2)(\lambda U_1 + U_2)\mathcal{D}(\lambda U_1 + U_2)^*(\lambda L_1 + L_2)^* ,$$

has unique critical point 0. Therefore the Conley index of the critical point 0 of φ equals the Conley index of the critical point 0 of the function

$$\widetilde{\varphi}(x) = (\widetilde{\mathcal{D}}x, x) ,$$

where

$$\widetilde{\mathcal{D}} = \begin{bmatrix} |\lambda_1| l_{11}^2 u_{11}^2 & & 0 \\ & \ddots & \\ 0 & & |\lambda_m| l_{NN}^2 u_{NN}^2 \end{bmatrix} \begin{bmatrix} \operatorname{sign} \lambda_1 & & 0 \\ & \ddots & \\ 0 & & \operatorname{sign} \lambda_N \end{bmatrix} ,$$

which, in turn, equals the Conley index of the critical point 0 of the function

$$\psi(x) = \psi(x_1, \ldots, x_N) = \sum_{i=1}^{r} x_i^2 - \sum_{i=r+1}^{N} x_i^2 , \tag{4.1.19}$$

where r is the number of positive eigenvalues of H (taking account of multiplicities).

The gradient system (4.1.16) for the function (4.1.19) has the form

$$\dot{x}_i = 2x_i \quad (i = 1, 2, \ldots, r) , \tag{4.1.20}$$

$$\dot{x}_i = -2x_i \quad (i = r+1, \ldots, N) . \tag{4.1.21}$$

Consequently,

$$p(t, x) = (x_1 e^{2t}, \ldots, x_r e^{2t}, x_{r+1} e^{-2t}, \ldots, x_N e^{-2t}) . \tag{4.1.22}$$

Thus the flow under consideration is the product of r flows of the form described by Eq. (4.1.20) and $N-r$ flows of the form described by Eq. (4.1.21). By Theorem 4.1.3, the Conley index of the flow (4.1.22) is the product of the Conley indices of these flows.

Consider the invariant set $\{0\}$ relative to the flow

$$p(t, x) = xe^{2t}$$

described by Eq. (4.1.20). The pair

$$\langle N_1, N_2 \rangle = \langle [-1, 1], \{-1, 1\} \rangle$$

is an index pair, and N_1/N_2 is homotopy equivalent to the one-dimensional sphere

$$S^1 = \{(x_1, x_2) \subset \mathbb{R}^2 \mid x_1^2 + x_2^2 = 1\}$$

with base point.

For the invariant set $\{0\}$ relative to the flow

$$p(t, x) = xe^{-2t} ,$$

the pair

$$\langle N_1, N_2 \rangle = \langle [-1, 1], \varnothing \rangle$$

is an index pair and N_1/N_2 is homotopy equivalent to the zero-dimensional sphere $S^0 = \{-1, 1\}$ with base point.

From Proposition 1.1.4 we have

$$S^p \wedge S^q \sim S^{p+q}$$

for all p, $q \geqslant 0$. We have thus proved the following assertion.

Theorem 4.1.6. *The Conley index of an isolated nondegenerate critical point x_0 of a function $f \in C^2$ is the homotopy type of a sphere S^r with base point, where r is the number of positive eigenvalues of the Hessian $\nabla^2 f(x_0)$ (taking account of multiplicities).*

Theorem 4.1.7. *Suppose that the family of functions $f(\cdot;\lambda)$ defined in a domain $\Omega \subset \mathbb{R}^N$ depends continuously on $\lambda \in [0,1]$. Suppose that the gradients of the functions depend continuously on λ and satisfy a Lipschitz condition. Assume that the point $x_0 \in \Omega$ is the unique critical point of these functions. If the Hessians $\nabla^2 f(x_0;0)$ and $\nabla^2 f(x_0;1)$ exist and are nondegenerate, then the matrices of the quadratic forms $(\nabla^2 f(x_0;0)\xi,\xi)$ and $(\nabla^2 f(x_0;1)\xi,\xi)$ have the same number of negative eigenvalues.*

This theorem follows from Theorems 4.1.5 and 4.1.6.

4.2 Conley Index in Infinite-Dimensional Problems

4.2.1 (E,H)-Regular Functionals

Let E be a real Banach space that is continuously and densely embedded in a Hilbert space H with inner product $(\cdot,\cdot)$. We say that a functional $f \in C^4$ defined on a ball $\mathring{B}(r)$ is (E,H)-regular if

$$|\langle \nabla f(x), h\rangle| \leqslant M\|h\|_H \quad (h \in E, x \in \mathring{B}(r))\,, \tag{4.2.1}$$

$$|\langle \nabla^2 f(x)g, h\rangle| \leqslant M\|g\|_H\|h\|_H \quad (h,g \in E,\ x \in \mathring{B}(r))\,, \tag{4.2.2}$$

$$\lim_{\|h\|_E\to 0} \sup_{\substack{x\in\mathring{B}(r),\, p,q\in E\\ \|p\|_H=\|q\|_H=1}} \langle(\nabla^2 f(x+h) - \nabla^2 f(x))p, q\rangle = 0\,. \tag{4.2.3}$$

By (4.2.1), the linear functional $\nabla f(x)$ admits a continuous extension from E to H, and by Riesz's theorem this extension can be represented in the form

$$\langle \nabla f(x), h\rangle = (\nabla_H f(x), h) \quad (h \in E,\ \nabla_H f(x) \in H)\,.$$

It follows from (4.2.2) that the bilinear form $\langle \nabla^2 f(x)g, h\rangle$ admits an extension to $H \times H$ which is representable as

$$\langle \nabla^2 f(x)g, h\rangle = (\nabla^2_H f(x)g, h) \quad (g,h \in E)\,,$$

where $\nabla^2_H f(x) : H \to H$ is a self-adjoint bounded operator.

Lemma 4.2.1. *The operator $\nabla_H f : \mathring{B}(r) \to H$ is uniformly and continuously Fréchet differentiable on $\mathring{B}(r)$ and the operator $\nabla^2_H f(x) : E \to H$ is its derivative at x.*

Proof. In order to prove that the operator $\nabla_H f : \mathring{B}(r) \to H$ is uniformly differentiable on $\mathring{B}(r)$, it suffices to establish the equality

$$\lim_{\|h\|_E \to 0} \sup_{x\in\mathring{B}(r)} \|h\|_E^{-1} \left\|\nabla_H f(x+h) - \nabla_H f(x) - \nabla_H^2 f(x)h\right\|_H = 0\,. \quad (4.2.4)$$

Now we have

$$\begin{aligned}
&\|h\|_E^{-1} \left\|\nabla_H f(x+h) - \nabla_H f(x) - \nabla_H^2 f(x)h\right\|_H \\
&\leqslant C\,\|h\|_H^{-1} \left\|\nabla_H f(x+h) - \nabla_H f(x) - \nabla_H^2 f(x)h\right\|_H \\
&= C\,\|h\|_H^{-1} \sup_{g\in H, \|g\|_H \leqslant 1} (\nabla_H f(x+h) - \nabla_H f(x) - \nabla_H^2 f(x)h, g) \\
&= C\,\|h\|_H^{-1} \sup_{g\in E, \|g\|_H \leqslant 1} \langle \nabla f(x+h) - \nabla f(x) - \nabla^2 f(x)h, g\rangle \\
&= C\,\|h\|_H^{-1} \sup_{g\in E, \|g\|_H \leqslant 1} \int_0^1 \langle (\nabla^2 f(x+\theta h) - \nabla^2 f(x))h, g\rangle\, d\theta \\
&= C \sup_{\substack{g\in E, \|g\|_H = 1\\ \theta\in[0,1]}} \langle (\nabla^2 f(x+\theta h) - \nabla^2 f(x))\,\|h\|_H^{-1}\, h, g\rangle.
\end{aligned}$$

This inequality and (4.2.3) imply (4.2.4).

Now we establish that the operator $\nabla_H^2 f : \mathring{B}(r) \to \mathcal{L}(H)$ is uniformly continuous (here and in the sequel, $\mathcal{L}(H)$ denotes the space of bounded linear operators on H).

Let $x, v \in \mathring{B}(r)$. Then, since E is dense in H, we have

$$\begin{aligned}
\left\|\nabla_H^2 f(x) - \nabla_H^2 f(v)\right\|_{\mathcal{L}(H)} &= \sup_{\substack{p,q\in H\\ \|p\|_H = \|q\|_H = 1}} ((\nabla_H^2 f(x) - \nabla_H^2 f(v))p, q) \\
&= \sup_{\substack{p,q\in E\\ \|p\|_H = \|q\|_H = 1}} \langle \left(\nabla^2 f(x) - \nabla^2 f(v)\right) p, q\rangle.
\end{aligned}$$

This inequality and (4.2.3) imply that $\nabla_H^2 f : \mathring{B}(r) \to \mathcal{L}(H)$ is uniformly continuous, and the lemma is proved. □

4.2.2 The Conley Index of Critical Points of (E, H)-Regular Functionals

In this section, we define the Conley index of an isolated critical point.

A finite-dimensional subspace L of E is said to be *admissible* for a functional f at x_* if, for some $m > 0$, the inequality

$$(\nabla_H^2 f(x_*)v, v) \geqslant m\|v\|_H^2 \tag{4.2.5}$$

holds for all $v \in L^\perp$, where

$$L^\perp \stackrel{\text{def}}{=} \{v : (u, v) = 0 \text{ for all } u \in L\} .$$

For an admissible subspace to exist, it is necessary and sufficient that the set

$$\sigma(\nabla_H^2 f(x_*)) \cap (-\infty, 0] \tag{4.2.6}$$

either be empty or consist of a finite number of regular eigenvalues of finite multiplicity (see, e.g., [125]).

We fix an arbitrary admissible subspace L and suppose that Q is the orthogonal projection operator onto $L^\perp$ in H. Below we assume that the following conditions hold:

(1) the operator $\nabla_H f(\cdot)$ maps $B(r)$ into E and is continuously Fréchet differentiable,

(2) for every $w \in E \cap L^\perp$, the linear equation

$$Q\nabla_H^2 f(x_*)x = w \tag{4.2.7}$$

has a solution x which lies in $E \cap L^\perp$.

Lemma 4.2.2. *The operator $Q : E \to E$ is bounded.*

Proof. Since the embedding of E into H is continuous, for some $\gamma > 0$ we have

$$\|x\|_H \leqslant \gamma\|x\|_E \quad (x \in E) . \tag{4.2.8}$$

From the properties of orthogonal projection operators,

$$\|Qx\|_H \leqslant \|x\|_H \quad (x \in H) , \tag{4.2.9}$$

$$\|(I - Q)x\|_H \leqslant \|x\|_H \quad (x \in H) , \tag{4.2.10}$$

where I denotes the identity operator. Since L is finite-dimensional, all norms on it are equivalent, and so for some $C > 0$ we have

$$\|\xi\|_E \leqslant C\|\xi\|_H \quad (\xi \in L) . \tag{4.2.11}$$

From (4.2.8)–(4.2.11) we find that

$$\begin{aligned}\|Qx\|_E = \|x - (I - Q)x\|_E &\leqslant \|x\|_E + \|(I - Q)x\|_E \\ &\leqslant \|x\|_E + C\|(I - Q)x\|_H \leqslant \|x\|_E + C\|x\|_H \\ &\leqslant (1 + C\gamma)\|x\|_E\end{aligned}$$

for all $x \in E$, as required. □

It follows from Lemma 4.2.2 that the set QE is closed in E.

Consider the operator equation

$$F(u,v) = Q\nabla_H f(x_* + u + v) = 0\,, \tag{4.2.12}$$

where F is a mapping from $L \times QE$ into QE. We have

$$\frac{\partial F}{\partial v}(0,0) = Q\nabla_H^2 f(x_*)Q\,.$$

The extension A of this operator from QE to $QH = L^\perp$ is a symmetric (and consequently self-adjoint) operator. By (4.2.5) and properties of the spectrum of self-adjoint operators, $0 \notin \sigma(A)$. Therefore A has a continuous inverse. Consequently, the operator

$$Q\nabla_H^2 f(x_*)Q : QE \mapsto QE$$

is injective; it follows also from condition (2) that it is surjective.

Thus we can apply the implicit function theorem to Eq. (4.2.12); it implies that for some neighborhood $\mathring{B}(r) \cap L$ there exists a C^3-mapping

$$\varphi : \mathring{B}(r) \cap L \to QE$$

satisfying the condition

$$Q\nabla_H f(x_* + u + \varphi(u)) = 0 \quad (u \in \mathring{B}(r) \cap L)\,. \tag{4.2.13}$$

Consider the functional

$$\Phi(u) = f(x_* + u + \varphi(u))\,. \tag{4.2.14}$$

We shall show later that 0 is an isolated critical point of Φ, so that the Conley index $h(0;\Phi)$ is defined. It turns out that $h(0;\Phi)$ is independent of the choice of the admissible subspace L. Therefore the Conley index of the critical point x_* of f can be defined by

$$h(x_*;f) = h(0;\Phi)\,.$$

In order to justify this definition of Conley index, we need an auxiliary result.

Lemma 4.2.3. *Let Ω be domain in $\mathbb{R}^N$ and g a C^1-function on Ω whose gradient vanishes only at the point $x_0 \in \Omega$. Suppose that $T : \Omega \to \Omega_0$ is a diffeomorphism to a domain Ω_0. Then the point $y_0 = T(x_0)$ is the unique critical point of the C^1-function*

$$g_0 = g \circ T^{-1} : \Omega_0 \to \mathbb{R}\,.$$

Moreover, the critical point x_0 of g and the critical point y_0 of g_0 have the same Conley index.

Proof. Let $p(t, x)$ be the shift operator along the trajectories of the system

$$\dot{x} = \nabla g(x) . \tag{4.2.15}$$

Now p determines a flow in the domain Ω. It follows from Theorem 4.1.1 that the Conley index of the invariant set $\{x_0\}$ relative to the flow p coincides with the Conley index of the invariant set $\{y_0\}$ relative to the flow q in the domain Ω_0 defined by

$$q(t, y) = T(p(t, T^{-1}(y))) . \tag{4.2.16}$$

Let $J(x)$ be the Jacobian of T. The gradient system for g_0 has the form

$$\dot{y} = \nabla g_0(y) = ((J(T^{-1}(y)))^{-1})^T \nabla g(T^{-1}(y)) . \tag{4.2.17}$$

Let $q_0(t, y)$ be the shift operator along the trajectories of the system (4.2.17) and set

$$p_0(t, x) = T^{-1}(q_0(t, T(x)) . \tag{4.2.18}$$

The flow (4.2.18) is a shift operator along the trajectories of the system

$$\dot{x} = (J^T(x)J(x))^{-1} \nabla g(x) . \tag{4.2.19}$$

It follows from (4.2.17) that the gradient of g_0 vanishes only for $y = y_0$. To complete the proof of the lemma, it suffices to show that the Conley indices of the invariant set $\{x_0\}$ relative to the flows p and p_0 are the same.

We set $G = J(x_0)$. We shall consider the systems (4.2.15) and (4.2.19) in a sufficiently small ball $\overset{\circ}{B}(\nu, x_0)$. Since the matrix G is nondegenerate, the linear deformation

$$\dot{x} = (\lambda(J^T(x)J(x))^{-1} + (1 - \lambda)(G^T G)^{-1}) \nabla g(x)$$

is nondegenerate. Therefore we can restrict attention to the system

$$\dot{x} = (G^T G)^{-1} \nabla g(x) \tag{4.2.20}$$

instead of (4.2.19). The linear deformation

$$\dot{x} = (\lambda I + (1 - \lambda)(G^T G)^{-1}) \nabla g(x)$$

is nondegenerate since the matrix $\lambda E + (1 - \lambda)(G^T G)^{-1}$ is nondegenerate for all $\lambda \in [0, 1]$; this follows from the inequality

$$(\lambda I + (1 - \lambda)G^{-1}(G^{-1})^T x, x) = \lambda(x, x) + (1 - \lambda)((G^{-1})^T x, (G^{-1})^T x) > 0$$

which holds for all $x \neq 0$. The lemma follows. □

The gradient $\nabla \Phi$ of the functional (4.2.14) is defined by

$$\langle \nabla \Phi(u), h \rangle = (\nabla_H f(x_* + u + \varphi(u)), h) \quad (h \in L) . \tag{4.2.21}$$

If $\nabla\Phi(u_1) = 0$ for some $u_1 \in L$, then by (4.2.21) and (4.2.13) we have

$$\nabla_H f(x_* + u_1 + \varphi(u_1)) = 0 .$$

It now follows from the definition of ∇_H that

$$\nabla f(x_* + u_1 + \varphi(u_1)) = 0 .$$

Therefore $u_1 = 0$, and thus 0 is an isolated critical point of Φ. This functional is defined on the finite-dimensional subspace L and the Conley index $h(0;\Phi)$ is defined to be the Conley index of the critical point 0 of the function

$$\Psi(x_1, \dots, x_N) = f\left(x_* + \sum_{i=1}^{N} x_i e_i + \varphi\left(\sum_{i=1}^{N} x_i e_i\right)\right) ,$$

where $N = \dim L$ and $\{e_1, \dots, e_N\}$ is any orthonormal basis in L. It follows from Lemma 4.2.3 that $h(0;\Phi)$ is well defined.

Lemma 4.2.4. *The Conley index $h(0;\Phi)$ is independent of the choice of the admissible subspace.*

Proof. Let L be an admissible subspace of E of dimension N and suppose that M is a subspace containing L and of dimension $N+1$. From the definition of an admissible subspace, M is also an admissible subspace. Let $\{e_1, \dots, e_{N+1}\}$ be an orthonormal basis of M such that $\{e_1, \dots, e_N\}$ is an orthonormal basis of L.

Arguing as above, we can construct functionals

$$\Phi_L(u) = f(x_* + u + \varphi_L(u)) \quad (u \in L) ,$$
$$\Phi_M(v) = f(x_* + v + \varphi_M(v)) \quad (v \in M) ,$$

where φ_L and φ_M are mappings such that

$$Q_L \nabla_H f(x_* + u + \varphi_L(u)) = 0 ,$$
$$Q_M \nabla_H f(x_* + v + \varphi_M(v)) = 0 .$$

We assume that Φ_L and Φ_M are defined on a fixed ball $\mathring{B}(\rho)$. We set

$$\Psi(x_1, \dots, x_N) = f\left(x_* + \sum_{i=1}^{N} x_i e_i + \varphi_L\left(\sum_{i=1}^{N} x_i e_i\right)\right) ,$$
$$\Psi_0(x_1, \dots, x_{N+1}) = f\left(x_* + \sum_{i=1}^{N+1} x_i e_i + \varphi_M\left(\sum_{i=1}^{N+1} x_i e_i\right)\right) .$$

By the implicit function theorem, φ_L and φ_M are uniquely defined, and so

$$\Psi(x_1, \dots, x_N) = \Psi_0(x_1, \dots, x_N, 0) .$$

We can assume that Ψ_0 is defined on a ball $\mathring{B}(\mu) \subset \mathbb{R}^{N+1}$. We have

$$\begin{aligned}
&\frac{\partial}{\partial x_i}\Psi_0(x_1,\dots,x_{N+1}) \\
&\quad = \left(\nabla_H f\left(x_* + \sum_{i=1}^{N+1} x_i e_i + \varphi_M\left(\sum_{i=1}^{N+1} x_i e_i\right)\right), e_i + \varphi_M'\left(\sum_{i=1}^{N+1} x_i e_i\right) e_i\right) \\
&\quad = \left(\nabla_H f\left(x_* + \sum_{i=1}^{N+1} x_i e_i + \varphi_M\left(\sum_{i=1}^{N+1} x_i e_i\right)\right), e_i\right).
\end{aligned}$$

If $x_{N+1} = 0$, then

$$\sum_{i=1}^{N+1} x_i e_i \in L$$

and consequently

$$\frac{\partial}{\partial x_{N+1}}\Psi_0(x_1,\dots,x_N,0) = 0\,. \tag{4.2.22}$$

We shall need the following easily verified result: if $\sigma(t)$ is a twice continuously differentiable real function on the interval $(-\mu,\mu)$ and $\sigma'(0) = 0$, then, for all $t \in (-\mu,\mu)$, we have

$$\sigma(t) = \sigma(0) + t^2 \int_0^1\!\!\int_0^1 \sigma''(\alpha\beta t)\alpha\, d\alpha d\beta\,. \tag{4.2.23}$$

It follows from (4.2.23) that

$$\begin{aligned}
\Psi_0(x_1,\dots,x_{N+1}) &= \Psi_0(x_1,\dots,x_N,0) \\
&\quad + x_{N+1}^2 \int_0^1\!\!\int_0^1 \frac{\partial^2}{\partial x_{N+1}^2}\Psi_0(x_1,\dots,x_N,\alpha\beta x_{N+1})\alpha\, d\alpha d\beta\,.
\end{aligned} \tag{4.2.24}$$

Since

$$\begin{aligned}
&\frac{\partial^2}{\partial x_i \partial x_j}\Psi_0(x_1,\dots,x_{N+1}) \\
&\quad = \left(\nabla_H^2 f\left(x_* + \sum_{i=1}^{N+1} x_l e_l + \varphi_M\left(\sum_{l=1}^{N+1} x_l e_l\right)\right) e_j, e_i\right)
\end{aligned}$$

for all i,j, it follows by (4.2.5) that

$$\frac{\partial^2}{\partial x_{N+1}^2}\Psi_0(0,\dots,0) > 0\,. \tag{4.2.25}$$

Consider the mapping $T : \mathbb{R}^{N+1} \to \mathbb{R}^{N+1}$ defined by

$$T(x_1, \dots, x_{N+1})$$
$$= \left(x_1, \dots, x_N, x_{N+1} \left(\int_0^1 \int_0^1 \frac{\partial^2}{\partial x_{N+1}^2} \Psi_0(x_1, \dots, x_N, \alpha\beta x_{N+1}) \alpha \, d\alpha d\beta \right)^{1/2} \right).$$

By (4.2.25), the Jacobian of this mapping at 0 is nondegenerate. Since we have $\Psi_0 \in C^3(\mathring{B}(\mu))$, the mapping T is a diffeomorphism in some neighborhood of 0. From (4.2.24) we find that

$$\Psi_0(T^{-1}(x_1, \dots, x_{N+1})) = \Psi_0(x_1, \dots, x_N, 0) + x_{N+1}^2 . \tag{4.2.26}$$

By Lemma 4.2.3, the Conley index of the critical point 0 of the function $\Psi_0(T^{-1}(x_1, \dots, x_{N+1}))$ coincides with the Conley index $h(0; \Phi_M)$ of the critical point 0 of $\Phi_M(v)$. The Conley index of the critical point 0 of the function appearing on the right-hand side of (4.2.26) is the product of the Conley index $h(0; \Phi_L)$ of the critical point 0 of $\Phi_L(u)$ and the Conley index of the critical point 0 of the real function $\chi(t) = t^2$. Therefore $h(0; \Phi_M) = h(0; \Phi_L)$.

Let L_1 and L_2 be arbitrary admissible subspaces. We set $L_3 = L_1 + L_2$. Write $h_k(0)$ for the Conley index corresponding to the subspace L_k for $k = 1, 2, 3$. Repeating the arguments above, we can show that

$$h_1(0) = h_3(0) \quad \text{and} \quad h_2(0) = h_3(0) .$$

Therefore

$$h_1(0) = h_2(0) ,$$

and the lemma is proved. □

4.2.3 Homotopy Invariance of the Conley Index

In this section, we establish that the Conley index is homotopy invariant under nondegenerate deformation.

Consider a family of functions $f(x; \lambda)$ $(0 \leqslant \lambda \leqslant 1)$ defined on an open set Ω in $\mathbb{R}^N$ and satisfying the following conditions:

(1) the gradient $\nabla f(\cdot; \lambda)$ satisfies a Lipschitz condition on Ω for every $\lambda \in [0, 1]$,

(2) $f(\cdot; \lambda)$ and $\nabla f(\cdot; \lambda)$ depend continuously on λ,

(3) for every $\lambda \in [0, 1]$ the gradient $\nabla f(\cdot; \lambda)$ vanishes only at some point $x_* \in \Omega$.

Under these conditions, Theorem 4.1.5 is applicable to the flow in Ω determined by the gradient system

$$\dot{x} = \nabla f(x; \lambda) .$$

Therefore we have the following result.

Theorem 4.2.1. *The Conley index of the isolated critical point x_* of the function $f(\cdot;\lambda)$ is the same for all $\lambda \in [0,1]$.*

Theorem 4.2.1 can be generalized to the infinite-dimensional case.

Suppose that E and H are a Banach space and a Hilbert space respectively, and that E is continuously and densely embedded into H. Consider a one-parameter family $f(\cdot;\cdot) : B(r,x_*) \times [0,1] \to \mathbb{R}$ of (E,H)-regular functionals whose only critical point on $B(r,x_*)$ is the point x_*. We assume that $f(\cdot;\lambda)$, $\nabla_H f(\cdot;\lambda)$, $\nabla_H^2 f(\cdot;\lambda)$ are continuous in λ uniformly with respect to $x \in B(r,x_*)$.

Theorem 4.2.2. *Suppose that the following conditions are satisfied:*

(1) *for every $\lambda \in [0,1]$ the operator $\nabla_H f(\cdot;\lambda)$ exists and is a continuously differentiable operator from E into E,*

(2) *for every $\lambda \in [0,1]$ there exists an admissible finite-dimensional subspace L_λ of E such that for every $w \in E \cap L_\lambda^\perp$ the solution x of the linear equation*

$$Q_\lambda \nabla_H^2 f(x_*;\lambda)x = w\ ,$$

where $Q(\lambda)$ is the orthogonal projection onto $L_\lambda^\perp$ in H, lies in E. Then, for all $\lambda \in [0,1]$, the Conley index $h(x_*; f(\cdot;\lambda))$ of the critical point x_* of $f(\cdot;\lambda)$ is the same.

Proof. We fix $\lambda_0 \in [0,1]$ and show that

$$h(x_*; f(\cdot;\lambda)) = h\left(x_*; f(\cdot;\lambda_0)\right)$$

for all points λ that are sufficiently close to λ_0.

Let L_{λ_0} be an admissible subspace for $f(\cdot;\lambda_0)$ satisfying condition (2). This means that we have

$$(\nabla_H^2 f(x_*;\lambda_0)v, v) \geqslant m\|v\|_H^2 \quad (v \in L_{\lambda_0}^\perp) \tag{4.2.27}$$

for some $m > 0$. Since $\nabla_H^2 f(x_*;\lambda)$ depends continuously on λ, we have

$$\|\nabla_H^2 f(x_*;\lambda) - \nabla_H^2 f(x_*;\lambda_0)\| \leqslant \tfrac{1}{2}m$$

if λ is sufficiently close to λ_0. Then we have

$$\begin{aligned}(\nabla_H^2 f(x_*;\lambda)v, v) &= (\nabla_H^2 f(x_*;\lambda_0)v, v) + ((\nabla_H^2 f(x_*;\lambda) - \nabla_H^2 f(x_*;\lambda_0))v, v)\\ &\geqslant m\|v\|_H^2 - \tfrac{1}{2}m\|v\|_H^2 = \tfrac{1}{2}m\|v\|_H^2\end{aligned} \tag{4.2.28}$$

for all $v \in L_{\lambda_0}^\perp$. By (4.2.28), the subspace L_{λ_0} is admissible if λ is sufficiently close to λ_0. The Conley index of the critical point x_* of $f(x;\lambda)$ is equal, by definition, to the Conley index of the critical point 0 of the functional

$$\Phi(u;\lambda) = f(x_* + u + \varphi(u;\lambda))\ .$$

Applying Theorem 4.2.1 to this functional we may now conclude that $h(x_*; f(\cdot;\lambda)) = h(x_*; f(\cdot;\lambda_0))$ if λ is sufficiently close to λ_0, and the argument used to prove Theorem 4.1.5 gives the assertion of Theorem 4.2.2. □

5 Applications

The theorems on the homotopy invariance of minimizers and on Conley index that we have proved in the preceding chapters provide effective means for tackling a wide variety of problems: proving inequalities, finding criteria for multilinear forms to be positive definite, investigating the stability of systems of differential equations, analysing whether degenerate critical points are minimizers, and investigating bifurcation points of solutions of nonlinear problems. We shall describe some applications of the homotopy method in this chapter.

5.1 Problems of Classical Analysis

5.1.1 Proving Inequalities (General Principles)

Many important inequalities have the form

$$f_1(x) \geqslant 0 \quad (x \in \mathbb{R}^N)\,, \tag{5.1.1}$$

where $f_1(0) = 0$ and f_1 is differentiable for $x \neq 0$. The homotopy method can be used to prove some well-known inequalities and certain new inequalities of this form. To apply the homotopy method it is often convenient to use the modification given below.

Theorem 5.1.1. *Let $f_0 : \mathbb{R}^N \to \mathbb{R}$ be a growing function and suppose that*

$$(1-\lambda)\nabla f_0(x) + \lambda \nabla f_1(x) \neq 0 \quad (x \neq 0,\ 0 \leqslant \lambda < 1)\,. \tag{5.1.2}$$

Then the inequality (5.1.1) *holds.*

Proof. By Theorem 2.1.6, the point 0 is a strict global minimizer of each function

$$f_{1-\varepsilon}(x) = f_1(x) + \varepsilon(f_0(x) - f_1(x)) \quad (0 < \varepsilon \leqslant 1)\,.$$

Therefore

$$f_{1-\varepsilon}(x) \geqslant \varepsilon f_0(0) \quad (x \in \mathbb{R}^N)\,,$$

and we obtain the result by passing to the limit as $\varepsilon \to 0$. □

The following result turns out to be useful for proving inequalities of form (5.1.1) in the case when f_1 is a positive homogeneous function (of some positive degree α).

Theorem 5.1.2. *Suppose that* $f(x;\lambda) : \mathbb{R}^N \times [0,1] \to \mathbb{R}$ *is a continuous function that is homogeneous of degree* $\alpha > 0$ *with respect to* x. *Suppose that the gradient* $\nabla f(\cdot;\cdot)$ *exists and is continuous on* $(\mathbb{R}^N \setminus \{0\}) \times [0,1)$. *Suppose, finally, that for each* $\lambda \in [0,1)$ *the function* $f(\cdot;\lambda)$ *has no nonzero critical points and that* 0 *is a minimizer of* $f(\cdot;0)$. *Then*

$$f(x;1) \geqslant 0 \quad (x \in \mathbb{R}^N) . \tag{5.1.3}$$

Proof. From Theorem 2.1.1 the point 0 is a local minimizer for $f(\cdot;\lambda)$ for each $\lambda \in [0,1)$. Since the function $f(\cdot;\lambda)$ is positive homogeneous, this minimizer is a global minimizer. Thus

$$f(x;\lambda) \geqslant 0 \quad (x \in \mathbb{R}^N,\ 0 \leqslant \lambda < 1) . \tag{5.1.4}$$

We now obtain (5.1.3) from (5.1.4) by taking the limit as $\lambda \to 0$. □

5.1.2 Sylvester's Criterion

Let

$$A = \begin{bmatrix} a_{11} & a_{12} & \dots & a_{1N} \\ a_{21} & a_{22} & \dots & a_{2N} \\ \dots & \dots & \dots & \dots \\ a_{N1} & a_{N2} & \dots & a_{NN} \end{bmatrix} \tag{5.1.5}$$

be a real symmetric matrix. We say that A is *positive (resp. nonnegative) definite* if all of its eigenvalues are positive (resp. nonnegative).

In many problems of analysis and its applications it is necessary to decide whether a symmetric matrix A is positive definite. Since the direct approach of solving the equation

$$\det(A - \lambda I) = 0 \tag{5.1.6}$$

for the eigenvalues of A is impractical for large N, it is important to have other methods for determining the signs of the eigenvalues. The most widely used of these is *Sylvester's criterion,* which we formulate in two separate theorems.

The *principal minors* of a symmetric $N \times N$ matrix $A = (a_{ij})$ are the determinants

$$\mathcal{D}_1 = a_{11},\ \mathcal{D}_2 = \begin{vmatrix} a_{11} & a_{12} \\ a_{21} & a_{22} \end{vmatrix}, \dots,\ \mathcal{D}_N = \begin{vmatrix} a_{11} & a_{12} & \dots & a_{1N} \\ a_{21} & a_{22} & \dots & a_{2N} \\ \dots & \dots & \dots & \dots \\ a_{N1} & a_{N2} & \dots & a_{NN} \end{vmatrix} . \tag{5.1.7}$$

Theorem 5.1.3. *A symmetric matrix is positive definite if and only if all of its principal minors are positive.*

Theorem 5.1.4. *A symmetric matrix has at least one negative eigenvalue if and only if at least one of its principal minors is negative.*

We shall give a deformational proof of Sylvester's criterion. We regard $\mathbb{R}^N$ as an inner product space with the standard inner product: thus for vectors $x = (x_1, \ldots, x_N)$ and $y = (y_1, \ldots, y_N)$ in $\mathbb{R}^N$ we define

$$(x, y) = \sum_{i=1}^{N} x_i y_i \,.$$

It is easy to see that a symmetric matrix A is positive definite if and only if the function

$$f(x) = (Ax, x) \tag{5.1.8}$$

takes positive values for all for nonzero $x \in \mathbb{R}^N$, and so if and only if 0 is a strict global minimizer for f on $\mathbb{R}^N$.

Let $A = (a_{ij})$ be an $N \times N$ symmetric matrix and suppose that

$$\mathcal{D}_k = \begin{vmatrix} a_{11} & \ldots & a_{1k} \\ \ldots & \ldots & \ldots \\ a_{k1} & \ldots & a_{kk} \end{vmatrix} < 0$$

for some k. Then the $k \times k$ matrix

$$D_k = \begin{bmatrix} a_{11} & \ldots & a_{1k} \\ \ldots & \ldots & \ldots \\ a_{k1} & \ldots & a_{kk} \end{bmatrix} \tag{5.1.9}$$

has at least one negative eigenvalue. Therefore there exists a nonzero vector

$$x_0 = (x_1, \ldots, x_k) \in \mathbb{R}^k \tag{5.1.10}$$

for which

$$(D_k x_0, x_0) < 0 \,,$$

and hence

$$(Ay_0, y_0) < 0$$

where

$$y_0 = (x_1, \ldots, x_k, 0, \ldots, 0) \in \mathbb{R}^N \,. \tag{5.1.11}$$

Consequently A has at least one negative eigenvalue.

If instead $\mathcal{D}_k = 0$, then the matrix D_k has a zero eigenvalue. Therefore

$$(D_k x_0, x_0) = 0$$

for some nonzero vector $x_0 = (x_1, \ldots, x_k) \in \mathbb{R}^k$, and hence we have

$$(Ay_0, y_0) = 0 ,$$

where $y_0 = (x_1, \ldots, x_k, 0, \ldots, 0) \in \mathbb{R}^N$. Consequently A is not positive definite.

Thus if A is positive definite then all of its principal minors must be positive. To complete the proof of Theorem 5.1.2, we must show that if all principal minors $\mathcal{D}_1, \ldots, \mathcal{D}_N$ of A are positive then A is positive definite. This is clear for $N = 1$ and so we assume that $N > 1$ and argue by induction.

We set

$$A(\lambda) = \begin{bmatrix} a_{11} & \ldots & a_{1N-1} & \lambda a_{1N} \\ a_{21} & \ldots & a_{2N-1} & \lambda a_{2N} \\ \ldots & \ldots & \ldots & \ldots \\ a_{N-1\,1} & \ldots & a_{N-1\,N-1} & \lambda a_{N-1\,N} \\ \lambda a_{N1} & \ldots & \lambda a_{NN-1} & \lambda^2 a_{NN} + (1-\lambda^2)\mathcal{D}_N/\mathcal{D}_{N-1} \end{bmatrix} .$$

Since

$$\nabla(A(\lambda)x, x) = 2A(\lambda)x \quad (x \in \mathbb{R}^N;\ 0 \leqslant \lambda \leqslant 1)$$

and

$$\det A(\lambda) = \mathcal{D}_N > 0 ,$$

the deformation $\varphi(x, \lambda) = (A(\lambda)x, x)$ of the function $\varphi(x, 0) = (A(0)x, x)$ into the function $\varphi(x, 1) = (A(1)x, x)$ is nondegenerate. It follows from Theorem 2.1.1 that A will be positive definite if $A(0)$ is positive definite. Clearly the eigenvalues of $A(0)$ are the eigenvalues of the matrix

$$\begin{bmatrix} a_{11} & \ldots & a_{1\,N-2} & a_{1\,N-1} \\ \ldots & \ldots & \ldots & \ldots \\ a_{N-2\,1} & \ldots & a_{N-2\,N-2} & a_{N-2\,N-1} \\ a_{N-1\,1} & \ldots & a_{N-1\,N-2} & a_{N-1\,N-1} \end{bmatrix}$$

together with $\mathcal{D}_N/\mathcal{D}_{N-1}$, and the principal minors of $A(0)$ are the principal minors $\mathcal{D}_1, \ldots, \mathcal{D}_{N-1}$ of the above matrix together with $\mathcal{D}_N$. Thus the eigenvalues of the above matrix are positive by induction, and hence the eigenvalues of $A(0)$ and of A are positive. Therefore Theorem 5.1.3 is proved. □

We supplement Theorem 5.1.3 with the following important observation: if all eigenvalues of A are nonnegative, then all principal minors of A are nonnegative. To prove this, we consider the family of matrices

$$A(\lambda) = A + \lambda I \quad (0 \leqslant \lambda \leqslant 1) ,$$

where I is the identity matrix. If the eigenvalues of A are nonnegative, then the eigenvalues of each matrix $A(\lambda)$ are strictly positive and, by Theorem

5.1.3, the principal minors of each matrix $A(\lambda)$ are positive. Therefore, we may conclude that the principal minors of $A(0) = A$ are nonnegative by taking the limit as $\lambda \to 0$.

The converse statement is not true, i.e., if A is a symmetric matrix all of whose minors are nonnegative then A need not be nonnegative definite. For example, the matrix

$$\begin{bmatrix} 0 & 0 \\ 0 & -1 \end{bmatrix}$$

is not nonnegative definite.

Now we prove Theorem 5.1.4. While proving Theorem 5.1.3 we established that if some minor $\mathcal{D}_k$ of A is negative then A has at least one negative eigenvalue. The reverse implication follows from the remark above. □

5.1.3 Young's Inequality

We say that two even convex functions $M, N : \mathbb{R} \to \mathbb{R}$ are *Young conjugate* functions if they are of the form

$$M(u) = \int_0^{|u|} p(\tau)d\tau\ , \quad N(v) = \int_0^{|v|} q(\tau)d\tau\ , \tag{5.1.12}$$

where p and q are nonnegative increasing continuous functions with $p(0) = 0$ such that q is the inverse of p.

In this section we prove Young's inequality

$$uv \leqslant M(u) + N(v) \quad (u, v \geqslant 0)\ , \tag{5.1.13}$$

for functions M, N satisfying these conditions.

We set

$$f_0(u, v) = M(u^2) + N(v^2), \quad f_1(u, v) = f_0(u, v) - u^2v^2\ .$$

We claim that the point $(0, 0)$ is a global minimizer for f_1. Since f_0 is growing, by Theorem 5.1.1 it suffices to show that

$$(1 - \lambda)\nabla f_0(u, v) + \lambda \nabla f_1(u, v) \neq 0 \tag{5.1.14}$$

for $\lambda \in [0, 1)$ and $u^2 + v^2 \neq 0$. In other words, it suffices to verify that the system of two equations

$$up(u^2) = \lambda uv^2, \quad vq(v^2) = \lambda u^2 v \tag{5.1.15}$$

has no nonzero solutions for $\lambda \in [0, 1)$. If for some $\lambda_0 \in [0, 1)$ there is such a solution (u_0, v_0), then u_0, $v_0 \neq 0$, and (5.1.15) yields that

$$p(u^2) = \lambda_0 v^2, \quad q(v^2) = \lambda_0 u^2\ ;$$

thus

$$p(u^2) < v^2, \quad q(v^2) < u^2 ,$$

contradicting the hypothesis that q is the inverse of p. Therefore Young's inequality is proved. □

In the mathematical literature, the inequality

$$xy \leqslant \frac{x^p}{p} + \frac{y^q}{q} , \tag{5.1.16}$$

where $x, y > 0$, $1 < p < \infty$ and $1/p + 1/q = 1$, is often called Young's inequality. It is easy to see that (5.1.16) is a consequence of (5.1.13). The homotopy method allows us to obtain the following strengthened version of (5.1.16):

$$\min\left(\frac{1}{p}, \frac{1}{q}\right)(x^{p/2} - y^{q/2})^2 + xy \leqslant \frac{x^p}{p} + \frac{y^q}{q} . \tag{5.1.17}$$

The proof of this inequality is similar to the proof of (5.1.13).

5.1.4 Minkowski's Inequality

This inequality asserts that

$$\left(\sum_{i=1}^{N} |x_i + y_i|^p\right)^{1/p} \leqslant \left(\sum_{i=1}^{N} |x_i|^p\right)^{1/p} + \left(\sum_{i=1}^{N} |y_i|^p\right)^{1/p} , \tag{5.1.18}$$

for all x_i, $y_i \in \mathbb{R}$ and $p > 1$.

For the proof we can clearly assume that x_i, $y_i \geqslant 0$. We set

$$x_i = u_i^2, \quad y_i = v_i^2 \quad (i = 1, \ldots, N) ,$$

$$f_0(u_1, \ldots, u_N; v_1, \ldots, v_N) = \left(\sum_{i=1}^{N} |u_i|^{2p}\right)^{1/p} + \left(\sum_{i=1}^{N} |v_i|^{2p}\right)^{1/p} , \tag{5.1.19}$$

$$\begin{aligned} &f_1(u_1, \ldots, u_N; v_1, \ldots, v_N) \\ &\quad = f_0(u_1, \ldots, u_N; v_1, \ldots, v_N) - \left(\sum_{i=1}^{N} (u_i^2 + v_i^2)^p\right)^{1/p} . \end{aligned} \tag{5.1.20}$$

The inequality (5.1.18) is equivalent to the inequality

$$f_1(u_1, \ldots, u_N; v_1, \ldots, v_N) \geqslant 0 . \tag{5.1.21}$$

Since the function f_0 is growing, to prove (5.1.21) it suffices by Theorem 5.1.1 to establish that, for $\lambda \in [0, 1)$, the system of $2N$ equations

$$(1-\lambda)\frac{\partial f_0(u_1,\ldots,u_N;v_1,\ldots,v_N)}{\partial u_i}+\lambda\frac{\partial f_1(u_1,\ldots,u_N;v_1,\ldots,v_N)}{\partial u_i}=0\,,$$

$$(1-\lambda)\frac{\partial f_0(u_1,\ldots,u_N;v_1,\ldots,v_N)}{\partial v_i}+\lambda\frac{\partial f_1(u_1,\ldots,u_N;v_1,\ldots,v_N)}{\partial v_i}=0$$

$$(i=1,\ldots,N)$$

has no nonzero solutions. Written out in full, this system becomes

$$\left(\sum_{j=1}^{N}|u_j|^{2p}\right)^{(1/p)-1}|u_i|^{2p-2}\,u_i=\lambda\left(\sum_{j=1}^{N}(u_j^2+v_j^2)^p\right)^{(1/p)-1}(u_i^2+v_i^2)^{p-1}u_i\,,$$

$$\left(\sum_{j=1}^{N}|v_j|^{2p}\right)^{(1/p)-1}|v_i|^{2p-2}\,v_i=\lambda\left(\sum_{j=1}^{N}(u_j^2+v_j^2)^p\right)^{(1/p)-1}(u_i^2+v_i^2)^{p-1}v_i$$

$$(i=1,\ldots,N).$$

Suppose that $u_1,\ldots,u_N,v_1,\ldots,v_N$ is a nonzero solution. If for some i we have $u_i=0$ and $v_i\neq 0$ then

$$\left(\sum_{j=1}^{N}|v_j|^{2p}\right)^{(1/p)-1}=\lambda\left(\sum_{j=1}^{N}(u_j^2+v_j^2)^p\right)^{(1/p)-1};$$

but then clearly $u_k=0$ for $k=1,\ldots,N$ and $\lambda=1$, a contradiction. This and a similar argument show that $u_i=0$ if and only if $v_i=0$. From the system of equations we now have

$$u_i^2=kv_i^2 \tag{5.1.22}$$

for $i=1,\ldots,N$, where

$$k=\left(\frac{\sum_{j=1}^{N}|u_j|^{2p}}{\sum_{j=1}^{N}|v_j|^{2p}}\right)^{1/p}.$$

Substituting (5.1.22) into the system of equations, we conclude again that $\lambda=1$, and this contradiction proves Minkowski's inequality.

5.1.5 Jensen's Inequality

Let $\varphi:\mathbb{R}\to\mathbb{R}$ be a convex function and let

$$0<p_1,\ldots,p_N<1 \quad\text{and}\quad p_1+\cdots+p_N=1\,.$$

Jensen's inequality asserts that

$$\varphi\left(\sum_{i=1}^{N}p_ix_i\right)\leqslant\sum_{i=1}^{N}p_i\varphi(x_i) \tag{5.1.23}$$

for all $x_1, \ldots, x_N \in \mathbb{R}$. We shall prove it by the homotopy method. We may assume, for simplicity, that the function φ is continuously differentiable. We set

$$x = (x_1, \ldots, x_N) \in \mathbb{R}^N ,$$

$$f_0(x) = \frac{1}{2} \sum_{i=1}^{N} p_i x_i^2 , \quad f_1(x) = \sum_{i=1}^{N} p_i \varphi(x_i) - \varphi\left(\sum_{i=1}^{N} p_i x_i\right) .$$

Since the function f_0 is growing, to prove Jensen's inequality it suffices by Theorem 5.1.1 to show that for $\lambda \in [0, 1)$ the vector equation

$$(1 - \lambda)\nabla f_0(x) + \lambda \nabla f_1(x) = 0 \tag{5.1.24}$$

has only the zero solution. This equation yields the system of scalar equations

$$(1 - \lambda)x_i + \lambda\varphi'(x_i) = \lambda\varphi'\left(\sum_{i=1}^{N} p_j x_j\right) \quad (i = 1, \ldots, N) . \tag{5.1.25}$$

Consider the function

$$\psi(\tau) = (1 - \lambda)\tau + \lambda\varphi'(\tau) \quad (-\infty < \tau < \infty) . \tag{5.1.26}$$

Since the function φ is convex, ψ is strictly monotonic and, consequently, has an inverse ψ^{-1}. It now follows from (5.1.26) that

$$x_i = \psi^{-1}\left(\lambda\varphi'\left(\sum_{j=1}^{N} p_i x_j\right)\right) \quad (i = 1, \ldots, N) ,$$

i.e.,

$$x_1 = \cdots = x_N .$$

Therefore the system (5.1.25) has the form

$$(1 - \lambda)x_i = 0 \quad (i = 1, \ldots, N) , \tag{5.1.27}$$

and since $\lambda \in [0, 1)$, it follows that

$$x_1 = \cdots = x_N = 0 ,$$

as required.

5.1.6 Cauchy's Inequality (Inequality of the Arithmetic and Geometric Means)

This section and the next one are devoted to one of the best-known of all inequalities.

Let $x_1, \ldots, x_N$ be nonnegative real numbers and $x = (x_1, \ldots, x_N)$. The *arithmetic mean* $M(x)$ and the *geometric mean* $\Gamma(x)$ of $x_1, \ldots, x_N$ are defined by

$$M(x) = \frac{1}{N}\sum_{i=1}^{N} x_i , \tag{5.1.28}$$

and

$$\Gamma(x) = \left(\prod_{i=1}^{N} x_i\right)^{1/N} . \tag{5.1.29}$$

The inequality of the arithmetic mean and geometric mean, due to Cauchy, asserts that

$$M(x) \geqslant \Gamma(x) . \tag{5.1.30}$$

In order to prove this inequality, we set $x_i = y_i^{2N}$ and consider the functions

$$f_0(y) = \frac{1}{N}\sum_{i=1}^{N} y_i^{2N}, \tag{5.1.31}$$

$$f_1(y) = \frac{1}{N}\sum_{i=1}^{N} y_i^{2N} - \prod_{i=1}^{N} y_i^2. \tag{5.1.32}$$

Clearly (5.1.30) is equivalent to the inequality

$$f_1(y) \geqslant 0 . \tag{5.1.33}$$

Since f_0 is growing, to prove (5.1.33) it suffices by Theorem 5.1.1 to show that for $\lambda \in [0, 1)$ the zero solution is the unique solution of the equation

$$(1-\lambda)\nabla f_0(y) + \lambda \nabla f_1(y) = 0 .$$

This equation can be written in the form

$$y_i^{2N-1} - \lambda y_i \prod_{j\neq i} y_j^2 = 0 \quad (i = 1, \ldots, N) . \tag{5.1.34}$$

It follows from (5.1.34) all numbers y_i have the same absolute value, a say. Thus the equations in (5.1.34) give

$$a^{2N-1}(1-\lambda) = 0 ,$$

and so $a = 0$, as required.

5.1.7 Improvements and Extensions

Cauchy's inequality (5.1.30) can be strengthened and generalized in various ways. Below we give some of the results (apparently hitherto unknown) that can be established using the homotopy method.

Write

$$\mathcal{D}(x) = \frac{1}{N}\sum_{i=1}^{N}\left(\sqrt{x_i} - \frac{1}{N}\sum_{j=1}^{N}\sqrt{x_j}\right)^2 .$$

We claim that

$$M(x) \geqslant \Gamma(x) + \mathcal{D}(x) , \qquad (5.1.35)$$

for all $x_1, \ldots, x_N$. To prove this, we set

$$y_i = x_i^{4N}, \quad y = (y_1, \ldots, y_N) ,$$

$$f_0(y) = \frac{1}{N}\sum_{i=1}^{N} y_i^{4N} ,$$

$$f_1(y) = \frac{1}{N}\sum_{i=1}^{N} y_i^{4N} - \prod_{i=1}^{N} y_i^4 - \frac{1}{N}\sum_{i=1}^{N}\left(y_i^{2N} - \frac{1}{N}\sum_{j=1}^{N} y_j^{2N}\right)^2 .$$

Since f_0 is growing, to prove the inequality (5.1.35) it suffices to show that for $\lambda \in [0, 1)$ the system of equations

$$y_i^{4N-1} = \lambda y_i^3 \prod_{j\neq i} y_j^4 + \lambda\left(y_i^{2N} - \frac{1}{N}\sum_{j=1}^{N} y_j^{2N}\right) y_i^{2N-1} \quad (i = 1, \ldots, N) \quad (5.1.36)$$

has no nonzero solutions. Suppose that $y_1, \ldots, y_N$ is a nonzero solution. Then Eq. (5.1.36), rewritten in the form,

$$(1-\lambda)y_i^{4N} + \left(\frac{\lambda}{N}\sum_{j=1}^{N} y_j^{2N}\right) y_i^{2N} - \lambda\prod_{j=1}^{N} y_j^4 = 0 \quad (i = 1, \ldots, N) ,$$

implies both that all numbers y_i^{2N} are non-zero and that they are equal to the unique positive root u_0 of the quadratic equation

$$u^2 + au + b = 0 ,$$

where

$$a = \frac{\lambda}{(1-\lambda)N}\sum_{j=1}^{N} y_j^{2N}, \quad b = -\frac{\lambda}{1-\lambda}\prod_{j=1}^{N} y_j^4 .$$

Replacing the numbers u_i^{2N} in (5.1.36) by u_0, we conclude that $\lambda = 1$, and this contradiction establishes the inequality (5.1.35).

For an N-tuple $x = (x_1, \dots, x_N)$ we write

$$\gamma(x) = N \left(\sum_{k=1}^{N} \frac{1}{x_k} \right)^{-1} ;$$

thus $\gamma(x)$ is the *harmonic mean* of the numbers x_i. The inequality

$$\alpha M(x) + (1-\alpha)\gamma(x) \geqslant \Gamma(x) , \tag{5.1.37}$$

where

$$\alpha = \frac{N^2}{N^2 + 4(N-1)} \quad (N \geqslant 2) ,$$

can easily be proved by the homotopy method using arguments like those above. Since the classical inequality

$$\Gamma(x) \geqslant \gamma(x)$$

is well known, (5.1.37) can be regarded as an improvement of the inequality (5.1.30).

5.2 Nonlinear Programming Problems

5.2.1 Extremals of Lipschitz Nonlinear Programming Problems

Let $g_i : \mathbb{R}^N \to \mathbb{R}$ $(i = 1, \dots, m)$ be locally Lipschitzian functions. We consider the Lebesgue sets

$$Q_i = \{x \in \mathbb{R}^N : g_i(x) \leqslant 0\} \quad (i = 1, \dots, m) \tag{5.2.1}$$

of these functions, and set

$$Q = \bigcap_{i=1}^{m} Q_i . \tag{5.2.2}$$

We shall study the minimization problem for a locally Lipschitzian function f on Q. It is customary to write this problem in the form

$$\begin{cases} f(x) \to \min, \\ g_i(x) \leqslant 0 \quad (i = 1, \dots, m) . \end{cases} \tag{5.2.3}$$

Problem (5.2.3) is called a *nonlinear* (or *mathematical*) *programming problem.* It is obviously equivalent to the problem

$$\begin{cases} f(x) \to \min , \\ g(x) \leqslant 0 , \end{cases} \tag{5.2.4}$$

where

$$g(x) = \max_{1 \leqslant i \leqslant m} g_i(x) . \tag{5.2.5}$$

A point $x_* \in Q$ is an *extremal* of problem (5.2.3), (5.2.4) if there exist nonnegative numbers μ, $y_1, \ldots, y_m$, not all zero, for which

$$0 \in \mu \partial f(x_*) + y_1 \partial g_1(x_*) + \cdots + y_m \partial g_m(x_*) \tag{5.2.6}$$

and

$$y_i g_i(x_*) = 0 \quad (i = 1, \ldots, m) . \tag{5.2.7}$$

We call x_* a *local minimizer* for problem (5.2.3) if $x_* \in Q$ and if

$$f(x) \geqslant f(x_*) \quad (x \in Q \cap B(\rho, x_*)) \tag{5.2.8}$$

for some $\rho > 0$. If x_* is a local minimizer for problem (5.2.3), then x_* is an extremal of this problem. In particular, every solution of problem (5.2.3) is an extremal. If

$$g_i(x_*) < 0 \tag{5.2.9}$$

for some i, then by (5.2.7) we have

$$y_i = 0.$$

If

$$g_i(x_*) = 0 \tag{5.2.10}$$

for some i, then the constraint in problem (5.2.3) corresponding to this index is said to be *active*. We write $J(x_*)$ for the set of indices corresponding to active constraints.

5.2.2 Extremals of Classical Nonlinear Programming Problems

When the functions f and g_i $(i = 1, \ldots, m)$ in problem (5.2.3) are continuously differentiable, (5.2.3) is a classical nonlinear programming problem.

The function

$$\mathcal{L}(x, y, \mu) = \mu f(x) + (y, G(x)) , \tag{5.2.11}$$

where

$$G(x) = \{g_1(x), \ldots, g_m(x)\} , \tag{5.2.12}$$

is known as the *Lagrangian function* of problem (5.2.3).

If x_* is a local minimizer for problem (5.2.3), then there exist simultaneously nonzero Lagrange multipliers $\mu > 0,\ y \in \mathbb{R}^m_+$ for which

$$\mathcal{L}_x(x_*, \mu, y) = 0 \tag{5.2.13}$$

and

$$y_i g_i(x_*) = 0 \quad (i = 1, 2, \ldots, m) . \tag{5.2.14}$$

This result is the *John–Kuhn–Tucker theorem* (see, e.g., [4]).

Condition (5.2.14) (just like its analog (5.2.7) for the case of Lipschitz problems) is called the condition of *complementary slackness*.

The definition of an extremal for a classical nonlinear programming problem is similar to the corresponding definition for the case of Lipschitz problems: we call x_* an *extremal* of problem (5.2.3) if there exist simultaneously nonzero Lagrange multipliers $\mu > 0$ and $y \in \mathbb{R}^m_+$ such that conditions (5.2.13), (5.2.14) hold.

We should point out that the definition of an extremal of problem (5.2.3) can be reformulated in terms of the normal cone N_Q to the set Q: a point x_* is an extremal of the Lipschitz problem (5.2.3) if

$$0 \in \partial f(x_*) + N_Q(x_*) \,. \tag{5.2.15}$$

If f is differentiable, then (5.2.15) assumes the form

$$0 \in \nabla f(x_*) + N_Q(x_*) \,. \tag{5.2.16}$$

5.2.3 The Deformation Theorem

Let f_0 and g_0 be locally Lipschitzian functions and consider the nonlinear programming problem

$$\begin{cases} f_0(x) \to \min \,, \\ g_0(x) \leqslant 0 \,. \end{cases} \tag{5.2.17}$$

We call a one-parameter family of problems

$$\begin{cases} f(x;\lambda) \to \min \,, \\ g(x;\lambda) \leqslant 0 \quad (0 \leqslant \lambda \leqslant 1) \end{cases} \tag{5.2.18}$$

a *nondegenerate deformation* of problem (5.2.17) into the problem

$$\begin{cases} f_1(x) \to \min \,, \\ g_1(x) \leqslant 0 \end{cases} \tag{5.2.19}$$

if

(1) the functions f, g are continuous on $\mathbb{R}^N \times [0,1]$ and locally Lipschitzian with respect to x for every $\lambda \in [0,1]$,

(2) the multivalued mappings

$$\partial_x f : \mathbb{R}^N \times [0,1] \to \mathbb{R}^N, \quad \partial_x g : Q(\lambda) \times [0,1] \to \mathbb{R}^N \,,$$

where

$$Q(\lambda) = \{x \in \mathbb{R}^N : g(x;\lambda) \leqslant 0\} \quad (0 \leqslant \lambda \leqslant 1) \,, \tag{5.2.20}$$

are upper semicontinuous,

(3) for every $\lambda \in [0,1]$ the constraint in problem (5.2.18) satisfies the following regularity conditions:

(a) for every x such that $g(x;\lambda) = 0$ and every $v \in \mathbb{R}^N$, the generalized derivative $g^0(x,v;\lambda)$ coincides with the classical derivative in the direction of v:

$$g^0(x,v;\lambda) = g'(x,v;\lambda) ,$$

(b)

$$0 \notin \partial_x g(x;\lambda) ,$$

(4) for every $\lambda \in [0,1]$, problem (5.2.18) has a unique extremal $x(\lambda)$ which continuously depends on the parameter $\lambda \in [0,1]$,

(5) $f(\cdot;0) = f_0$, $g(\cdot;0) = g_0$, $f(\cdot;1) = f_1$ and $g(\cdot;1) = g_1$.

Theorem 5.2.1. *Suppose that there exists a nondegenerate deformation of problem* (5.2.17) *into problem* (5.2.19). *If the extremal* $x_0 = x(0)$ *is a local minimizer for problem* (5.2.17) *then the extremal* $x_1 = x(1)$ *is a local minimizer for problem* (5.2.19).

Proof. We assume for simplicity that

$$f(0;\lambda) = g(0;\lambda) = 0 \quad (0 \leqslant \lambda \leqslant 1)$$

and

$$x(\lambda) = 0 \quad (0 \leqslant \lambda \leqslant 1) .$$

We set

$$d_{Q(\lambda)}(x) = \min_{y \in Q(\lambda)} |x - y| \quad (x \in \mathbb{R}^N; 0 \leqslant \lambda \leqslant 1) , \tag{5.2.21}$$

$$F(x,k,\lambda) = f(x,\lambda) + k d_{Q(\lambda)}(x) \quad (x \in \mathbb{R}^N; 0 \leqslant \lambda \leqslant 1) . \tag{5.2.22}$$

We claim that for sufficiently large k the family of functions $F(\cdot,k,\lambda)$ satisfies the hypothesis of Theorem 2.2.1 on some ball $B(\rho)$ with $\rho > 0$. It suffices to show that for large k the point 0 is a local minimizer of $F(\cdot,k,0)$ and is an extremal of the function $F(\cdot,k,\lambda)$ that is uniformly isolated with respect to $\lambda \in [0,1]$.

Let $k > k_1$, where k_1 is the Lipschitz constant of f_0 on the unit ball. Then, for $\rho < 1$, the minimizers of $F(\cdot,k,0)$, regarded as a function defined on $B(\rho)$, lie in the set $Q_0 \cap B(\rho)$. However, 0 is the unique minimizer of $F(\cdot,k,0)$ on $Q_0 \cap B(\rho)$ for small $\rho > 0$. Thus 0 is a local minimizer of $F(\cdot,k;0)$.

It remains to show that for large k the function $F(\cdot,k;\lambda)$ has no nonzero critical points of small norm.

First consider the function $F(\cdot,k;\lambda)$ on $Q(\lambda)$. If

$$x_0 \in \partial Q(\lambda), \quad x_0 \neq 0, \quad |x_0| \ll 1$$

and

$$0 \in \partial_x F(x_0,k;\lambda) ,$$

then

$$0 \in \partial_x(f(x_0;\lambda) + kd_{Q(\lambda)}(x_0)) \subset \partial_x f(x_0;\lambda) + k\partial_x d_{Q(\lambda)}(x_0)$$
$$\subset \partial_x f(x_0) + \bigcup_{\mu \geqslant 0} \mu\partial_x g(x_0;\lambda) = \partial_x f(x_0;\lambda) + N_{Q(\lambda)}(x_0) .$$

Therefore x_0 is a nonzero extremal of problem (5.2.18) of small norm, and this is a contradiction.

If

$$x_0 \in \mathring{Q}(\lambda), \quad x_0 \neq 0, \quad |x_0| \ll 1$$

and

$$0 \in \partial_x F(x_0, k; \lambda) ,$$

then

$$0 \in \partial_x f(x_0;\lambda) ,$$

and we again find that x_0 is a nonzero extremal of problem (5.2.18) of small norm, and this contradicts the hypothesis of the theorem.

Thus $F(\cdot, k; \lambda)$ has no nonzero critical points of small norm on $Q(\lambda)$.

We shall show now that for large k the function $F(\cdot, k; \lambda)$ has no nonzero critical points of small norm on $\mathbb{R}^N \setminus Q(\lambda)$.

Since

$$0 \notin \partial_x g(0;\lambda)$$

by the regularity condition (b), the set

$$M = \operatorname{co}\left\{\left(\bigcup_{\mu \geqslant 0} \mu\partial_x g(0;\lambda)\right) \cap \partial B\right\}$$

does not contain zero. Consequently,

$$0 \notin U(\varepsilon_0, M) = \{x \in \mathbb{R}^N : |y - x| \leqslant \varepsilon_0, y \in M\}$$

for some $\varepsilon > 0$. Since for some $\varepsilon_1 > 0$ the multivalued mapping

$$\partial_x d_{Q(\lambda)} : \mathbb{R}^N \to \mathbb{R}^N$$

is upper semicontinuous for $|x| \leqslant \varepsilon_1$, we have

$$\partial_x d_{Q(\lambda)}(x) \subset V(\varepsilon_0, \partial d_{Q(\lambda)}(0)), \tag{5.2.23}$$

and

$$\partial_x d_{Q(\lambda)}(0) = \left(\bigcup_{\mu \geqslant 0} \mu\partial_x g(0;\lambda)\right) \cap B . \tag{5.2.24}$$

On the other hand,

$$\partial_x d_{Q(\lambda)}(x) = \operatorname{co}\{y \in \mathbb{R}^N,\ \nabla_x d_{Q(\lambda)}(x_k) \to y,\\ x_k \to x, x_k \in \Lambda(d_{Q(\lambda)})\} \tag{5.2.25}$$

and

$$|\nabla_x d_{Q(\lambda)}(x)| = 1 \quad (x \in \mathbb{R}^N \setminus Q(\lambda) \cap \Lambda(d_{Q(\lambda)}))\ , \tag{5.2.26}$$

where ∇ is the ordinary gradient and $\Lambda(d_{Q(\lambda)})$ is the set of points at which $d_{Q(\lambda)}$ is differentiable.

It follows from (5.2.24)–(5.2.25) that

$$\partial_x d_{Q(\lambda)}(x) \subset \mathring{U}(\varepsilon_0, M) \quad (x \in \mathbb{R}^N \setminus Q(\lambda), |x| \leqslant \varepsilon_1) \tag{5.2.27}$$

for all sufficiently small $\varepsilon_1 > 0$.

We set

$$a = \min_{y \in U(\varepsilon_0, M)} |y|, \quad A = \max_{y \in \partial_x f(x;\lambda), x \in B} |y|, \quad k_2 = A/a\ .$$

Then by (5.2.27), for all $k > k_2$, $x \in Q(\lambda)$ and $|x| < \varepsilon_1$, we have

$$\begin{aligned}\min_{y \in \partial_x F(x,k;\lambda)} |y| &= \min_{y \in \partial_x (f(x;\lambda) + k d_{Q(\lambda)}(x))} |y| \\ &\geqslant \min_{y \in (\partial_x f(x;\lambda) + k \partial_x d_{Q(\lambda)}(x))} |y| \\ &\geqslant \min_{y \in (B(A) + kU(\varepsilon_0, M))} |y| \geqslant ka - A > 0\end{aligned}$$

so that

$$0 \notin \partial_x F(x,k;\lambda)\ .$$

Thus, for $k \geqslant \max\{k_1, k_2\}$, the critical point 0 of the function $F(\cdot, k; \lambda)$ is isolated in $\mathbb{R}^N$ uniformly with respect to λ. This follows from the upper semicontinuity of the mapping

$$\partial_x g(\cdot;\cdot) : \mathbb{R}^N \times [0,1] \to \mathbb{R}^N\ .$$

We have now verified all conditions of Theorem 2.2.1, and we can conclude that the point 0 is a local minimizer for the function

$$F(\cdot, k; 1) \quad (k \geqslant \max\{k_1, k_2\})$$

and thus is a local minimizer for problem (5.2.19), as required. □

Consider now the classical nonlinear programming problem

$$\begin{cases} f_0(x) \to \min, \\ g_i^0(x) \leqslant 0 \quad (i = 1, \ldots, m)\ . \end{cases} \tag{5.2.28}$$

The deformation method can also be applied to problem (5.2.28). The procedure is similar to our applications of the deformation method above: if we can deform problem (5.2.28) to some standard problem in such a way that the corresponding extremal remains isolated in the process of deformation, then the extremal of the problem under study is a minimizer if and only if the corresponding extremal of the standard problem is a minimizer. We explain the details below.

A one-parameter family of problems

$$\begin{cases} f(x;\lambda) \to \min\ , \\ g_i^0(x;\lambda) \leqslant 0 \quad (i=1,\dots,m, 0 \leqslant \lambda \leqslant 1) \end{cases} \tag{5.2.29}$$

is a *nondegenerate deformation* of problem (5.2.28) into the problem

$$\begin{cases} f_1(x) \to \min\ , \\ g_i^1(x) \leqslant 0 \quad (i=1,\dots,m) \end{cases} \tag{5.2.30}$$

if the following conditions hold:

(1) the functions $f(x;\lambda)$, $g_i(x;\lambda)$ $(i=1,\dots,m)$ and their gradients ∇f, ∇g_i $(i=1,\dots,m)$ are continuous on $\mathbb{R}^N \times [0,1]$,

(2) (*regularity condition*) if $x \in \partial Q(\lambda)$, where

$$Q(\lambda) = \{x \in \mathbb{R}^N : g_i(x;\lambda) \leqslant 0,\ i=1,\dots,m\}\ ,$$

then the vectors

$$\nabla_x g_i(x;\lambda) \quad (i \in J(x;\lambda))$$

are positively linearly independent; here $J(x;\lambda)$ is the set of indices corresponding to active constraints, i.e., $i \in J(x;\lambda)$ if and only if $g_i(x;\lambda)=0$,

(3) for every $\lambda \in [0,1]$, problem (5.2.29) has a unique extremal $x(\lambda)$ which continuously depends on λ,

(4) $f(\cdot;0)=f_0$, $f(\cdot;1)=f_1$, $g_i(\cdot;0)=g_i^0$ and $g_i(\cdot;1)=g_i^1$ $(i=1,\dots,m)$.

Theorem 5.2.2. *Suppose that the family of nonlinear programming problems* (5.2.29) *is a nondegenerate deformation of problem* (5.2.28) *into problem* (5.2.30). *If the extremal x_0 is a local minimizer for problem* (5.2.28) *then the extremal x_1 is a local minimizer for problem* (5.2.30).

Proof. We may assume for simplicity that

$$f(0;\lambda)=0 \quad (0 \leqslant \lambda \leqslant 1)\ ,$$

$$g_i(0;\lambda)=0 \quad (0 \leqslant \lambda \leqslant 1;\ i=1,\dots,m)\ ,$$

$$x(\lambda)=0 \quad (0 \leqslant \lambda \leqslant 1)\ .$$

We set

$$g(x;\lambda) = \max_{1 \leqslant i \leqslant m} g_i(x;\lambda)\ .$$

Then problems (5.2.28), (5.2.29), and (5.2.30) are equivalent, respectively, to the problems

$$\begin{cases} f_0(x) \to \min , \\ g_0(x) \leqslant 0 , \end{cases} \tag{5.2.31}$$

$$\begin{cases} f(x;\lambda) \to \min , \\ g(x;\lambda) \leqslant 0 \quad (0 \leqslant \lambda \leqslant 1) , \end{cases} \tag{5.2.32}$$

$$\begin{cases} f_1(x) \to \min , \\ g_1(x) \leqslant 0 . \end{cases} \tag{5.2.33}$$

We shall show that problems (5.2.31)–(5.2.33) satisfy the hypothesis of Theorem 5.2.1. For this purpose, we must show that 0 is the unique extremal of each problem (5.2.32), that the multivalued mapping

$$\partial_x g : Q(\lambda) \times [0,1] \to \mathbb{R}^N$$

is upper semicontinuous, and that for every $\lambda \in [0,1]$ the constraint in problem (5.2.32) satisfies the regularity conditions (a) and (b).

Since the point 0 is an extremal of problem (5.2.29), there exist simultaneously nonzero Lagrange multipliers

$$\mu \in \mathbb{R}_+, \quad y = \{y_1, \dots, y_m\} \in \mathbb{R}^m_+$$

for which

$$\mu \nabla f(0;\lambda) + \sum_{i=1}^{m} y_i \nabla g_i(0;\lambda) = 0 . \tag{5.2.34}$$

Since the vectors $\nabla g_i(0;\lambda)$ $(i = 1, \dots, m)$ are positively linearly independent, it follows that $\mu \neq 0$. Therefore we can rewrite (5.2.34) in the form

$$\nabla f(0;\lambda) + \sum_{i=1}^{m} \tau_i \nabla g_i(0;\lambda) = 0 , \tag{5.2.35}$$

where $\tau_i = y_i/\mu$. However,

$$N_{Q(\lambda)}(0) = \left\{ \sum_{i=1}^{m} \mu_i \nabla g_i(0;\lambda) : \mu_i \geqslant 0, \quad i = 1, \dots, m \right\} , \tag{5.2.36}$$

and therefore (5.2.35) implies that

$$0 \in \nabla f(0;\lambda) + N_{Q(\lambda)}(0) ,$$

i.e., 0 is an extremal of problem (5.2.32).

Now assume that problem (5.2.32) has a nonzero extremal $x_* \in Q(\lambda)$ for some $\lambda \in [0,1]$. Then

$$0 \in \nabla f(x_*;\lambda) + N_{Q(\lambda)}(x_*) . \tag{5.2.37}$$

However,

$$N_{Q(\lambda)}(x_*) = \left\{ \sum_{i \in J(x_*;\lambda)} \mu_i \nabla g_i(x_*;\lambda) : \mu_i \geqslant 0 \right\}. \tag{5.2.38}$$

It follows from (5.2.37) and (5.2.38) that

$$\nabla f(x_*;\lambda) + \sum_{i \in J(x_*;\lambda)} \mu_i^* \nabla g_i(x_*;\lambda) = 0$$

for some positive reals μ_i^* $(i \in J(x_*;\lambda))$. Therefore x_* is an extremal of the nonlinear programming problem (5.2.29). Hence $x_* = 0$, and this is a contradiction.

We have thus shown that, for every λ, problem (5.2.32) has a unique extremal, namely 0.

Furthermore, since

$$\partial_x g(x;\lambda) = \mathrm{co}\{\nabla g_i(x;\lambda) : i \in J(x;\lambda)\} \tag{5.2.39}$$

for $x \in Q(\lambda)$ and since J is upper semicontinuous on $Q(\lambda) \times [0,1]$, the mapping $\partial_x g$ is upper semicontinuous on $Q(\lambda) \times [0,1]$.

Finally, since the functions $g_i(x;\lambda)$ are continuously differentiable the function g is regular, and since the vectors $\nabla g_i(x;\lambda)$ $(i \in J(x;\lambda))$ are positively linear independent, (5.2.39) implies that

$$0 \notin \partial_x g(x;\lambda)$$

for $x \in \partial Q(\lambda)$.

Thus all conditions of Theorem 5.2.1 are satisfied. We conclude that 0 is a minimizer for problem (5.2.34), and hence also for problems (5.2.30) and (5.2.31), and the theorem follows. □

5.2.4 Linear Deformations of Nonlinear Programming Problems and Invariance of Global Minimizers

Under the conditions of Theorem 5.2.2, the point x_1 is in general only a local minimizer. Below we consider conditions under which x_1 is a global minimizer for problem (5.2.30).

We call the family of problems (5.2.29) a *nondegenerate linear deformation* of problem (5.2.28) into problem (5.2.30) if

$$f(x;\lambda) = (1-\lambda) f_0(x) + \lambda f_1(x),$$
$$g_i(x;\lambda) = g_i^0(x) \quad (x \in \mathbb{R}^N;\ 0 \leqslant \lambda \leqslant 1;\ i = 1,\dots,m),$$

and, for every $\lambda \in [0,1]$, the point 0 is the unique extremal of problem (5.2.29).

Theorem 5.2.3. *Suppose that there exists a nondegenerate linear deformation of problem* (5.2.28) *into problem* (5.2.30). *Write*

$$Q_0 = \{x \in \mathbb{R}^N : g_i^0(x) \leqslant 0,\ i = 1, \dots, m\}$$

and suppose that for every nonzero $x \in Q_0$ *there exists a vector* $s(x) \in \mathbb{R}^N$ *such that*

$$(\nabla g_i^0(x), s(x)) < 0 \quad (i \in J(0, x)) . \tag{5.2.40}$$

Suppose, finally, that

$$\lim_{|x|\to\infty, x\in Q_0} f_0(x) = \infty . \tag{5.2.41}$$

Then 0 *is a global minimizer for problem* (5.2.30).

Proof. We fix $c > 0$ and consider the problem

$$\begin{cases} f_1(x) \to \min , \\ g_i^0(x) \leqslant 0 \quad (i = 1, \dots, m) , \\ f_0(x) - c \leqslant 0 . \end{cases} \tag{5.2.42}$$

By condition (5.2.41), the intersection of the set Q_0 and the set

$$G = \{x \in \mathbb{R}^N : f_0(x) \leqslant c\}$$

is compact. Therefore problem (5.2.42) has a solution x_0. Let us show that $x_0 = 0$.

By the John–Kuhn–Tucker theorem, we have

$$\mu_0 \nabla f_1(x_0) + \nu_0 \nabla f_0(x_0) + \sum_{i=1}^m y_i^0 \nabla g_i^0(x_0) = 0 , \tag{5.2.43}$$

$$y_i^0 g_i^0(x_0) = 0 \quad (i = 1, \dots, m) , \tag{5.2.44}$$

$$\nu_0(f_0(x) - c) = 0 \tag{5.2.45}$$

for some μ_0, $\nu_0 \in \mathbb{R}_+$ and $y_0 = \{y_1^0, \dots, y_m^0\} \in \mathbb{R}_+^m$ which are simultaneously nonzero. Multiplying (5.2.43) by $s(x_0)$, we obtain that

$$\mu_0(\nabla f_1(x_0), s(x_0)) + \nu_0(\nabla f_0(x_0), s(x_0)) + \sum_{i=1}^m y_i^0(\nabla g_i^0(x_0), s(x_0)) = 0 .$$

If $\mu_0 + \nu_0 = 0$ we have a contradiction to (5.2.40). Therefore

$$\mu_0 + \nu_0 > 0 ,$$

and we can rewrite (5.2.43) and (5.2.44) in the form

$$\frac{\mu_0}{\mu_0 + \nu_0} \nabla f_1(x_0) + \frac{\nu_0}{\mu_0 + \nu_0} \nabla f_0(x_0) + \sum_{i=1}^m \frac{y_i^0}{\mu_0 + \nu_0} \nabla g_i^0(x_0) = 0 ,$$

$$\frac{y_i^0}{\mu_0 + \nu_0} g_i^0(x_0) = 0 \quad (i = 1, \ldots, m) .$$

Therefore x_0 is an extremal of the problem

$$\begin{cases} \frac{\mu_0}{\mu_0 + \nu_0} f_1(x) + \frac{\nu_0}{\mu_0 + \nu_0} f_0(x) \to \min , \\ g_i^0(x) = 0 \quad (i = 1, \ldots, m) . \end{cases}$$

Consequently,

$$x_0 = 0 .$$

Since the constant c is arbitrary and (5.2.41) holds, 0 is a global minimizer for problem (5.2.30), as required. □

5.2.5 Sufficient Conditions for Minimizers in Nonlinear Programming Problems

In this section and the next, we indicate some applications of the results on the homotopy invariance of minimizers established in the preceding sections.

Let $f, g_i : \mathbb{R}^N \to \mathbb{R}$ $(i = 1, \ldots, m)$ be continuously differentiable functions. Consider the nonlinear programming problem

$$\begin{cases} f(x) \to \min, \\ g_i(x) \leqslant 0 \quad (i = 1, \ldots, m) . \end{cases} \tag{5.2.46}$$

Let x_* be an extremal of problem (5.2.46), i.e.,

$$x_* \in Q = \{x \in \mathbb{R}^N : g_i(x) \leqslant 0, \; i = 1, \ldots, m) ,$$

and suppose that for simultaneously nonzero Lagrange multipliers $\mu > 0$, $y \in \mathbb{R}_+^m$ we have

$$\mathcal{L}_x(x_*, \mu, y) = 0 , \tag{5.2.47}$$

$$(y, G(x_*)) = 0 , \tag{5.2.48}$$

where

$$\mathcal{L}(x, \mu, y) = \mu f(x) + \sum_{i=1}^{m} y_i g_i(x) \tag{5.2.49}$$

is the Lagrangian and

$$G(x) = \{g_1(x), \ldots, g_m(x)\} . \tag{5.2.50}$$

When investigating nonlinear programming problems, one first finds their extremals and then subjects them to further analysis to establish which of them are minimizers. Some of the requirements can be investigated by the homotopy method.

We begin by finding first-order sufficient conditions for a minimizer.

Theorem 5.2.4. *Let x_* be an extremal of problem* (5.2.46) *and suppose that*

$$m = N \ . \tag{5.2.51}$$

Suppose that the vectors $\nabla g_1(x_*), \dots, \nabla g_m(x_*)$ are linearly independent and that the Lagrange multipliers $\mu = 1, y_1, \dots, y_m$ corresponding to the extremal x_* are positive. Then x_* is a local minimizer for problem (5.2.51).

Proof. We assume for simplicity that

$$x_* = 0, \quad f(0) = 0, \quad g_i(0) = 0 \quad (i = 1, \dots, m) \ .$$

Consider the family of problems

$$\begin{cases} (1-\lambda)f(x) + \lambda(\nabla f(0), x) \to \min \ , \\ (1-\lambda)g_i(x) + \lambda(\nabla g_i(0), x) \leqslant 0 \quad (i = 1, \dots, N) \end{cases} \tag{5.2.52}$$

depending on the parameter $\lambda \in [0,1]$. For every λ the point 0 is an extremal of the problem corresponding to λ in (5.2.52). Below we shall establish that this extremal is uniformly isolated, relative to $\lambda \in [0,1]$. It will then follow from Theorem 5.2.2 that 0 is a local minimizer for problem (5.2.46) if and only if it is a local minimizer for the linear programming problem

$$\begin{cases} (f(0), x) \to \min \ , \\ (\nabla g_i(0), x) \leqslant 0 \quad (i = 1, \dots, N) \ . \end{cases} \tag{5.2.53}$$

Therefore, since we can write (5.2.47) in the form

$$\nabla f(0) = -\sum_{i=1}^{N} y_i \nabla g_i(0) \ ,$$

we can conclude that

$$(\nabla f(0), x) = -\sum_{i=1}^{N} y_i (\nabla g_i(0), x) \ ,$$

and consequently that 0 is a solution of problem (5.2.53), so that the assertion of the theorem will follow.

It remains to show that the extremal 0 is uniformly isolated relative to $\lambda \in [0,1]$ in problem (5.2.52). If this is false, then there exists a sequence (λ_k) in $[0,1]$ that converges to some λ_0, where each λ_k is associated with some nonzero extremal x_k of problem (5.2.52) for $\lambda = \lambda_k$ such that $x_k \to 0$. By the definition of extremals, we have equations

$$\begin{aligned} &\mu_k \left((1-\lambda_k)\nabla f(x_k) + \lambda_k \nabla f(0)\right) \\ &\qquad + \sum_{i=1}^{N} y_i^k \left((1-\lambda_k)\nabla g_i(x_k) + \lambda_k \nabla g_i(0)\right) = 0 \ , \end{aligned} \tag{5.2.54}$$

$$y_i^k\left((1-\lambda-k)g_i(x_k)+\lambda_k(\nabla g_i(0),x_k)\right)=0 \quad (i=1,\dots,N)\,, \tag{5.2.55}$$

where μ_k, $y_1^k,\dots,y_N^k$ are nonnegative and not all zero. Since $x_k \to 0$, the vectors

$$b_i^k=(1-\lambda_k)\nabla g_i(x_k)+\lambda_k\nabla g_i(0) \quad (i=1,\dots,N)$$

are linearly independent for sufficiently large k, and therefore the numbers μ_k are positive for these values of k. Consequently, for each i the sequence y_i^k/μ_k converges to y_i, and hence the numbers y_i^k are positive for large k. Therefore the equations (5.2.55) imply that

$$(1-\lambda_k)g_i(x_k)+\lambda_k(\nabla g_i(0),x_k)=0 \quad (i=1,\dots,N),$$

and these equations can be rewritten in the form

$$|x_k|^{-1}(1-\lambda_k)\left(g_i(x_k)-(\nabla g_i(0),x_k)\right)+\left(\nabla g_i(0),|x_k|^{-1}x_k\right)=0\,.$$

Passing to the limit here as $k\to\infty$ and assuming that $|x_k|^{-1}x_k \to x_0$, we conclude that

$$(\nabla g_i(0),x_0)=0 \quad (i=1,\dots,N)$$

and hence, since the vectors $\nabla g_i(0)$ are linearly independent, that

$$x_0=0\,.$$

On the other hand,

$$|x_0|=1\,.$$

This contradiction completes the proof of the theorem. □

We now consider second-order sufficient conditions for a minimizer in problem (5.2.46).

Let x_* be an extremal of problem (5.2.46). We assume that $m\leqslant N$ and that the vectors $\nabla g_i(x_*)$ $(i=1,\dots,m)$ are linearly independent. We denote by Π the subspace of vectors in $\mathbb{R}^N$ orthogonal to all of the vectors $\nabla g_i(x_*)$ $(i=1,\dots,m)$.

Theorem 5.2.5. *Let*

$$\nabla f(x_*)\neq 0$$

and suppose that the Lagrange multipliers μ, $y_1,\dots,y_m$ corresponding to the extremal x_* are positive. If

$$(\mathcal{L}_{xx}(x_*,\mu,y)h,h)\geqslant a\,|h|^2 \quad (a>0,\ h\in\Pi)\,, \tag{5.2.56}$$

then the extremal x_* is a local minimizer for problem (5.2.46).

Proof. Just as in the proof of Theorem 5.2.4, it is convenient to assume that

$$x_* = 0 \text{ and } \mu = 1 .$$

Then (as pointed out above) condition (5.2.56) has the form

$$\nabla f(0) + \sum_{i=1}^{m} y_i \nabla g_i(0) = 0 . \tag{5.2.57}$$

Let P be the orthogonal projection onto Π and consider the family of problems

$$\begin{cases} \lambda f(x) - (1-\lambda)\sum_{i=1}^{m} y_i g_i(x) \\ \quad + \frac{1}{2}(1-\lambda)\,(\mathcal{L}_{xx}(0,1,y)Px, Px) \to \min , \\ g_i(x) \leqslant 0 \quad (i = 1, \ldots, m) \end{cases} \tag{5.2.58}$$

depending on the parameter $\lambda \in [0,1]$.

For each problem (5.2.58), the point 0 is an extremal. We show below that this extremal is uniformly isolated, with respect to $\lambda \in [0,1]$, in the set

$$Q = \{x \in \mathbb{R}^N : g_i(x) \leqslant 0,\ i = 1, \ldots, m\} .$$

Therefore Theorem 5.2.2 implies that the assertion that we must prove holds if and only if 0 is a local minimizer for problem (5.2.58) for $\lambda = 0$.

Now for $\lambda = 0$ problem (5.2.58) is the problem

$$\begin{cases} \frac{1}{2}(\mathcal{L}_{xx}(0,1,y_0)Px, Px) - \sum_{i=1}^{m} y_i^0 g_i(x) \to \min , \\ g_i(x) \leqslant 0 \quad (i = 1, \ldots, m) . \end{cases} \tag{5.2.59}$$

It follows from (5.2.56) and from the fact that the multipliers $y_1^0, \ldots, y_m^0$ are positive that the extremal 0 is a minimizer for the problems (5.2.59). It remains to prove that this extremal is uniformly isolated with respect to $\lambda \in [0,1]$ in the set Q.

Suppose the contrary. Then there exists a sequence (λ_k) in $[0,1]$, which converges to some λ_0, such that each λ_k is associated with a nonzero extremal $x_k \in Q$ of problem (5.2.59) for $\lambda = \lambda_k$ with the property that $x_k \to 0$ and $x_k/|x_k| \to x_0$, where $|x_0| = 1$. By the definition of extremals, we have

$$\mu_k\Big(\lambda_k \nabla f(x_k) - (1-\lambda_k)\sum_{i=1}^{m} y_i \nabla g_i(x_k) + (1-\lambda_k)P\mathcal{L}_{xx}(0,1,y)Px_k\Big) + \sum_{i=1}^{m} y_i^k \nabla g_i(x_k) = 0 , \tag{5.2.60}$$

$$y_i^k g_i(x_k) = 0 \quad (i = 1, \ldots, m) \tag{5.2.61}$$

for every k, and the numbers μ_k, $y_1^k, \dots, y_m^k$ are nonnegative and not all zero.

Arguing as in the proof of Theorem 5.2.4, we can show that $x_0 \in \Pi$ and

$$(\mathcal{L}_{xx}(0,1,y)x_0, x_0) = 0 \,.$$

This equality contradicts condition (5.2.56), and the theorem is proved. □

Theorem 5.2.5 gives a sufficient condition for a second-order minimum. Another sufficient second-order condition is contained in the following result.

Theorem 5.2.6. *Let x_* be an extremal of problem* (5.2.46),

$$\nabla f(x_*) = 0 \,,$$

let $m \leqslant N$, and suppose that

$$(\nabla^2 f(x_*)h, h) > 0$$

for all $h \neq 0$ in the cone

$$K = \{h \in \mathbb{R}^N : (\nabla g_i(x_*), h) \leqslant 0, \; i = 1, \dots, m\} \,.$$

Then x_* is a local minimizer for problem (5.2.46).

Proof. We assume for simplicity that

$$x_* = 0, \quad g_i(0) = 0 \quad (i = 1, \dots, m) \,.$$

Consider the one-parameter family of problems

$$\begin{cases} \lambda f(x) + \frac{1}{2}(1-\lambda)(\nabla^2 f(0)x, x) \to \min \,, \\ \lambda g_i(x) + (1-\lambda)(\nabla g_i(0), x) \leqslant 0 \quad (i = 1, \dots, m) \end{cases} \quad (0 \leqslant \lambda \leqslant 1) \,. \tag{5.2.62}$$

The point 0 is an extremal of each problem of this family. We shall show that this extremal is isolated uniformly with respect to $\lambda \in [0,1]$.

If the contrary holds, then there exist sequences of points $x_k \neq 0$, numbers $\lambda_k \in [1,0]$, and nonzero vectors $\{\mu_k, y_1^k, \dots, y_m^k\}$ with nonnegative components, such that

$$\lim_{k\to\infty} |x_k| = 0 \,, \tag{5.2.63}$$

$$\lambda_k g_i(x_k) + (1-\lambda_k)(\nabla g_i(0), x_k) \leqslant 0 \quad (i = 1, \dots, m; \; k = 1, 2, \dots) \,, \tag{5.2.64}$$

$$\begin{aligned} &\mu_k(\lambda_k \nabla f(x_k) + (1-\lambda_k)\nabla^2 f(0)x_k) \\ &\quad + \sum_{i=1}^m y_i^k(\lambda_k \nabla g_i(x_k) + (1-\lambda_k)\nabla g_i(0)) = 0 \quad (k = 1, 2, \dots) \,, \end{aligned} \tag{5.2.65}$$

$$\sum_{i=1}^m y_i^k(\lambda_k g_i(x_k) + (1-\lambda_k)(\nabla g_i(0), x_k)) = 0 \quad (k = 1, 2, \dots) \,. \tag{5.2.66}$$

We can rewrite the equations (5.2.65) and (5.2.66) as

$$\mu_k \nabla^2 f(0) x_k + \sum_{i=1}^{m} y_i^k (\nabla g_i(0) + O(|x_k|)) = 0 \quad (k = 1, 2, \ldots) \tag{5.2.67}$$

and

$$\sum_{i=1}^{m} y_i^k ((\nabla g_i(0), x_k) + o(|x_k|)) = 0 \quad (k = 1, 2, \ldots) . \tag{5.2.68}$$

Since the vectors $\nabla g_i(0)$ $(i = 1, \ldots, m)$ are linearly independent, it follows from (5.2.67) that the numbers μ_k are nonzero for large k. Therefore, for large k, the equations (5.2.67) can be rewritten in the form

$$\nabla^2 f(0) x_k + \sum_{i=1}^{m} \frac{y_i^k}{\mu_k} (\nabla g_i(0) + O(|x_k|)) = 0 . \tag{5.2.69}$$

In turn, these equations give the inequality

$$0 \leqslant \frac{y_i^k}{\mu_k} \leqslant C|x_k| . \tag{5.2.70}$$

Multiplying the equations (5.2.69) by the scalars $x_k/|x_k|^2$ we have

$$\left(\nabla^2 f(0) \frac{x_k}{|x_k|}, \frac{x_k}{|x_k|}\right) + \sum_{k=1}^{m} \frac{y_i^k}{\mu_k |x_k|^2} ((\nabla g_i(0), x_k) + o(|x_k|)) . \tag{5.2.71}$$

Now it follows from the equations (5.2.68) and the inequalities (5.2.70) that

$$\sum_{i=1}^{m} \frac{y_i^k}{\mu_k |x_k|^2} (\nabla g_i(0), x_k) = O(|x_k|) . \tag{5.2.72}$$

However, then from (5.2.70)–(5.2.72) we have

$$\left(\nabla^2 f(0) \frac{x_k}{|x_k|}, \frac{x_k}{|x_k|}\right) = O(|x_k|) . \tag{5.2.73}$$

We can assume without loss of generality that the sequence $(x_k/|x_k|)$ converges to some element $h_0 \in K$ with

$$|h_0| = 1 .$$

Passing to the limit as $k \to \infty$ in (5.2.73), we find that

$$(\nabla^2 f(0) h_0, h_0) = 0 ,$$

and this contradicts the hypothesis of the theorem.

Thus the one-parameter family of problems (5.2.62) is a nondegenerate deformation of the original problem

$$\begin{cases} f(x) \to \min , \\ g_i(x) \leqslant 0 \quad (i = 1, \dots, m) \end{cases}$$

into the problem

$$\begin{cases} \frac{1}{2}(\nabla^2 f(0)x, x) \to \min , \\ (\nabla g_i(0), x) \leqslant 0 \end{cases}$$

for which the point 0 is obviously a minimizer. The result now follows from an appeal to Theorem 5.2.2. □

5.2.6 Inequalities with Constraints

In Sects. 5.1.6 and 5.1.7 we illustrated the homotopy method by strengthening and generalizing the inequality

$$M(x) \geqslant \Gamma(x) , \tag{5.2.74}$$

where $x = (x_1, \dots, x_N)$ is a vector with nonnegative components, and where

$$M(x) = \frac{1}{N} \sum_{i=1}^{N} x_i \quad \text{and} \quad \Gamma(x) = \left(\prod_{i=1}^{N} x_i \right)^{1/N} .$$

The inequality (5.2.74) is strict unless $x_1 = x_2 = \cdots = x_N$. Assume that $x_1, \dots, x_N$ satisfy the additional constraint

$$\sum_{k=1}^{N} a_k x_k \leqslant 0 , \tag{5.2.75}$$

where

$$\sum_{k=1}^{N} a_k > 0. \tag{5.2.76}$$

We shall show that, under the constraints (5.2.75), (5.2.76), the inequality (5.2.74) can be improved to

$$M(x) \geqslant k_0 \Gamma(x) , \tag{5.2.77}$$

where

$$k_0 = \left(\prod_{k=1}^{N} (1 + t_0 a_k) \right)^{1/N} \tag{5.2.78}$$

and t_0 is the least positive root of the equation

$$\sum_{k=1}^{N} \frac{1}{1+a_k t} - N = 0 . \tag{5.2.79}$$

If $a_k > 0$ for each k, then the constraints

$$\sum_{k=1}^{N} a_k x_k \leqslant 0, \quad x_1 \geqslant 0, \ldots, x_N \geqslant 0$$

have only the point $0 \in \mathbb{R}^N$ as a solution, and (5.2.77) holds trivially. Therefore we may assume that some of the numbers a_k are negative; since the inequality depends continuously on $a_1, \ldots, a_N$ we may also assume without loss of generality that

$$a_1 < a_2 < \cdots < a_N \quad \text{and} \quad a_k \neq 0 \ \text{ for each } k .$$

In order to prove inequality (5.2.77), we shall use Theorem 5.2.3. We fix some $\varepsilon > 0$ and consider the family of problems

$$\begin{cases} \dfrac{1}{N}\displaystyle\sum_{k=1}^{N} x_k - \lambda k_0 \left(\prod_{k=1}^{N} x_k + \varepsilon\right)^{1/N} \to \min , \\ \displaystyle\sum_{k=1}^{N} a_k x_k \leqslant 0 , \\ x_k \geqslant 0 \quad (k = 1, \ldots, N;\ 0 \leqslant \lambda \leqslant 1) . \end{cases} \tag{5.2.80}$$

This family is a linear deformation of the problem

$$\begin{cases} \dfrac{1}{N}\displaystyle\sum_{k=1}^{N} x_k \to \min , \\ \displaystyle\sum_{k=1}^{N} a_k x_k \leqslant 0 , \\ x_k \geqslant 0 \quad (k = 1, \ldots, N) \end{cases} \tag{5.2.81}$$

into the problem

$$\begin{cases} \dfrac{1}{N}\displaystyle\sum_{k=1}^{N} x_k - k_0 \left(\prod_{k=1}^{N} x_k + \varepsilon\right)^{1/N} \to \min , \\ \displaystyle\sum_{k=1}^{N} a_k x_k \leqslant 0 , \\ x_k \geqslant 0 \quad (k = 1, \ldots, N) . \end{cases} \tag{5.2.82}$$

We shall show that this linear deformation is nondegenerate.

Suppose that this is false and choose $\lambda_0 \in [0,1]$ for which some nonzero point

$$x_0 = (x_1^0, \dots, x_N^0)$$

is an extremal of problem (5.2.80). Thus for some nonnegative numbers μ_0, $y_0^0, y_1^0, \dots, y_N^0$ which are not all equal to zero we have

$$\mathcal{L}_x(x_0, \mu_0, y_0) = 0 , \tag{5.2.83}$$

$$y_0^0 \sum_{k=1}^{N} a_k x_k^0 = 0 , \tag{5.2.84}$$

$$x_k^0 y_k^0 = 0 \quad (k = 1, \dots, N) , \tag{5.2.85}$$

where $\mathcal{L}(x, \mu, y)$ is the Lagrangian of problem (5.2.80) for $\lambda = \lambda_0$:

$$\mathcal{L}(x, \mu, y) = \mu \left(\frac{1}{N} \sum_{k=1}^{N} x_k - \lambda_0 k_0 \left(\prod_{k=1}^{N} x_k + \varepsilon \right) \right) + y_0 \sum_{k=1}^{N} a_k x_k - \sum_{k=1}^{N} y_k x_k .$$

It follows from (5.2.83) that $\mu_0 \neq 0$, since for $\mu_0 = 0$ the equations (5.2.83) assume the form

$$y_0^0 a_k - y_k^0 = 0 \quad (k = 1, \dots, N),$$

and so it follows from (5.2.85) that $y_0^0 = 0$; but then $y_k^0 = 0$ for $k = 1, \dots, N$ and so all Lagrange multipliers are zero.

Since $\mu_0 \neq 0$, we can rewrite (5.2.83)–(5.2.85) in the form

$$1 - \lambda_0 k_0 \left(\prod_{k=1}^{N} x_k^0 + \varepsilon \right)^{(1-N)/N} \prod_{i \neq k} x_i^0 - t_0 a_k - t_k = 0 \quad (k = 1, \dots, N) , \tag{5.2.86}$$

$$t_0 \sum_{k=1}^{N} a_k x_k^0 = 0 , \tag{5.2.87}$$

$$x_k^0 t_k = 0 \quad (k = 1, \dots, N) , \tag{5.2.88}$$

where

$$t_k = \frac{y_k^0 N}{\mu_0} \quad (k = 0, 1, \dots, N) .$$

It follows from (5.2.86) that $t_0 \neq 0$. Indeed, if $t_0 = 0$, then (5.2.86) and (5.2.88) imply that

$$x_k^0 = \lambda_0 k_0 \left(\prod_{k=1}^{N} x_k^0 + \varepsilon \right)^{(1-N)/N} \prod_{k=1}^{N} x_k^0 ,$$

i.e.,

$$x_1^0 = \cdots = x_N^0 .$$

Therefore

$$x_1^0 = \cdots = x_N^0 = 0 .$$

Furthermore, (5.2.86)–(5.2.88) imply that

$$x_k^0(1+t_0a_k) = \lambda_0 k_0 \left(\prod_{k=1}^N x_k^0 + \varepsilon\right)^{(1-N)/N} \prod_{k=1}^N x_k^0 \quad (k=1,\ldots,N) \tag{5.2.89}$$

and

$$\sum_{k=1}^N x_k^0 = \lambda_0 N k_0 \left(\prod_{k=1}^N x_k^0 + \varepsilon\right)^{(1-N)/N} \prod_{k=1}^N x_k^0 . \tag{5.2.90}$$

It follows from (5.2.90) that

$$x_k^0 > 0 \quad (k=1,\ldots,N) ,$$

and hence by (5.2.89) we have

$$1+t_0a_k > 0 \quad (k=1,\ldots,N) \tag{5.2.91}$$

and

$$x_k^0 = (1+t_0a_k)^{-1}\lambda_0 k_0 \left(\prod_{k=1}^N x_k^0 + \varepsilon\right)^{(1-N)/N} \prod_{k=1}^N x_k^0 . \tag{5.2.92}$$

Multiplying the above equations together, we obtain

$$\prod_{k=1}^N x_k^0 = \prod_{k=1}^N (1+t_0a_k)^{-1}\lambda_0^N k_0^N \left(\prod_{k=1}^N x_k^0 + \varepsilon\right)^{1-N} \left(\prod_{k=1}^N x_k^0\right)^N ,$$

whence we find that

$$\lambda_0 = k_0^{-1}\left(1+\varepsilon\left(\prod_{k=1}^N x_k^0\right)^{-1}\right)^{(N-1)/N} \left(\prod_{k=1}^N (1+t_0a_k)\right)^{1/N} .$$

Therefore

$$\lambda_0 > k_0^{-1}\left(\prod_{k=1}^N (1+t_0a_k)\right)^{1/N} . \tag{5.2.93}$$

Adding the equations (5.2.92) together, we get

$$\sum_{k=1}^N x_k^0 = \lambda_0 k_0 \left(\prod_{k=1}^N x_k^0 + \varepsilon\right)^{(1-N)/N} \prod_{k=1}^N x_k^0 \sum_{k=1}^N (1+t_0a_k)^{-1} .$$

This and (5.2.90) imply that

$$\sum_{k=1}^{N} \frac{1}{1+t_0 a_k} = N \, . \tag{5.2.94}$$

Thus λ_0 satisfies the inequality (5.2.93), where t_0 is a positive root of Eq. (5.2.94) satisfying the constraints (5.2.91).

The smallest positive root of Eq. (5.2.79) has these properties. Indeed, Eq. (5.2.79) is equivalent to a polynomial equation of degree N, and so it has most N real roots; since it also has at least N real roots, the number of roots is equal to N. A positive root of Eq. (5.2.79) satisfying (5.2.91) must lie in the interval $(-a_N^{-1}, -a_0^{-1})$. However, there is just one positive root of Eq. (5.2.79) in this interval, namely, the smallest positive root. Thus from (5.2.78) and (5.2.93) we have

$$\lambda_0 > 1 \, .$$

This is a contradiction, and we conclude that the linear deformation (5.2.80) must be nondegenerate.

Now we verify that condition (5.2.40) holds. For this purpose, we divide the boundary S of the domain $\mathcal{D}$ of points in $\mathbb{R}^N$ satisfying the inequalities

$$\sum_{k=1}^{N} a_k x_k < 0, \quad x_k > 0 \quad (k = 1, \dots, N)$$

into two parts, S_0 and S_1:

$$S_0 = \left\{ x = \{x_1, \dots, x_N\} \in S : \sum_{k=1}^{N} a_k x_k < 0 \right\} ,$$

$$S_1 = \left\{ x = \{x_1, \dots, x_N\} \in S : \sum_{k=1}^{N} a_k x_k = 0 \right\} .$$

If $x \in S_0$ and $x \neq 0$, then the set $J_0(x)$ of indices corresponding to the active constraints in problem (5.2.81) fails to contain at least one index k in $\{1, \dots, N\}$. In this case, we define $s(x) \in \mathbb{R}^N$ by

$$s(x) = \{\varepsilon, \dots, \varepsilon, -1, \varepsilon, \dots, \varepsilon\} ,$$

where -1 is the kth component and $\varepsilon = 1/N$.

If instead $x \in S_1$ and $x \neq 0$, then the set $J_0(x)$ of indices corresponding to the active constraints among the constraints $x_k \geqslant 0$ ($k = 1, \dots, N$) fails to contain at least two indices from $\{1, \dots, N\}$. In this case, $a_l > 0$ for at least one $l \notin J(x)$. Define

$$s(x) = \{\varepsilon, \dots, \varepsilon, -1, \varepsilon, \dots, \varepsilon\}$$

where the lth component is equal to -1 and

$$\varepsilon = \min\left\{\frac{1}{N}, \frac{a_l}{\sum_{k=1}^N |a_k|}\right\} .$$

Thus condition (5.2.40) of Theorem 5.2.3 holds. Condition (5.2.41) also holds since

$$\lim_{x\in\mathbb{R}_+^N,\, |x|\to\infty} \frac{1}{N}\sum_{k=1}^{N} x_k = \infty .$$

We have shown that the point 0 is a global minimizer for problem (5.2.82); equivalently, we have established the inequality

$$\frac{1}{N}\sum_{k=1}^{N} x_k \geqslant k_0\left(\prod_{k=1}^{N} x_k + \varepsilon\right)^{1/N} - k_0\varepsilon^{1/N} \tag{5.2.95}$$

for nonnegative $x_1, \ldots, x_N$ under the constraints (5.2.75), (5.2.76). We now obtain the inequality (5.2.77) by letting $\varepsilon \to 0$ in (5.2.95).

It is easy to verify that the constant k_0 in (5.2.50) exceeds unity.

5.2.7 Bernstein's Inequality

The inequalities established in the preceding sections were obtained using smooth functions, and the smooth version of the deformation principle for minimizers was used for proving them. In this section we prove Bernstein's well-known inequality connecting the norms in $C[0,2\pi]$ of a trigonometric polynomial and its derivative. Since the norm function on $C[0,2\pi]$ is not smooth, we use the Lipschitz version of the deformation principle.

Consider the trigonometric polynomial

$$x(t) = x_0 + \sum_{k=1}^{n}(x_k \sin kt + x_{k+n}\cos kt) . \tag{5.2.96}$$

Bernstein's inequality bounds the norm of the derivative $\dot{x}$ of x in the space $C[0,2\pi]$ in terms of the norm of x: namely, we have

$$\|\dot{x}\|_{C[0,2\pi]} \leqslant n\|x\|_{C[0,2\pi]} . \tag{5.2.97}$$

Let us prove (5.2.97). Since the norms of the functions $x(t)$ and $\dot{x}(t)$ in $C[0,2\pi]$ coincide, respectively, with the norms of $x(t+\tau)$ and $\dot{x}(t+\tau)$ for any constant τ, it suffices to prove the inequality

$$\dot{x}(0) \leqslant n\|x\|_{C[0,2\pi]}$$

or, equivalently, the inequality

$$\sum_{k=1}^{n} kx_k \leqslant n\varphi(x) , \tag{5.2.98}$$

where

$$x = (x_0, \ldots, x_{2n}) ,$$

$$\varphi(x) = \max_{0 \leqslant t \leqslant 2\pi} \varphi(x,t)$$

and

$$\varphi(x,t) = \left| x_0 + \sum_{k=1}^{n} (x_k \sin kt + x_{k+n} \cos kt) \right| .$$

By Theorem 2.2.1, in order to prove (5.2.98) it suffices to show that, for all $\lambda \in (0, 1/n)$, the point 0 is the unique extremal of the function $f(\cdot; \lambda)$ of the one-parameter family

$$f(x;\lambda) = \varphi(x) - \lambda \sum_{k=1}^{n} kx_k .$$

Suppose that this is false. Then we have

$$0 \in \partial_x f(x_*; \lambda_*) \tag{5.2.99}$$

for some

$$x_* = (x_0^*, \ldots, x_{2n}^*) \neq 0, \quad \lambda_* \in (0, 1/n) .$$

We set

$$x_*(t) = x_0^* + \sum_{k=1}^{n} (x_k^* \sin kt + x_{k+n}^* \cos kt)$$

and

$$\mathfrak{M} = \{\tau \in [0, 2\pi] : |x_*(\tau)| = \|x_*(\cdot)\|_{C[0,2\pi]}\} .$$

The points of $\mathfrak{M}$ are zeros of the nonzero polynomial x_*, and so there are at most $2n$ of them; let

$$\mathfrak{M} = \{\tau_1, \ldots, \tau_l\} .$$

Then, by Proposition 1.5.4, we have

$$\partial\varphi(x_*) = \mathrm{co}\{\nabla\varphi(x_*, \tau_1), \ldots, \nabla\varphi(x_*)\tau_l)\} .$$

From this and (5.2.99) we must have

$$\sum_{k=1}^{l} \alpha_k \operatorname{sign} x_*(\tau_k) = 0 ,$$
$$\sum_{k=1}^{l} \alpha_k \operatorname{sign} x_*(\tau_k) \sin \tau_k = \lambda_* ,$$
$$\dots\dots\dots\dots\dots\dots\dots\dots\dots$$
$$\sum_{k=1}^{l} \alpha_k \operatorname{sign} x_*(\tau_k) \sin n\tau_k = n\lambda_* ,$$
$$\sum_{k=1}^{l} \alpha_k \operatorname{sign} x_*(\tau_k) \cos \tau_k = 0 ,$$
$$\dots\dots\dots\dots\dots\dots\dots\dots\dots$$
$$\sum_{k=1}^{l} \alpha_k \operatorname{sign} x_*(\tau_k) \cos n\tau_k = 0$$

for some $\alpha_1, \dots, \alpha_l \in [0, 1]$ with $\alpha_1 + \cdots + \alpha_l = 1$. Multiplying these equations by y_0, $y_1, \dots$, y_{2n} respectively and adding, we conclude that

$$\sum_{k=1}^{l} \alpha_k \operatorname{sign} x_*(\tau_k) y(\tau_k) = \lambda_* \dot{y}(0) \tag{5.2.100}$$

for each polynomial

$$y(t) = y_0 + \sum_{k=1}^{n} (y_k \sin kt + y_{k+n} \cos kt) . \tag{5.2.101}$$

It follows from (5.2.100) that $l = 2n$. Indeed, if $l < 2n$, then there exists a polynomial $y_*(t)$ of the form (5.2.101) that vanishes at $2n$ points including $\tau_1, \dots, \tau_l$ and 0. It now follows from (5.2.100) that 0 is a double root of $y_*(t)$, and this is impossible.

We set

$$M = \max_{0 \leqslant t \leqslant 2\pi} |x_*(t)| .$$

Then the polynomials $M^2 - x_*^2(t)$ and $\dot{x}_*^2(t)$ have double roots at $\tau_1, \dots, \tau_{2n}$. Therefore these polynomials are proportional, and so

$$\dot{x}_*^2(t) = n^2(M^2 - x_*^2(t)) .$$

It follows that

$$x_*(t) = \pm M \sin(nt + \gamma) .$$

Substituting the numbers $x_*(\tau_k)$ for $y(\tau_k)$ in (5.2.100), we obtain

$$A = \pm \lambda_* A n \cos \gamma .$$

This is impossible, and so we have a contradiction. This completes the proof of Bernstein's inequality. □

5.3 Multicriteria Problems

5.3.1 Definitions

We consider the space $\mathbb{R}^M$, with norm $|\cdot|$. Let K be the cone of vectors in $\mathbb{R}^M$ with nonnegative coordinates. This cone defines a partial order in $\mathbb{R}^M$: we write $y \leqslant z$ if

$$z - y \in K \,. \tag{5.3.1}$$

If $z - y \in K$ and $z \neq y$, then we write $y < z$.

Consider a mapping $F : \mathbb{R}^N \to \mathbb{R}^M$ whose components

$$f_i(x) = f_i(x_1, \ldots, x_N) \quad (i = 1, \ldots, M)$$

are continuously differentiable functions. A point $x_* \in \mathbb{R}^N$ is said to be a *locally Pareto optimal point* of F if x_* has a neighborhood in $\mathbb{R}^N$ which contains no points x such that

$$F(x) < F(x_*) \,. \tag{5.3.2}$$

If there are no points $x \in \mathbb{R}^N$ for which this inequality holds, then the point x_* is said to be *Pareto optimal.*

We shall be interested in the problem of finding and investigating Pareto optimal or locally Pareto optimal points for the mapping F. For brevity, we shall call this problem the P-optimum problem (or local P-optimum problem) and denote it by

$$F(x) \to \text{P-opt} \,. \tag{5.3.3}$$

A point $x_* \in \mathbb{R}^N$ is called a *locally optimal point* of F if

$$F(x_*) \leqslant F(x) \quad (x \in V) \tag{5.3.4}$$

for all x in some neighborhood V of x_*. If

$$F(x_*) < F(x) \quad (x \in V, x \neq x_*) \,,$$

then x_* is said to be a *strictly locally optimal point* of F.

The notions of an optimal point and a strictly optimal point of F are defined similarly. We shall denote the problem of finding optimal (locally optimal) points of F by

$$F(x) \to \text{opt} \,. \tag{5.3.5}$$

Note that optimal (locally optimal) points of F are P-optimal (locally P-optimal). The converse statement is in general not true.

We call x_* a *critical point* of problems (5.3.3) and (5.3.5) if

$$0 \in \mathrm{co}\{\nabla f_i(x_*);\ i = 1, \dots, M\} .$$

It is well known (see [65]) that locally P-optimal points (and consequently also locally optimal points) are critical points.

In this section, we study the homotopy invariance of solutions to problems (5.3.3) and (5.3.5).

Consider a one-parameter family of problems

$$F(x; \lambda) \to \text{P-opt} \quad (x \in \mathbb{R}^N,\ 0 \leqslant \lambda \leqslant 1) . \tag{5.3.6}$$

We say that (5.3.6) is a *nondegenerate deformation on the ball* $B(R)$ (or $B(R)$*-nondegenerate deformation*) of the problem

$$F_0(x) \to \text{P-opt} \tag{5.3.7}$$

into the problem

$$F_1(x) \to \text{P-opt} \tag{5.3.8}$$

if

(1) the components $f_i(x; \lambda)$ $(i = 1, \dots, m)$ of the mapping F and their derivatives with respect to x are continuous,

(2) for every $\lambda \in [0, 1]$, problem (5.3.7) has on the ball $B(R)$ a unique critical point $x(\lambda) \in \mathring{B}(R)$ which continuously depends on the parameter λ and for which

$$0 \notin \partial \max_{1 \leqslant i \leqslant M} (f_i(x; \lambda) - f_i(x(\lambda); \lambda))$$

for all $x \in B(R) \setminus \{x(\lambda)\}$.

(3) $F(\cdot; 0) = F_0$, $F(\cdot; 1) = F_1$.

5.3.2 The Deformation Theorem

We shall prove the following result.

Theorem 5.3.1. *Suppose that there exists a* $B(R)$*-nondegenerate deformation of problem* (5.3.7) *into problem* (5.3.8). *Let* $x_0 = x(0)$ *be a locally* P-*optimal point for problem* (5.3.7). *Then* $x_1 = x(1)$ *is a locally* P*-optimal point of problem* (5.3.8). *If* x_1 *is the unique extremal of problem* (5.3.8) *in the ball* $B(R)$, *then this point is locally optimal.*

Proof. We shall show that Theorem 5.3.1 follows from Theorem 3.4.1.

We assume, without loss of generality, that

$$x(\lambda) = 0 \quad (0 \leqslant \lambda \leqslant 1)$$

and

$$F(0; \lambda) = 0 \quad (0 \leqslant \lambda \leqslant 1) .$$

Let

$$f(x;\lambda) = \max_{1 \leqslant i \leqslant M} f_i(x;\lambda) .$$

We set

$$J(x;\lambda) = \{j = 1, \ldots, M : f_j(x;\lambda) = f(x;\lambda)\} .$$

Since

$$\partial_x f(x;\lambda) = \operatorname{co}\{\nabla f_i(x;\lambda) : i \in J(x;\lambda)\}$$

and since the mappings

$$\nabla f_i : B(R) \times [0,1] \to \mathbb{R}^N \quad (i = 1, \ldots, M)$$

are continuous and the mapping

$$J : B(R) \times [0,1] \to \{1, \ldots, M\}$$

is upper semicontinuous, it follows that the mapping

$$\partial_x f : B(R) \times [0,1] \to \mathbb{R}^N$$

is upper semicontinuous. In order to prove the first assertion of the theorem, it suffices to show that 0 is a strict local minimizer of f_1, and this follows from Theorem 3.4.1 since 0 is a local minimizer of f_0.

Now we prove the second assertion of the theorem.

Since F_1 has no critical points in the set $B(R) \setminus \{0\}$, the function $l(F_1(x))$ has no critical points in $B(R) \setminus \{0\}$, where l is any nonzero linear functional from the dual cone

$$K^* = \{l \in (\mathbb{R}^M)^* : l(u) \geqslant 0,\ u \in K\} .$$

Consider the set $\mathcal{L}$ of functionals

$$l = \{l_1, \ldots, l_M\} \in K^*$$

for which

$$|l| = \sum_{i=1}^{M} l_i = 1 .$$

For each function $l(F_1(x))$ we construct a set of splines $\mathfrak{M}_l$ using the procedure in Sect. 3.4.3. We have

$$\max_{1 \leqslant i \leqslant M} f_1^i(x) = \max_{l \in \mathcal{L}} l(F_1(x)) , \tag{5.3.9}$$

which implies that there is a ball $B(\rho)$ such that for any point $x \in B(\rho) \setminus \{0\}$ and any $l \in \mathcal{L}$ there exists a spline $x_l \in \mathfrak{M}_l$ for which $x_l(0) = x$ and which converges to zero at infinity. It follows that $l(F_1(x)) > 0$ for all $l \in \mathcal{L}$, i.e., $F_1(x) > 0$. This completes the proof of the theorem. □

5.3.3 Multicriteria Problems with Constraints

In the preceding section we considered deformations of multicriteria problems without constraints. Here we shall study problems of the form

$$\begin{cases} F(x) \to \text{P-opt} , \\ g_\nu(x) = 0 \quad (\nu = 1, \dots, n) , \\ h_j(x) \leqslant 0 \quad (j = 1, \dots, m) , \\ x \in C , \end{cases} \tag{5.3.10}$$

where $F : \mathbb{R}^N \to \mathbb{R}^M$ is the mapping considered in Sect. 5.3.1, $g_\nu : \mathbb{R}^N \to \mathbb{R}$, $h_j : \mathbb{R}^N \to \mathbb{R}$ are locally Lipschitzian functions, and C is a convex compact set in $\mathbb{R}^N$. For problem (5.3.10) we may define in an obvious way the notions of locally Pareto optimal points and Pareto optimal points.

A set D of points $x \in \mathbb{R}^N$ satisfying the constraints of problem (5.3.10) is said to be an *admissible* set. We call $x_* \in D$ an *extremal* of problem (5.3.10) if

$$0 \in \mathrm{co}\{\nabla f_i(x_*),\ i = 1, \dots, M\} + \sum_{\nu=1}^{n} r_\nu \partial g_\nu(x_*) + \sum_{j=1}^{m} s_j \partial h_j(x_*) + \alpha \partial d_C(x_*) ,$$

where $r_\nu \in \mathbb{R}$ $(\nu = 1, \dots, n)$, $s_j \geqslant 0$ $(j = 1, \dots, m)$, $\alpha \geqslant 0$, d_C is the usual distance function to C defined by

$$d_C(x) = \min_{y \in C} |x - y| , \tag{5.3.11}$$

and the conditions of complementary slackness

$$s_i h_j(x_*) = 0 \quad (j = 1, \dots, m) \tag{5.3.12}$$

are fulfilled.

Consider a one-parameter family of problems

$$\begin{cases} F(x; \lambda) \to \text{P-opt} , \\ g_\nu(x; \lambda) = 0 \quad (\nu = 1, \dots, n) , \\ h_j(x; \lambda) \leqslant 0 \quad (j = 1, \dots, m) , \\ x \in C(\lambda) \quad (0 \leqslant \lambda \leqslant 1) . \end{cases} \tag{5.3.13}$$

Suppose that for every $\lambda \in [0, 1]$ the mapping $F(\cdot; \lambda)$ and the functions $g_\nu(\cdot; \lambda)$ $(\nu = 1, \dots, n)$, $h_j(\cdot; \lambda)$ $(j = 1, \dots, m)$ satisfy the requirements listed above, that the mappings

$$\nabla f_i : \mathbb{R}^N \times \mathbb{R} \to \mathbb{R}^N \quad (i = 1, \dots, M)$$

are continuous in the collection of variables, and the mappings

$$\partial g_\nu : \mathbb{R}^N \times \mathbb{R} \to \mathbb{R}^N \quad (\nu = 1, \dots, n) ,$$

$$\partial h_j : \mathbb{R}^N \times \mathbb{R} \to \mathbb{R}^N \quad (j = 1, \dots, m) ,$$

$$\partial d_{C(\cdot)}(\cdot) : \mathbb{R}^N \times \mathbb{R} \to \mathbb{R}^N$$

are upper semicontinuous.

Suppose that for all $\lambda \in [0,1]$ and for $\nu = 1,\dots,n$ (resp. $j = 1,\dots,m$) the functions g_ν (resp. h_j) are regular at the points x at which $g_\nu(x;\lambda) = 0$ (resp. $h_j(x;\lambda) = 0$).

We call the family of problems (5.3.13) a $B(R)$-*nondegenerate deformation* of the problem

$$\begin{cases} F_0(x) \to \text{P-opt}\ , \\ g_\nu^0(x) = 0 \quad (\nu = 1,\dots,n)\ , \\ h_j^0(x) \leqslant 0 \quad (j = 1,\dots,m)\ , \\ x \in C_0 \end{cases} \tag{5.3.14}$$

into the problem

$$\begin{cases} F_1(x) \to \text{P-opt}\ , \\ g_\nu^1(x) = 0 \quad (\nu = 1,\dots,n)\ , \\ h_j^1(x) \leqslant 0 \quad (j = 1,\dots,m)\ , \\ x \in C_1 \end{cases} \tag{5.3.15}$$

if, for every λ, there is an extremal $x_*(\lambda) \in \mathring{B}(R)$ of problem (5.3.13) such that

$$\begin{aligned} 0 \notin \partial &\max_{1\leqslant i\leqslant M} (f_i(x;\lambda) - f_i(x_*(\lambda);\lambda)) \\ &+ \sum_{\nu=1}^{n} r_\nu \partial g_\nu(x;\lambda) + \sum_{j=1}^{m} s_j \partial h_j(x;\lambda) + \alpha \partial d_C(x)\ , \end{aligned} \tag{5.3.16}$$

where

$$x \in B(R),\quad x \neq x_*(\lambda),\quad 0 \leqslant \lambda \leqslant 1,\quad r_\nu \in \mathbb{R}\ \ (\nu = 1,\dots,n)\ ,$$
$$\alpha > 0,\quad s_j \geqslant 0 \quad (j = 1,\dots,m)$$

and

$$s_j h_j(x_*(\lambda);\lambda) = 0 \quad (j = 1,\dots,m)\ , \tag{5.3.17}$$

with

$$\begin{aligned} &F(\cdot;0) = F_0,\quad F(\cdot;1) = F_1\ , \\ &g_\nu(\cdot;0) = g_\nu^0,\quad g_\nu(\cdot;1) = g_\nu^1 \quad (\nu = 1,\dots,n)\ , \\ &h_j(\cdot;0) = h^0 j,\quad h_j(\cdot;1) = h_j^1 \quad (j = 1,\dots,m)\ , \\ &C(0) = C_0,\quad C(1) = C_1\ . \end{aligned}$$

Theorem 5.3.2. *Suppose that there exists a $B(R)$-nondegenerate deformation of problem* (5.3.14) *into problem* (5.3.15). *Suppose that the regularity condition*

$$0 \notin \sum_{\nu=1}^{n} r_\nu \partial g_\nu(x_*(\lambda);\lambda) + \sum_{j=1}^{M} s_j \partial h_j(x_*(\lambda);\lambda) + \alpha \partial d_{C(\lambda)}(x_*(\lambda)) \tag{5.3.18}$$

holds for all $r_\nu \in \mathbb{R}$, $s_j \geqslant 0$, $\alpha \geqslant 0$ *and that*

$$\sum_{\nu=1}^{n} |r_\nu| + \sum_{j=1}^{m} s_j > 0 . \tag{5.3.19}$$

Suppose finally that x_0 *is a locally* P*-optimal point for problem* (5.3.14) *for* $\lambda = 0$. *Then* x_1 *is locally* P*-optimal for problem* (5.3.15). *If in addition the condition*

$$\begin{aligned} 0 \notin \operatorname{co}\{\nabla f_1(x;1), \ldots, \nabla f_M(x;1)\} &+ \sum_{\nu=1}^{n} r_\nu \partial g_\nu(x;1) \\ &+ \sum_{j=1}^{m} s_j \partial h_j(x;1) + \alpha \partial d_{C_1}(x) \end{aligned} \tag{5.3.20}$$

holds for all

$$r_\nu \in \mathbb{R},\ s_j \geqslant 0,\ \alpha \geqslant 0,\ x \in B(R) \cap D_1 \setminus \{x_1\},\ \sum_{j=1}^{m} s_j h_j(x,1) = 0 , \tag{5.3.21}$$

then the point x_1 *is locally optimal.*

Proof. We can assume without loss of generality that

$$x(\lambda) = 0 \quad (0 \leqslant \lambda \leqslant 1), \qquad F(0;\lambda) = 0 \quad (0 \leqslant \lambda \leqslant 1) .$$

Consider the functions defined by

$$\begin{aligned} g_{1\nu}(x;\lambda) &= \max(0, g_\nu(x;\lambda)) , \\ g_{2\nu}(x;\lambda) &= \max(0, -g_\nu(x;\lambda)) , \\ h_j^+(x;\lambda) &= \max(0, h_j(x;\lambda)) , \end{aligned}$$

$$\begin{aligned} \varphi(x;\lambda;\alpha) = &\sum_{\nu=1}^{n} (g_{1\nu}(x;\lambda) + g_{2\nu}(x;\lambda)) \\ &+ \sum_{j=1}^{m} h_j^+(x;\lambda) + \alpha d_{C(\lambda)}(x) \quad (\alpha > 0) . \end{aligned} \tag{5.3.22}$$

It follows from (5.3.22) that

$$\varphi(x;\lambda;\alpha) \geqslant 0$$

and

$$D(\lambda) = \{x \in \mathbb{R}^N : \varphi(x;\lambda;\alpha) = 0\} .$$

Therefore, in order to prove the first assertion of Theorem 5.3.2, it suffices to show that for sufficiently large α and k the family of functions

$$\Psi(x; \lambda; a, k) = f(x; \lambda) + k\varphi(x; \lambda, \alpha) ,$$

where

$$f(x; \lambda) = \max_{i=1,\dots,M} f_i(x; \lambda) ,$$

satisfies the conditions of Theorem 3.4.1, and, for $\lambda = 0$, the function $\Psi(x; 0; \alpha, k)$ has a local minimum at 0. We fix $\lambda \in [0, 1]$. Let

$$x_1 \notin C(\lambda), \quad y \in \partial d_{C(\lambda)}(x_1) .$$

We shall show that

$$|y|_{(\mathbb{R}^N)^*} = 1 .$$

Note first that by Theorem 1.7.13 the distance function d_C satisfies a Lipschitz condition with coefficient 1, and if C is convex, then d_C is a convex function.

Let $v \in \mathbb{R}^N$ and

$$d_{C(\lambda)}(x_1 + v) = 0 ,$$
$$|v| = d_{C(\lambda)}(x_1) .$$

Then we have

$$d_{C(\lambda)}((1 - t)x_1 + t(x_1 + \nu)) \leqslant (1 - t)d_{C(\lambda)}(x_1) \tag{5.3.23}$$

for $0 < t < 1$. Since any convex function is regular, i.e., its classical derivative in each direction exists and is equal to the generalized derivative, it follows from (5.3.23) that

$$d^0_{C(\lambda)}(x_1; \nu) \leqslant -d_{C(\lambda)}(x_1) = -|\nu| . \tag{5.3.24}$$

This implies that

$$|y|_{(\mathbb{R}^N)^*} \geqslant 1 .$$

Therefore we have

$$|y|_{(\mathbb{R}^N)^*} = 1 .$$

By the upper semicontinuity of the mappings ∂g_ν, ∂h_j, $\partial d_{C(\lambda)}$, the mappings $\partial g_{1,\nu}, \partial g_{2,\nu}, \partial h_j^+$ are also upper semicontinuous on $B(R) \times [0, 1]$. Therefore we have

$$\max_{\substack{y \in \sum_{\nu=1}^n (\partial g_{1\nu}(x;\lambda) + \partial g_{2\nu}(x;\lambda)) + \sum_{j=1}^m \partial h_j^+(x;\lambda) + \partial d_{C(\lambda)}(x) \\ x \in B(R),\ \lambda \in [0,1]}} |y|_{(\mathbb{R}^N)^*} = M_1 < \infty . \tag{5.3.25}$$

It follows from (5.3.24) and (5.3.25) that

$$\inf_{y\in\partial\varphi(x;\lambda;\alpha),\ x\in B(R)\setminus C(\lambda),\ \lambda\in[0,1]} |y|_{(\mathbb{R}^N)^*} \geqslant b_1 \tag{5.3.26}$$

for $b_1 = \alpha - M_1 > 0$. Let

$$\begin{aligned} &G_{1\nu}(\lambda) = \{x : g_{1\nu}(x;\lambda) \leqslant 0\}\,,\\ &G_{2\nu}(\lambda) = \{x : g_{2\nu}(x;\lambda) \leqslant 0\}\,,\\ &H_j(\lambda) = \{x : h_j^+(x;\lambda) \leqslant 0\}\,,\\ &(\nu = 1,\dots,n;\quad j = 1,\dots,m)\,. \end{aligned}$$

Then we have

$$\begin{aligned} B(r)\setminus D(\lambda) = \left(\bigcup_\nu (B(r)\setminus G_{1\nu}(\lambda))\right) \cup \left(\bigcup_\nu (B(r)\setminus G_{2\nu}(\lambda))\right)\\ \cup \left(\bigcup_j (B(r)\setminus H_j(\lambda))\right) \cup (B(r)\setminus C(\lambda))\,. \end{aligned} \tag{5.3.27}$$

It follows from (5.3.26), (5.3.27) and the upper semicontinuity of the mappings $\partial g_{1\nu}$, $\partial g_{2\nu}$, ∂h_j^+, $\partial d_{C(\lambda)}$ that there exists some $\rho > 0$ such that

$$\inf_{y\in\partial\varphi(x;\lambda;\alpha),\ x\in B(\rho)\setminus D(\lambda),\ \lambda\in[0,1]} |y|_{(\mathbb{R}^N)^*} = m_1 > 0 \tag{5.3.28}$$

for $\alpha = M_1 + b_1$. Let

$$\max_{y\in\partial f(x;\lambda),x\in B(R),\lambda\in[0,1]} |y|_{(\mathbb{R}^N)^*} = M_2\,. \tag{5.3.29}$$

We shall show that for $\alpha = M_1 + b_1,\ k > 3M_2/m_1$ the conditions of Theorem 3.4.1 hold for the functions $\Psi(x;\lambda,\alpha,k)$.

Suppose that the point 0 is a minimizer of the function f_0 on $\mathcal{D}_0 \cap B(r)$, where $2r < \rho < R$. Let $\mathfrak{N}$ be the set of splines constructed for the function $\varphi(x) = \varphi(x;0,\alpha)$ and the set $B(2r)\setminus\mathcal{D}_0$ by the same procedure as the set of splines $\mathfrak{M}$ described in Sect. 3.4.3. Recall that the conditions

$$\dot{x}(t) = v_{i-1} \quad (t_{i-1} \leqslant t \leqslant t_i;\ |v_{i-1}| = 1)\,, \tag{5.3.30}$$

$$\varphi^0(x(t);v_{i-1}) \leqslant -\tfrac{1}{2}\mu\left(\varphi;\tfrac{3}{4}|x(t)|\right)\,, \tag{5.3.31}$$

where

$$\mu(\varphi;s) = \inf_{y\in\partial\varphi(x),s\leqslant|x|\leqslant R,x\notin\mathcal{D}_0} |y|_{(\mathbb{R}^N)^*}\,,$$

are satisfied on the ith link of a continuous spline $x \in \mathfrak{N}$, and for this link of the spline, (5.3.30) and (5.3.31) imply that

$$\Psi^0(x;0;\alpha;k;v_{i-1}) \leqslant f^0(x;0;v_{i-1}) + k\varphi^0(x;v_{i-1}) < 0\,.$$

Then the functions

$$\gamma(t) = \Psi(x(t); 0; \alpha; k), \quad \eta(t) = \varphi(x(t)) ,$$

satisfy almost everywhere the inequalities

$$\dot{\gamma}(t) < 0 , \tag{5.3.32}$$

$$\dot{\eta}(y) \leqslant -\tfrac{1}{2}\mu\left(\varphi; \tfrac{3}{4}|x(t)|\right) . \tag{5.3.33}$$

If the spline x begins at the point

$$x_0 \in B(r) \setminus \mathcal{D}_0 ,$$

then two cases are possible, namely, either there exists a T_0 (finite or infinite) such that

$$x(T_0) \in B(2r) \cap \mathcal{D}_0 ,$$

or there exists a T_1 such that

$$|x(T_1)| = 2r .$$

In the first case, (5.3.32) implies that

$$\Psi(x_0; 0; \alpha; k) > f(0; 0) , \tag{5.3.34}$$

and in the second case it follows from (5.3.33) that

$$k\varphi(x_0) > k_1 r \quad (k_1 > M_2) . \tag{5.3.35}$$

This implies that

$$k\varphi(x_0) > k_1 d_{\mathcal{D}_0}(x_0) , \tag{5.3.36}$$

where $d_{\mathcal{D}_0}$ is the distance function to the subset $\mathcal{D}_0$. Since the minimum of the function $f(\cdot; 0) + k_1 d_{\mathcal{D}_0}(\cdot)$ on the ball $B(R)$ is attained at 0, it follows from (5.3.35) and (5.3.36) that

$$\Psi(x_0; 0; \alpha; k) > f(0, 0) ,$$

i.e., for $\alpha = M_1 + b_1$, $k > 3M_2/m_1$ the function $\Psi(\cdot; 0; \alpha; k)$ attains its minimum on the ball $B(R)$ at the point $x = 0$.

Let $\lambda \in [0, 1]$. From (5.3.28) and (5.3.29), we have

$$\min_{y \in \partial\Psi(x;\lambda;\alpha;k)} |y|_{(\mathbb{R}^N)^*} \geqslant \min_{y \in \partial f(x;\lambda) + k\partial\varphi(x;\lambda;\alpha)} |y|_{(\mathbb{R}^N)^*} \geqslant km_1 - M_2 > 0$$

for all $x \in B(\rho) \setminus \mathcal{D}(\lambda)$. Now if

$$x \in B(\rho) \cap \mathcal{D}(\lambda) \setminus \{0\} ,$$

then, by the hypothesis of the theorem, we have

$$0 \in \partial\Psi(x;\lambda;\alpha;k)\,.$$

Thus the conditions of Theorem 3.4.1 hold, and the first assertion of Theorem 5.3.2 follows. In order to prove the second assertion, we consider the function

$$l(F(\cdot;1)) + k\varphi(\cdot;1;\alpha)\,,$$

where $\alpha = M_1 + b_1,\ k > 3M_2/m_1,\ l \in K^*,\ |l| = 1$. Applying for this function the method used to prove the second assertion of Theorem 5.3.1, we find a ball $B(\rho_1)$ whose nonzero points satisfy

$$l(F(x;1)) + k\varphi(x;1;\alpha) > 0\,.$$

Therefore

$$F(x;1) > F(0;1)$$

for

$$\varphi(x;1;\alpha) = 0, \quad x \neq 0\,,$$

and the theorem follows. □

5.4 Problems in the Calculus of Variations

5.4.1 One-Dimensional Problems

We shall consider the simplest functional of the calculus of variations,

$$f(x) = \int_0^T F(t, x(t), x'(t))\, dt \tag{5.4.1}$$

on the Hilbert space $\overset{\circ}{W}{}_2^1[0,T]$ of absolutely continuous functions x on $[0,T]$ satisfying

$$x(0) = x(T) = 0 \tag{5.4.2}$$

and with square-summable derivatives. We define the inner product in $\overset{\circ}{W}{}_2^1[0,T]$ by

$$(x,y) = \int_0^T x'(t)y'(t)\, dt\,.$$

Lemma 5.4.1. *Suppose that the Lagrangian $F(t,x,p)$ of f is continuous on $[0,T]\times\mathbb{R}\times\mathbb{R}$ and twice continuously differentiable with respect to x,p. Suppose that*

$$|F| + |F'_x| + |F''_{xx}| \leqslant c(1+p^2)\,, \tag{5.4.3}$$

$$|F_p'| + |F_{xp}''| \leqslant c(1 + |p|) , \tag{5.4.4}$$

$$a \leqslant F_{pp}'' \leqslant c , \tag{5.4.5}$$

where a and c are positive constants. Then the functional f is $\mathring{W}_2^1[0,T]$-regular.

We shall only sketch the proof this lemma.

The inequalities (5.4.3), (5.4.4), and the right-hand inequality in (5.4.5) ensure that f is differentiable on $\mathring{W}_2^1[0,T]$, and we have

$$\nabla f(x) = \int_0^t (F_p'(s,x(s),x'(s)) - \int_0^s F_x'(\tau,x(\tau),x'(\tau))d\tau)\, ds$$
$$- \frac{t}{T}\int_0^T (F_p'(s,x(s),x'(s)) - \int_0^s F_x'(\tau,x(\tau),x'(\tau))d\tau)\, ds .$$

Direct verification shows that the gradient satisfies a local Lipschitz condition on $\mathring{W}_2^1[0,T]$.

In the proof that the gradient ∇f has the $(S)_+$-property, we use the left-hand inequality (5.4.5) and the compactness of the embedding of $\mathring{W}_2^1[0,T]$ in $C[0,T]$.

Consider the one-parameter family of functionals

$$f(x,\lambda) = \int_0^T F(t,x(t),x'(t);\lambda)\, dt \quad (0 \leqslant \lambda \leqslant 1) \tag{5.4.6}$$

on $\mathring{W}_2^1[0,T]$. Suppose that the Lagrangian $F(t,x,p;\lambda)$ and its first and second derivatives with respect to x,p are continuous on $[0,T] \times \mathbb{R} \times \mathbb{R} \times [0,1]$ and that $F(t,x,p;\lambda)$ satisfies (5.4.3)–(5.4.5) for every $\lambda \in [0,1]$. Lemma 5.4.1 and Theorem 3.1.1 imply

Theorem 5.4.1. *Suppose that for all $\lambda \in [0,1]$ the Euler equation*

$$\frac{d}{dt}\frac{\partial F(t,x(t),x'(t);\lambda)}{\partial p} - \frac{\partial F(t,x(t),x'(t);\lambda)}{\partial x} = 0 \tag{5.4.7}$$

for the functional (5.4.6) *has a unique solution $x(t,\lambda)$ satisfying the boundary conditions*

$$x(0,\lambda) = x(T,\lambda) = 0 . \tag{5.4.8}$$

If the function $x_0 = x(\cdot,0)$ is a local minimizer of the functional $f_0 = f(\cdot,0)$ on $\mathring{W}_2^1[0,T]$ then $x_1 = x(\cdot,1)$ is a local minimizer of $f_1 = f(\cdot,1)$ on $\mathring{W}_2^1[0,T]$.

5.4.2 Multidimensional Integral Functionals

Let Ω be a bounded domain in $\mathbb{R}^N$ with smooth boundary, and let $\overset{\circ}{W}{}_2^m(\Omega)$ be the Sobolev space of functions u on Ω that vanish on the boundary $\partial\Omega$ and have generalized square-summable derivatives up to order m together with derivatives up to order $m-1$ (see Sect. 1.1.6).

Consider the integral functional

$$f(u) = \int_\Omega F(x, u(x), Du(x), \ldots, D^m u(x))\, dx \tag{5.4.9}$$

on $\overset{\circ}{W}{}_2^m(\Omega)$. Here

$$D^k u(x) = \{\mathcal{D}^\alpha u(x) : |\alpha| = k\} \quad (k = 1, \ldots, m)\,.$$

Suppose that the integrand

$$F(x,\xi) \quad (x \in \Omega,\ \xi = \{\xi_\alpha : |\alpha| \leqslant m\} \in \mathbb{R}^M)$$

and its first and second derivatives with respect to $\xi \in \mathbb{R}^M$ are continuous in $\Omega \times \mathbb{R}^m$. Suppose in addition that

$$\left|\frac{\partial^2 F}{\partial\xi_\alpha \partial\xi_\beta}\right| \leqslant C\left(1 + \sum_{m-\frac{N}{2}\leqslant|\gamma|\leqslant m} |\xi_\gamma|^{p_\gamma}\right)^{p_{\alpha\beta}}\,. \tag{5.4.10}$$

Here

p_γ is an arbitrary positive number if $\gamma = m - N/2$,
$p_\gamma = 2N/(N - 2(m - |\gamma|))$ if $m - N/2 < |\gamma| < m$,
$p_{\alpha\beta} = 1 - p_\alpha^{-1} - p_\beta^{-1}$ if $|\alpha| = |\beta| = m$,
$p_{\alpha\beta} = 1 - p_\alpha^{-1}$ if $m - N/2 \leqslant |\alpha| \leqslant m$,
$p_{\alpha\beta} = 1$ if $|\alpha|, |\beta| < m - N/2$,
$0 < p_{\alpha\beta} < 1 - p_\alpha^{-1} - p_\beta^{-1}$ if $|\alpha|, |\beta| > m - N/2$ and $|\alpha| + |\beta| < 2m$.

Then f is Fréchet differentiable on $\overset{\circ}{W}{}_2^m(\Omega)$ and its gradient ∇f satisfies a Lipschitz condition on every ball of $\overset{\circ}{W}{}_2^m(\Omega)$. If in addition

$$\sum_{|\alpha|=|\beta|=m} \frac{\partial^2 F}{\partial\xi_\alpha \partial\xi_\beta}\eta_\alpha\eta_\beta \geqslant c \sum_{|\alpha|=m} \eta_\alpha^2 \tag{5.4.11}$$

$$(c > 0,\ \ x \in \Omega,\ \ \xi \in \mathbb{R}^M,\ \ \eta = \{\eta_\alpha : |\alpha| = m\})\,,$$

then ∇f has the $(S)_+$-property. The proof of these assertions is not difficult but cumbersome: it makes use of the embedding theorem (see Sect. 1.1.6) and the equation $F(x,\xi) - F(x_*,\xi_*) = \langle F'_x(x_*,\xi_*), x - x_*\rangle + R(x,\xi,x_*,\xi_*)$, and the determination of bounds for the integrals of the remainder term

$R(x, \xi, x_*, \xi_*)$ and the variation $|F(x,\xi)-F(x_*,\xi_*)|$. In this case, the gradient $\nabla f : \overset{\circ}{W}{}_2^m(\Omega) \to \overset{\circ}{W}{}_2^m(\Omega)$ of f is defined by

$$\nabla f(u) = \Delta^{-m} \sum_{|\alpha| \leqslant m} (-1)^{m+|\alpha|} \mathcal{D}^\alpha \left(\frac{\partial F}{\partial \xi_\alpha} \right) ,$$

where $\Delta^m : \overset{\circ}{W}{}_2^m(\Omega) \to W_2^{-m}(\Omega)$ is the mth power of the Laplace operator Δ and $\Delta^{-m} : W_2^{-m}(\Omega) \to \overset{\circ}{W}{}_2^m(\Omega)$ is the inverse of Δ^m.

By definition, the critical points of f on $\overset{\circ}{W}{}_2^m(\Omega)$ are solutions in $\overset{\circ}{W}{}_2^m(\Omega)$ of the equation

$$\Delta^{-m} \sum_{|\alpha| \leqslant m} (-1)^{m+|\alpha|} \mathcal{D}^\alpha \left(\frac{\partial F}{\partial \xi_\alpha} \right) = 0 .$$

It is easy to see that the definition of a critical point is equivalent to the definition of an extremal point used in the calculus of variations, based on the notion of a generalized solution of the Euler equation for the functional (5.4.9):

$$\sum_{|\alpha| \leqslant m} (-1)^\alpha \mathcal{D}^\alpha \frac{\partial F}{\partial \xi_\alpha} (x, u(x), \ldots, D^m u(x)) = 0 , \qquad (5.4.12)$$

$$\mathcal{D}^\alpha u(x) \big|_{\partial\Omega} = 0 \quad (|\alpha| \leqslant m-1) . \qquad (5.4.13)$$

Recall that a function $u_* \in \overset{\circ}{W}{}_2^m(\Omega)$ is a *generalized solution* of problem (5.4.12), (5.4.13) if

$$\int_\Omega \sum_{|\alpha| \leqslant m} \frac{\partial F}{\partial \xi_\alpha} (x, u_*(x), \ldots, D^m u_*(x)) \mathcal{D}^\alpha v(x) dx = 0$$

for all $v \in \overset{\circ}{W}{}_2^m(\Omega)$.

5.4.3 The Deformation Theorem

Consider a one-parameter family of integral functionals

$$f(u; \lambda) = \int_\Omega F(x, u(x), \ldots, D^m u(x); \lambda)\, dx \qquad (5.4.14)$$

$$(u \in \overset{\circ}{W}{}_2^m(\Omega), 0 \leqslant \lambda \leqslant 1)$$

on $\overset{\circ}{W}{}_2^m(\Omega)$. We assume that the integrand $F(x, \xi; \lambda)$ and its first and second derivatives with respect to ξ are continuous on $\Omega \times \mathbb{R}^M \times [0,1]$. Suppose that for every $\lambda \in [0,1]$ the integrand $F(x, \xi; \lambda)$ satisfies (5.4.10), (5.4.11).

Theorem 3.1.1 implies

Theorem 5.4.2. *Suppose that for every $\lambda \in [0,1]$ the Dirichlet problem for the Euler equation of the functional* (5.4.14)

$$\sum_{|\alpha|=m} (-1)^{|\alpha|}\mathcal{D}^\alpha \frac{\partial F}{\partial \xi_\alpha}(x, u(x), \ldots, D^m u(x); \lambda) = 0 ,$$

$$\mathcal{D}^\alpha u(x)\big|_{\partial\Omega} = 0, \quad (|\alpha| \leqslant m-1)$$

has a unique generalized solution $u(\cdot;\lambda) \in \mathring{W}_2^m(\Omega)$. If $u_0 = u(\cdot;0)$ is a local minimizer of the functional $f_0 = f(\cdot;0)$ on $\mathring{W}_2^m(\Omega)$ then $u_1 = u(\cdot;1)$ is a local minimizer of $f_1 = f(\cdot;1)$ on $\mathring{W}_2^m(\Omega)$.

Theorem 5.4.2 establishes the deformation principle for minimizers for integral functionals on the Hilbert space $\mathring{W}_2^m(\Omega)$. A similar result holds for integral functionals on the Banach spaces $\mathring{W}_p^m(\Omega)$ $(2 \leqslant p < \infty)$; Theorem 3.2.1 implies

Theorem 5.4.3. *Suppose that the inequalities*

$$\left|\frac{\partial^2 F(x,\xi;\lambda)}{\partial\xi_\alpha\partial\xi_\beta}\right| \leqslant g_1\left(\sum_{|\gamma|<m-N/p} |\xi_\gamma|\right)\left(1 + \sum_{m-N/p\leqslant\gamma\leqslant m} |\xi_\gamma|^{p_\gamma}\right)^{p_{\alpha\beta}}$$

hold, where

g_1 is a continuous positive nondecreasing function,
p_γ is an arbitrary positive number for $|\gamma| = m - N/p$,
$p_\gamma = Np/(N-(m-|\gamma|)p)$ for $m - N/p < |\gamma| \leqslant m$,
$p_{\alpha\beta} = p_{\beta\alpha}$ and these numbers are defined by

$$p_{\alpha\beta} = 1 - \frac{1}{p_\alpha} - \frac{1}{p_\beta} \quad \text{for } |\alpha| = |\beta| = m ,$$

$$p_{\alpha\beta} = 1 - \frac{1}{p_\alpha} \quad \text{for } m - \frac{N}{p} \leqslant \alpha \leqslant m, \quad |\beta| < m - \frac{N}{p} ,$$

$$p_{\alpha\beta} = 1 \quad \text{for } |\alpha|, |\beta| < m - \frac{N}{p} ,$$

$$0 < p_{\alpha\beta} < 1 - \frac{1}{p_\alpha} - \frac{1}{p_\beta} \quad \text{for } |\alpha|, |\beta| \geqslant m - \frac{N}{p}, \quad |\alpha| + \beta < 2m .$$

Suppose in addition that

$$\sum_{|\alpha|=|\beta|=m} \frac{\partial^2 F(x,\xi;\lambda)}{\partial\xi_\alpha\partial\xi_\beta}\eta_\alpha\eta_\beta$$
$$\geqslant g_2\left(\sum_{|\gamma|<m-N/p} |\xi_\gamma|\right) \sum_{|\gamma|=m} |\xi_\gamma|^{p-2} \sum_{|\alpha|=m} \eta_\alpha^2 ,$$

where $x \in \overline{\Omega}$, $\xi \in \mathbb{R}^M$, *and the function* g_2 *is continuous, positive, and nonincreasing. Suppose that for every* $\lambda \in [0,1]$ *the Dirichlet problem*

$$\sum_{|\alpha|=m} (-1)^{|\alpha|} \mathcal{D}^\alpha \frac{\partial F}{\partial \xi_\alpha}(x, u(x), \ldots, D^m u(x); \lambda) = 0 ,$$

$$\mathcal{D}^\alpha u(x)\,|_{\partial\Omega} = 0 \quad (|\alpha| \leqslant m-1)$$

has a unique generalized solution $u(\cdot;\lambda) \in \mathring{W}_p^m(\Omega)$. *If* $u_0 = u(\cdot;0)$ *is a local minimizer of the functional* $f_0 = f(\cdot;0)$ *on* $\mathring{W}_p^m(\Omega)$, *then* $u_1 = u(\cdot;1)$ *is a local minimizer of* $f_1 = f(\cdot;1)$ *on* $\mathring{W}_p^m(\Omega)$.

5.4.4 Deformations of Integral Functionals in the Problem of Weak Minimizers

So far in this section we have studied integral functionals of the calculus of variations on Sobolev spaces. The requirement of Fréchet differentiability for these functionals imposes rather severe constraints on the rate of increase of the integrands and their partial derivatives. Moreover, the fact that the simple integral functional

$$f(u) = \int_\Omega F(x, u(x), \nabla u(x))\, dx \tag{5.4.15}$$

has two successive Fréchet derivatives on its natural domain of definition $\mathring{W}_2^1(\Omega)$ implies that this is a quadratic functional with respect to the derivatives (see [214]).

In this section, we study the functional (5.4.15) on the space $\mathring{C}^1(\overline{\Omega})$ of continuously differentiable functions u on $\overline{\Omega}$ that vanish on $\partial\Omega$. The norm in $\mathring{C}^1(\overline{\Omega})$ is defined by

$$\|u\|_{\mathring{C}^1(\overline{\Omega})} = \max_{x \in \overline{\Omega}} |\nabla u(x)| .$$

For this space, the smoothness of the functional (5.4.15) is related only to the smoothness of its integrand $F(x,u,p)$ and it does not depend on conditions on the rate of increase of the integrand.

The problem of finding the local minimizers of the functional (5.4.15) on $\mathring{C}^1(\overline{\Omega})$ is called the *weak minimizer* problem.

We shall now study the homotopy invariance of weak local minimizers of integral functionals of the form (5.4.15). We assume that the integrands of these functionals are three-times continuously differentiable with respect to the collection of variables, and that Ω is a connected bounded domain whose boundary is sufficiently smooth. (For the validity of the assertions proved below it is sufficient that $\partial\Omega \in C^{2+\alpha}$ for some $\alpha > 0$.)

Consider the integral functionals

$$f_0(u) = \int_\Omega F_0(x, u(x), \nabla u(x))\, dx\ , \tag{5.4.16}$$

$$f_1(u) = \int_\Omega F_1(x, u(x), \nabla u(x))\, dx\ . \tag{5.4.17}$$

We assume that the functions u_0 and u_1 are the unique critical points of f_0 and f_1 respectively, i.e., they are classical solutions of the Dirichlet problem for the Euler equations of the functionals f_0 and f_1:

$$\sum_{i=1}^N \frac{d}{dx_i}\frac{\partial F_0(x, u_0(x), \nabla u_0(x))}{\partial p_i} - \frac{\partial F_0(x, u_0(x), \nabla u_0(x))}{\partial u} = 0\ , \tag{5.4.18}$$

$$u_0(x)\,|_{\partial\Omega} = 0\ , \tag{5.4.19}$$

$$\sum_{i=1}^N \frac{d}{dx_i}\frac{\partial F_1(x, u_1(x), \nabla u_1(x))}{\partial p_i} - \frac{\partial F_1(x, u_1(x), \nabla u_1(x))}{\partial u} = 0\ , \tag{5.4.20}$$

$$u_1(x)\,|_{\partial\Omega} = 0\ . \tag{5.4.21}$$

We call a one-parameter family of functionals

$$f(u;\lambda) = \int_\Omega F(x, u(x), \nabla u(x);\lambda)\, dx \quad (u \in \mathring{C}^1(\overline{\Omega}), 0 \leqslant \lambda \leqslant 1) \tag{5.4.22}$$

a *nondegenerate deformation* of f_0 into f_1 if

(1) the integrand $F(x, u, p; \lambda)$ and its rth order derivatives with respect to x, u, p for $r \leqslant 3$ depend continuously on λ,

(2) for every value of the parameter λ, the functional (5.4.22) has a unique critical point $u(\lambda) = u(\cdot;\lambda)$, i.e., $u(\cdot;\lambda)$ is a solution of the Dirichlet problem for the Euler equation

$$\sum_{i=1}^N \frac{d}{dx_i}\frac{\partial F(x, u(x;\lambda), \nabla u(x;\lambda);\lambda)}{\partial p_i} - \frac{\partial F(x, u(x;\lambda), \nabla u(x;\lambda);\lambda)}{\partial u} = 0\ ,$$

$$u(x;\lambda)|_{\partial\Omega} = 0\ ,$$

and this critical point depends continuously on λ in the norm $\mathring{C}^1(\overline{\Omega})$,

(3) the strengthened Legendre condition

$$\sum_{i,j=1}^N \frac{\partial^2 F(x, u, p;\lambda)}{\partial p_i \partial p_j}\xi_i\xi_j \geqslant c\sum_{i=1}^N \xi_i^2$$

$$(c > 0,\quad x \in \overline{\Omega},\quad u, p \in \mathbb{R},\quad \lambda \in [0,1],\quad \xi \in \mathbb{R}^N)$$

holds,

(4) $f(\cdot;0) = f_0, f(\cdot;1) = f_1$.

Theorem 5.4.4. *Suppose that there exists a nondegenerate deformation* $f(\cdot;\lambda)$ *of the functional* f_0 *into the functional* f_1. *If* $u_0 = u(\cdot;0)$ *is a weak local minimizer of* f_0 *then* $u_1 = u(\cdot;1)$ *is a weak local minimizer of* f_1.

For the proof we need several lemmas. For simplicity, we shall as usual assume that

$$f(0;\lambda) = 0 \quad (0 \leqslant \lambda \leqslant 1), \quad u(x;\lambda) = 0 \quad (0 \leqslant \lambda \leqslant 1)$$

and also that the function $u_0(x) = 0$ is an absolute minimizer of f_0 on the unit ball B in $\mathring{C}^1(\overline{\Omega})$.

Let $C^2(\overline{\Omega})$ be the Banach space of three-times continuously differentiable functions u on $\overline{\Omega}$ with the norm

$$\|u\|_{C^2(\overline{\Omega})} = \max_{x\in\overline{\Omega}}(|u(x)| + |\nabla u(x)| + |\nabla^2 u(x)|) .$$

Here $\nabla^2 u$ is the Hessian of u and $|\nabla^2 u|$ is its matrix norm defined by the Euclidean norm $|\cdot|$.

Lemma 5.4.2. *Let*

$$u \in \mathring{C}^1(\overline{\Omega}) \cap C^2(\overline{\Omega}) .$$

Then

$$\|u\|^{N+2}_{\mathring{C}^1(\overline{\Omega})} \leqslant K\|u\|^N_{C^2(\overline{\Omega})}\|u\|^2_{\mathring{W}^1_2(\Omega)} \tag{5.4.23}$$

for some constant K.

Proof. Since the boundary $\partial\Omega$ of Ω is smooth, we have

$$\inf_{x\in\overline{\Omega},r\in[0,1]} \operatorname{mes}\{\Omega \cap B(r,x)\} \geqslant ar^N \tag{5.4.24}$$

for some $a > 0$. Since Ω is connected, for any points $x, y \in \Omega$ there exists a smooth curve $\varphi(t) : [0,1] \to \Omega$ such that

$$\varphi(0) = x, \ \varphi(1) = y,$$

with length $l(\varphi)$ satisfying

$$l(\varphi) \leqslant A|x - y| \tag{5.4.25}$$

for some constant A that is independent of x and y.

Let x_0 be a point in $\overline{\Omega}$ for which

$$|\nabla u(x_0)| = \|u\|_{\mathring{C}^1(\overline{\Omega})} .$$

If $|\nabla u(x_0)| = 0$, then (5.4.23) is clear. Suppose that

$$|\nabla u(x_0)| = \alpha \neq 0 .$$

We set

$$\Omega_0 = \{x \in \overline{\Omega} : |\nabla u(x)| \geqslant \tfrac{1}{2}\alpha\}$$

and

$$\Omega_1 = \overline{\Omega} \cap B\left(\frac{3\alpha}{8A\beta}, x_0\right) ,$$

where

$$\beta = \max_{x\in\overline{\Omega}} |\nabla^2 u(x)| . \tag{5.4.26}$$

Since

$$\max_{x\in\overline{\Omega}} |\nabla u(x)| \leqslant A \max_{x\in\overline{\Omega}} |\nabla^2 u(x)| ,$$

it follows that

$$\frac{3\alpha}{8A\beta} \leqslant \frac{3}{8} .$$

We claim that $\Omega_1 \subset \Omega_0$. Indeed, if $x \in \Omega_1$, then there exists a curve $\varphi(t) : [0,1] \to \overline{\Omega}$ for which

$$\varphi(0) = x,\ \varphi(1) = x_0 ,$$

of length $l(\varphi)$ satisfying

$$l(\varphi) \leqslant \frac{3\alpha}{8\beta} . \tag{5.4.27}$$

Consider the real function

$$\psi(t) = \tfrac{1}{2}|\nabla u(\varphi(t))|^2 \quad (0 \leqslant t \leqslant 1) .$$

By (5.4.26), we have

$$\begin{aligned}\tfrac{1}{2}(\alpha^2 - |\nabla u(x)|^2) = \psi(1) - \psi(0) &= \int_0^1 \psi'(t)\, dt \\ &= \int_0^1 (\nabla^2 u(\varphi(t))\nabla u(\varphi(t)), \varphi'(t))\, dt \\ &\leqslant \beta \int_0^1 |\nabla u(\varphi(t))| \cdot |\varphi'(t)|\, dt.\end{aligned} \tag{5.4.28}$$

We change variable in the last integral by setting

$$s = \int_0^t |\varphi'(\tau)|\, d\tau .$$

The inequality (5.4.27) yields that

$$\int_0^1 |\nabla u(\varphi(t))| \cdot |\varphi'(t)|\, dt = \int_0^{l(\varphi)} |\nabla u(\varphi(t^{-1}(s)))|\, ds \leqslant \frac{3\alpha^2}{8\beta} . \tag{5.4.29}$$

From (5.4.27)–(5.4.29) we get

$$|\nabla u(x)| \geqslant \tfrac{1}{2}\alpha .$$

Therefore $x \in \Omega_0$ and it follows that $\Omega_1 \subset \Omega_0$.

Now by (5.4.24) we have

$$\|u\|^2_{\overset{\circ}{W}{}^1_2(\Omega)} \geqslant \int_{\Omega_1} |\nabla u(x)|^2\, dx \geqslant \tfrac{1}{4}\alpha^2 \operatorname{mes} \Omega_1 \geqslant \tfrac{1}{4} a\alpha^2 \left(\frac{3\alpha}{8A\beta}\right)^N$$

and, consequently,

$$\|u\|^{N+2}_{\overset{\circ}{C}{}^1(\overline{\Omega})} \leqslant \frac{4}{a}\left(\frac{8A}{3}\right)^N \max_{x\in\overline{\Omega}} |\nabla^2 u(x)|^N \|u(\cdot)\|^2_{\overset{\circ}{W}{}^1_2(\Omega)} .$$

This and the inequality

$$\max_{x\in\overline{\Omega}} |\nabla^2 u(x)| \leqslant \|u(\cdot)\|_{C^2(\overline{\Omega})}$$

together imply (5.4.23), with the constant

$$K = \frac{4}{a}\left(\frac{8A}{3}\right)^N ,$$

and the lemma follows. □

Consider a one-parameter family of functionals

$$f_0(u;\lambda) = \int_\Omega F_0(x, u(x), \nabla u(x); \lambda)\, dx . \tag{5.4.30}$$

We assume that the integrands $F_0(x, u, p; \lambda)$ are three-times continuously differentiable with respect to x, u, p. Suppose in addition that

$$|F_0| \leqslant C_0(1 + u^2 + |p|^2) , \tag{5.4.31}$$

$$\left|\frac{\partial F_0}{\partial u}\right| + \sum_{i=1}^N \left(\left|\frac{\partial F_0}{\partial p_i}\right| + \left|\frac{\partial^2 F_0}{\partial u \partial x_i}\right|\right) + \sum_{i,j=1}^N \left|\frac{\partial^2 F_0}{\partial p_i \partial x_j}\right| \leqslant C_0(1+|u|+|p|) , \tag{5.4.32}$$

$$\left|\frac{\partial^2 F_0}{\partial u^2}\right| + \sum_{i=1}^N \left|\frac{\partial^2 F_0}{\partial u \partial p_i}\right| + \sum_{i,j=1}^N \left|\frac{\partial^2 F_0}{\partial p_i \partial p_j}\right| \leqslant C_0 , \tag{5.4.33}$$

$$\sum_{i,j=1}^{N} \frac{\partial^2 F_0}{\partial p_i \partial p_j} \xi_i \xi_j \geqslant c_0 \sum_{i=1}^{N} \xi_i^2 , \tag{5.4.34}$$

$$(x \in \overline{\Omega}, \quad u \in \mathbb{R}, \quad p, \xi \in \mathbb{R}^N, \quad \lambda \in [0,1], \quad C_0, c_0 > 0) .$$

Finally, assume that

$$F(x,u,p;\lambda) = F_0(x,u,p;\lambda) \quad (x \in \overline{\Omega}, \ |u| + |p| \leqslant a_0, \ \lambda \in [0,1]) \tag{5.4.35}$$

for some $a_0 > 0$.

Lemma 5.4.3. *For every $\lambda \in [0,1]$ the functional $f_0(\cdot;\lambda)$ coincides with the functional $f(\cdot;\lambda)$ on the ball $B(r_*)$ in $\overset{\circ}{C}{}^1(\overline{\Omega})$ of radius*

$$r_* = a_0(1 + A \operatorname{diam} \Omega)^{-1} ,$$

where A is the constant from (5.4.25).

Proof. The result follows directly from the inequality

$$\max_{x \in \overline{\Omega}} |u(x)| \leqslant A \operatorname{diam} \Omega \|u\|_{\overset{\circ}{C}{}^1(\overline{\Omega})}$$

and (5.4.35). □

It follows from this lemma that the function 0 is an extremal of $f_0(\cdot;\lambda)$ for all $\lambda \in [0,1]$.

The inequalities (5.4.31)–(5.4.34) imply that the functionals $f_0(\cdot;\lambda)$ are defined and differentiable on the space $\overset{\circ}{W}{}_2^1(\Omega)$ and that their gradients $\nabla f_0(\cdot;\lambda)$ satisfy a Lipschitz condition on $\overset{\circ}{W}{}_2^1(\Omega)$ and have the $(S)_+$-property. Therefore the functionals $f_0(\cdot;\lambda)$ are $\overset{\circ}{W}{}_2^1(\Omega)$-regular.

Lemma 5.4.4. *The critical point 0 of the functionals $f_0(\cdot;\lambda)$ is isolated on $\overset{\circ}{W}{}_2^1(\Omega)$ uniformly with respect to $\lambda \in [0,1]$.*

Proof. If the contrary holds, then there exist a sequence (λ_n) in $[0,1]$ and a sequence (u_n) with u_n a nonzero critical point of $f_0(\cdot;\lambda_n)$ such that

$$\lim_{n \to \infty} \|u_n\|_{\overset{\circ}{W}{}_2^1(\Omega)} = 0 . \tag{5.4.36}$$

Each critical point u_n is a generalized solution from $\overset{\circ}{W}{}_2^1(\Omega)$ of the Dirichlet problem for the Euler equation of the functional $f_0(u;\lambda_n)$:

$$\sum_{i=1}^{N} \frac{d}{dx_i} \frac{\partial F_0(x, u_n(x), \nabla u_n(x); \lambda_n)}{\partial p_i} - \frac{\partial F_0(x, u_n(x), \nabla u_n(x); \lambda_n)}{\partial u} = 0 ,$$

$$u(x)\,|_{\partial\Omega} = 0 .$$

Since the integrands $F_0(x,u,p;\lambda)$ of the functionals $f_0(\cdot;\lambda)$ satisfy (5.4.31)–(5.4.34) with constants c_0 and C_0 independent of λ, it follows (see [153]) that

$$\operatorname{vrai}\max_{x\in\Omega}|u_n(x)| \leqslant M_n\,, \tag{5.4.37}$$

where the constants M_n depend only on c_0, C_0, Ω, and $\|u_n\|_{\mathring{W}_2^1(\Omega)}$. Therefore (5.4.36) gives the general inequality

$$\operatorname{vrai}\max_{x\in\Omega}|u_n(x)| \leqslant M_0\,.$$

Thus (see [227]) we have

$$u_n(x)\in \mathring{W}_2^2(\Omega)$$

and

$$\operatorname{vrai}\max_{x\in\Omega}|\nabla u_n(x)| \leqslant N_0\,.$$

Consequently (see [153]),

$$u_n\in C^2(\overline{\Omega}),\quad \|u_n\|_{C^2(\overline{\Omega})}\leqslant K_0\quad (n=1,2,\dots)\,,$$

where the constant K_0 depends only on the constants c_0, C_0, M_0, N_0, and the domain Ω. Therefore, by (5.4.23) we have

$$\|u_n\|^{N+2}_{\mathring{C}^1(\overline{\Omega})}\leqslant KK_0^N\|u_n\|^2_{\mathring{W}_2^1(\Omega)}\,.$$

This bound and (5.4.36) yield that

$$\lim_{n\to\infty}\|u_n\|_{\mathring{C}^1(\overline{\Omega})}=0\,.$$

Since the functionals $f(\cdot;\lambda)$ coincide with the functionals $f_0(\cdot;\lambda)$ on the ball $B(R_*)$ by Lemma 5.4.3, for large n the functions u_n are nonzero critical points of the functionals $f(\cdot;\lambda_n)$. This is a contradiction, and the lemma follows. □

Lemma 5.4.5. *The critical point* 0 *of the functional* $f_0(\cdot;0)$ *is a local minimizer on the space* $\mathring{W}_2^1(\Omega)$.

Proof. Assume the contrary and consider the functional $f_0(\cdot;0)$ on the sequence of balls

$$B_n=\{u\in\mathring{W}_2^1(\Omega):\|u\|_{\mathring{W}_2^1(\Omega)}\leqslant n^{-1}\}\quad (n=1,2,\dots)\,.$$

Since the functional $f_0(\cdot;0)$ is $\mathring{W}_2^1(\Omega)$-regular, it is lower semicontinuous on $\mathring{W}_2^1(\Omega)$ by Lemma 3.1.1. Therefore, since the balls B_n are weakly compact, for every n there exists a nonzero function $u_n\in B_n$ such that

$$f_0(u_n;0)=\inf_{u\in B_n}f_0(u;0)<0\,.$$

We claim that

$$u_n \in \partial B_n \quad \text{for large } n.$$

Indeed, if

$$\|u_{n_k}\|_{\overset{\circ}{W}{}^1_2(\Omega)} < 1/n_k \quad (k = 1, 2, \ldots) \tag{5.4.38}$$

for some sequence u_{n_k}, then

$$\nabla f_0(u_{n_k}; 0) = 0 \quad (k = 1, 2, \ldots)\,. \tag{5.4.39}$$

Consequently, the points u_{n_k} are critical points of $f_0(\cdot; 0)$; but then the inequalities (5.4.38) contradict the fact that 0 is an isolated critical point of $f_0(\cdot; 0)$ in $\overset{\circ}{W}{}^1_2(\Omega)$. Our claim follows.

Thus, by Theorem 1.7.7, the equations

$$\nabla f_0(u_n; 0) + \lambda_n u_n = 0\,, \tag{5.4.40}$$

in which $\lambda_n > 0$, hold for large n. Therefore the functions u_n are critical points of the functionals

$$g_n(u) = f_0(u; 0) + \tfrac{1}{2}\lambda_n (u, u)_{\overset{\circ}{W}{}^1_2(\Omega)}\,.$$

Since the gradient $\nabla f_0(\cdot; 0)$ of $f_0(\cdot; 0)$ satisfies a Lipschitz condition, by (5.4.31)–(5.4.34), equations (5.4.40) give the inequality

$$\sup_n \lambda_n \leqslant L < \infty\,.$$

Therefore the integrands $G_n(x, u, p)$ of the functionals g_n satisfy the inequalities

$$|G_n| \leqslant C_1(1 + u^2 + |p|^2)\,,$$

$$\left|\frac{\partial G_n}{\partial u}\right| + \sum_{i=1}^{N}\left(\left|\frac{\partial G_n}{\partial p_i}\right| + \left|\frac{\partial^2 G_n}{\partial u \partial x_i}\right|\right) + \sum_{i,j=1}^{N}\left|\frac{\partial^2 G_n}{\partial p_i \partial x_j}\right| \leqslant C_1(1 + |u| + |p|)\,,$$

$$\left|\frac{\partial^2 G_n}{\partial u \partial u}\right| + \sum_{i=1}^{N}\left|\frac{\partial^2 G_n}{\partial u \partial p_i}\right| + \sum_{i,j=1}^{N}\left|\frac{\partial^2 G_n}{\partial p_i \partial p_j}\right| \leqslant C_1\,,$$

$$\sum_{i,j=1}^{N}\frac{\partial^2 G_n}{\partial p_i \partial p_j}\xi_i\xi_j \geqslant c_0 \sum_{i=1}^{N}\xi_i^2$$

$$(x \in \overline{\Omega}, u \in \mathbb{R}, p, \xi \in \mathbb{R}^N, c_0, C_1 > 0),$$

where the constants c_0, C_1 are independent of n. Then, just as in the proof of Lemma 5.4.4, we find that

$$u_n \in \overset{\circ}{C}{}^1(\overline{\Omega})$$

for large n and

$$\lim_{n\to\infty} \|u_n\|_{\mathring{C}^1(\overline{\Omega})} = 0 \,. \tag{5.4.41}$$

By Lemma 5.4.3, the functionals $f_0(\cdot;0)$ and f_0 coincide on balls $B(\rho)$ in $\mathring{C}^1(\overline{\Omega})$ of small radius ρ. Consequently,

$$f_0(u_n;0) = f_0(u_n) < f_0(0) \tag{5.4.42}$$

for large n. The equations (5.4.41), (5.4.42) contradict the assumption that 0 is a weak local minimizer of f_0, and the lemma is proved. □

Now we prove Theorem 5.4.3.

Since the zero critical point of the family of functionals $f_0(\cdot;\lambda)$ is isolated on $\mathring{W}_2^1(\Omega)$ uniformly with respect to $\lambda \in [0,1]$ and is a local minimizer of $f_0(\cdot;0)$, it follows, by Theorem 3.1.1 (see the remark on this theorem at the end of Sect. 3.1.4), that 0 is a local minimizer of $f_0(\cdot;1)$ on $\mathring{W}_2^1(\Omega)$. Since the embedding of $\mathring{C}^1(\overline{\Omega})$ into $\mathring{W}_2^1(\Omega)$ is continuous, the function $u(x) = 0$ is a weak local minimizer of $f_0(\cdot;1)$, and since the functionals $f_0(\cdot;1)$ and f_1 coincide on balls $B(\rho)$ in $\mathring{C}^1(\overline{\Omega})$ of small radius ρ, the function is a weak local minimizer of f_1. The is completes the proof of the theorem. □

Remark. An analog of Theorem 5.4.3 holds in the situation where the critical point $u(\lambda)$ of the functional (5.4.22) is not unique: we require merely that this critical point is isolated in $\mathring{C}^1(\overline{\Omega})$ uniformly with respect to $\lambda \in [0,1]$. In this case, the strengthened Legendre condition does not need to be satisfied globally but only for the critical point $u(\lambda)$.

5.4.5 Functional Inequalities

We shall now prove some functional inequalities.

Let Ω be a bounded domain in $\mathbb{R}^N$ with a smooth boundary. Consider the space $\mathring{W}_2^1(\Omega)$. The *Poincaré–Friedrichs inequality* is the inequality

$$\int_\Omega |\nabla u(x)|^2\,dx \geqslant \lambda_1 \int_\Omega u^2(x)\,dx \,, \tag{5.4.43}$$

where λ_1 is the first eigenvalue of the boundary-value problem

$$-\Delta u = \lambda u \,, \tag{5.4.44}$$

$$u\big|_{\partial\Omega} = 0 \,. \tag{5.4.45}$$

We claim that this inequality holds for all $u \in \mathring{W}_2^1(\Omega)$. In order to prove this, it suffices to consider the one-parameter family of functionals

$$f(u;\lambda) = \int_\Omega (|\nabla u(x)|^2 - \lambda u^2(x))\,dx \,. \tag{5.4.46}$$

The critical points of this family are solutions of the Dirichlet problem for the Euler equation (5.4.44). If $\lambda \in [0, \lambda_1)$, then problem (5.4.44)–(5.4.45) has only the zero solution. We obtain the inequality (5.4.43) now by appealing to Theorem 3.3.3.

Here is one of the possible improvements of the Poincaré–Friedrichs inequality. We fix $\mu \in (0, \lambda_1)$ and $e \in \overset{\circ}{W}{}_2^1(\Omega)$. Then the inequality

$$\int_\Omega |\nabla u(x)|^2 \, dx \geqslant \mu \int_\Omega u^2(x) \, dx + \varkappa(\mu, e) \left(\int_\Omega u(x) e(x) \, dx \right)^2 \tag{5.4.47}$$

holds, where

$$\varkappa(\mu, e) = \left(\int_{\Omega \times \Omega} G(x, y, \mu) e(x) e(y) \, dx dy \right)^{-1} \tag{5.4.48}$$

and $G(x, y, \mu)$ is the Green's function of the differential operator $-(\Delta + I)$ under zero boundary conditions. The proof of this inequality follows the same pattern as the proof of the inequality (5.4.43).

The eigenvalue λ_1 of the boundary-value problem (5.4.44)–(5.4.45) is simple. Let u_1 be the nontrivial solution of norm 1 for problem (5.4.44)–(5.4.45) for $\lambda = \lambda_1$ (this solution is unique up to multiplication by -1 since λ_1 is a simple eigenvalue) and let e be some function from $\overset{\circ}{W}{}_2^1(\Omega)$ which is orthogonal to $u_1(x)$. We can show that

$$\lim_{\mu \to \lambda_1 - 0} \varkappa(\mu, e) = \varkappa_0(e) > 0 \, .$$

Therefore the Poincaré–Friedrichs inequality can be improved to

$$\int_\Omega |\nabla u(x)|^2 \, dx \geqslant \lambda_1 \int_\Omega u^2(x) \, dx + \varkappa_0(e) \left(\int_\Omega u(x) e(x) \, dx \right)^2 . \tag{5.4.49}$$

In some cases, the constant $\varkappa_0(e)$ and the eigenvalue λ_1 can be determined. Thus, for instance, if

$$\Omega = (0, T), \quad e(x) = \cos \frac{\pi}{T} x \, ,$$

then

$$\varkappa_0(e) = \frac{8\pi^2}{T^3}, \quad \lambda_1 = \frac{\pi^2}{T^2} \, .$$

Therefore, in the one-dimensional case, for

$$e(x) = \cos \frac{\pi}{T} x \, ,$$

the inequality (5.4.49) assumes the form

$$\int_0^T |u'(x)|^2\,dx \geqslant \frac{\pi^2}{T^2}\int_0^T u^2(x)\,dx + \frac{8\pi^2}{T^3}\left(\int_0^T u(x)\cos\frac{\pi}{T}x\,dx\right)^2 . \tag{5.4.50}$$

We now give another inequality related to the Poincaré–Friedrichs inequality.

Let

$$e \in \mathring{W}_2^1(\Omega), \quad \|e\| = 1$$

and let E be the orthogonal complement of e in $\mathring{W}_2^1(\Omega)$. If e is equal to the eigenfunction u_1 of problem (5.4.44)–(5.4.45) corresponding to the first eigenvalue, then the inequality

$$\int_\Omega |\nabla u(x)|^2\,dx \geqslant \lambda_2 \int_\Omega u^2(x)\,dx \quad (u \in E) \tag{5.4.51}$$

holds, where λ_2 is the second eigenvalue of problem (5.4.44)–(5.4.45). Using Theorem 3.3.3, we can find the exact constant $\varkappa_1(e)$ in the inequality

$$\int_\Omega |\nabla u(x)|^2\,dx \geqslant \varkappa_1(e) \int_\Omega u^2(x)\,dx \quad (u \in E) \tag{5.4.52}$$

in the case where e does not equal u_1: it is equal to the smallest root τ_* of the equation

$$1 + \tau + \tau^2 \int_{\Omega\times\Omega} G(x,y,\tau)e(x)e(y)\,dxdy = 0$$

lying in the interval (λ_1, λ_2).

5.4.6 Solvability of Boundary Value Problems and Criteria for Minimizers of Integral Functionals

We consider an integral functional

$$f(u) = \int_\Omega F(x, u(x), \nabla u(x))\,dx, \quad u(x)\big|_{\partial\Omega} = 0\,, \tag{5.4.53}$$

and write $\mathfrak{M}$ for the set of extremals of f, i.e., the set of solutions of the Dirichlet problem for the Euler equation

$$\sum_{i=1}^N \frac{d}{dx_i}\frac{\partial F(x,u(x),\nabla u(x))}{\partial p_i} - \frac{\partial F(x,u(x),\nabla u(x))}{\partial u} = 0\,, \tag{5.4.54}$$

$$u(x)\big|_{\partial\Omega} = 0\,. \tag{5.4.55}$$

We assume that $\mathfrak{M}$ is nonempty. In applications it is important to know whether f has a minimizer in $\mathfrak{M}$. In particular, if $\mathfrak{M}$ consists of one point u_*, then it is important to know whether u_* is a minimizer. Since it is generally difficult to find u_* explicitly, it is desirable to obtain this information just in terms of properties of the integrand F. The information can be used to justify various approximation procedures (for instance, gradient methods, projection methods, factor-procedures, etc.), to analyze the stability of extremals with respect to small perturbations of the integrand, to evaluate the number of extremals of the functional (5.4.53), etc.

In this section we show that under the conditions of many well-known theorems on the solvability of problem (5.4.54)–(5.4.55), the corresponding integral functional does have at least one minimizer.

We consider first the one-dimensional functional

$$f(x) = \int_0^T F(t, x(t), x'(t))\,dt, \quad x(0) = x(T) = 0\,. \tag{5.4.56}$$

We assume that the Lagrangian $F(t, x, p)$ is sufficiently smooth and satisfies the strengthened Legendre condition

$$\frac{\partial^2 F}{\partial p^2} \geqslant a > 0 \quad (0 \leqslant t \leqslant T, |x|, |p| < \infty)\,.$$

In his classical paper [15], Bernstein gives conditions guaranteeing the existence of an extremal for the functional (5.4.56).

Theorem 5.4.5 (Bernstein). *Suppose that the Euler equation for the functional* (5.4.56) *has the normal form*

$$x'' = \varphi(t, x, x')\,, \tag{5.4.57}$$

where

$$\varphi(t, x, p) = \left(\frac{\partial^2 F}{\partial p^2}\right)^{-1}\left(\frac{\partial F}{\partial x} - \frac{\partial^2 F}{\partial t \partial p} - \frac{\partial^2 F}{\partial x \partial p}\right). \tag{5.4.58}$$

Let

$$|\varphi(t, x, p)| \leqslant A(t)(1 + p^2)\,, \tag{5.4.59}$$

$$\frac{\partial \varphi}{\partial x} \geqslant \varepsilon > 0\,. \tag{5.4.60}$$

Then the functional (5.4.56) *has a unique critical point* $x_*(t)$.

The following result complements Bernstein's theorem.

Theorem 5.4.6. *Under the conditions of Bernstein's theorem, the critical point* $x_*(t)$ *is a strong minimizer for the functional* (5.4.56).

Proof. The inequalities (5.4.59), (5.4.60) give an *a priori* estimate

$$\|x_*(t)\|_{\mathring{C}^1[0,T]} < C \tag{5.4.61}$$

in which the constant C depends only on the function $A(t)$ and the numbers ε, T.

We shall consider the one-parameter family of functionals

$$f(x;\lambda) = \int_0^T F(\lambda t, x(t), \lambda^{-1}x'(t))\,dt \quad (x \in \mathring{C}^1[0,T],\ 0 < \lambda \leqslant 1) \tag{5.4.62}$$

and show that each functional $f(\cdot;\lambda)$ has a unique critical point $x_*(\cdot;\lambda)$. Indeed, the critical points $x_*(\cdot;\lambda)$ of the functionals (5.4.62) are related to the critical points $y_*(\cdot;\lambda)$ of the functionals

$$g(y;\lambda) = \int_0^{\lambda T} F(t, y(t), y'(t))\,dt \quad (y \in \mathring{C}^1[0,\lambda T],\ 0 < \lambda \leqslant 1) \tag{5.4.63}$$

by the equation

$$x_*(t;\lambda) = y_*(\lambda t;\lambda)\,.$$

However, by Bernstein's theorem each functional $g(\cdot;\lambda)$ has a unique critical point $y_*(\cdot;\lambda)$ satisfying (5.4.61). Therefore, for every $\lambda \in (0,1]$, the functional (5.4.62) has a critical point $x_*(\cdot;\lambda)$.

We shall show that for small $\lambda > 0$ the critical point $x_*(\cdot;\lambda)$ is a weak local minimizer for $f(\cdot;\lambda)$. To do this, it suffices to show that for small $\lambda > 0$ the critical point $y_*(\cdot;\lambda)$ is a local minimizer of $g(\cdot;\lambda)$ in $\mathring{C}^1[0,\lambda T]$.

Let $h \in \mathring{C}^1[0,\lambda T]$ and

$$\|h\|_{\mathring{C}^1[0,\lambda T]} \leqslant 1\,.$$

Then

$$g(y_* + h;\lambda) - g(y_*;\lambda)$$
$$= \frac{1}{2}\int_0^{\lambda T}\left(\frac{\partial^2 \widetilde{F}}{\partial x^2}h^2(t) + 2\frac{\partial^2 \widetilde{F}}{\partial x \partial p}h(t)h'(t) + \frac{\partial^2 \widetilde{F}}{\partial p^2}(h'(t))^2\right)dt\,.$$

It follows from (5.4.61) that

$$\left|\frac{\partial^2 \widetilde{F}}{\partial x^2}\right| \leqslant C_0, \quad \left|\frac{\partial^2 \widetilde{F}}{\partial x \partial p}\right| \leqslant C_0\,. \tag{5.4.64}$$

These inequalities, Legendre's condition, the Cauchy–Schwarz–Bunyakovskii inequality, and Wirtinger's inequality yield the chain of inequalities

$$\int_0^{\lambda T} \left(\frac{\partial^2 \widetilde{F}}{\partial x^2} h^2(t) + 2 \frac{\partial^2 \widetilde{F}}{\partial x \partial p} h(t)h'(t) + \frac{\partial^2 \widetilde{F}}{\partial p^2} (h'(t))^2 \right) dt$$

$$\geqslant a_0 \int_0^{\lambda T} |h'(t)|^2 \, dt - 2C_0 \int_0^{\lambda T} |h(t)||h'(t)| \, dt - C_0 \int_0^{\lambda T} h^2(t) \, dt$$

$$\geqslant \frac{a_0 \pi^2}{\lambda^2 T^2} \int_0^{\lambda T} h^2(t) \, dt - 2C_0 \left(\int_0^{\lambda T} h^2(t) \, dt \right)^{1/2} \cdot \left(\int_0^{\lambda T} |h'(t)|^2 \, dt \right)^{1/2} - C_0 \int_0^{\lambda T} h^2(t) \, dt$$

$$\geqslant \left(\frac{a_0 \pi^2}{\lambda^2 T^2} - \frac{2C_0 \pi}{\lambda T} - C_0 \right) \int_0^{\lambda T} h^2(t) \, dt \, .$$

Consequently we have

$$g(y_* + h; \lambda) - g(y_*; \lambda) > 0$$

for small $\lambda > 0$. However, for small $\lambda > 0$, the critical point $x_*(\cdot; \lambda)$ is a weak local minimizer for $f(\cdot; \lambda)$. By Theorem 5.4.4, the critical point $x_* = x_*(\cdot; 1)$ is a weak local minimizer of $f = f(\cdot; 1)$. Since the strengthened Legendre condition holds, it follows (see, e.g., [2]) that x_* is a strong local minimizer of f, and the theorem is proved. □

Results similar to Theorem 5.4.6 can be proved under the conditions of many well-known solvability criteria for two-point boundary-value problems for the Euler equations of integral functionals, e.g., those of the theorems of Picard [187], Nagumo [180], Lettenmeyer [160], Krasnosel'skii [151], Semenov [204].

Now we consider criteria for the existence of a weak minimizer for the integral functional (5.4.53) in the multidimensional case.

Let Ω be a convex bounded domain in $\mathbb{R}^N$ where $N \geqslant 3$, with $\partial\Omega \in C^{2+\alpha}$. We rewrite the Euler equation for the functional (5.4.53) in the form

$$\sum_{i,j=1}^{N} \frac{\partial^2 F}{\partial p_i \partial p_j} \frac{\partial^2 u(x)}{\partial x_i \partial x_j} + a(x, u(x), \nabla u(x)) = 0 \, , \tag{5.4.65}$$

$$u(x)\big|_{\partial\Omega} = 0. \tag{5.4.66}$$

Assume that the functions

$$\frac{\partial^2 F}{\partial p_i \partial p_j}$$

are independent of u and

$$\frac{\partial a}{\partial u} \leqslant \varepsilon < 0 . \tag{5.4.67}$$

The authors of [153] give the following criterion for the solvability of the boundary-value problem (5.4.65), (5.4.66): if we have the inequalities

$$\nu(1+|p|)^{m-2}\sum_{i=1}^{N}\xi_i^2 \leqslant \sum_{i,j=1}^{N}\frac{\partial^2 F}{\partial p_i \partial p_j}\xi_i\xi_j \leqslant \mu(1+|p|)^{m-2}\sum_{i=1}^{N}\xi_i^2 ,$$

$$\left|\frac{\partial F}{\partial u}\right| + \sum_{i=1}^{N}\left(\left|\frac{\partial^2 F}{\partial u \partial p_i}\right| + \left|\frac{\partial F}{\partial p_i}\right|\right)(1+|p|) + \sum_{i,j=1}^{N}\left|\frac{\partial^2 F}{\partial p_i \partial x_j}\right| \leqslant \mu(1+|p|)^m ,$$

where μ, ν are positive constants and $m > 1$, then problem (5.4.65), (5.4.66) has a unique solution

$$u_* \in C^{2+\alpha}(\overline{\Omega}) .$$

Using the method of proof of Theorem 5.4.6, we can easily establish that under the above conditions the extremal u_* is a weak local minimizer of the functional (5.4.53).

5.4.7 Investigation of Critical Points for a Minimizer

The deformation theorems establishing invariance of minimizers give a general method for determining which critical points are minimizers. We shall describe a modification which can be used for investigating the functional (5.4.53).

Let u_* be a critical point of the functional (5.4.53) under investigation. We construct a one-parameter family of functionals

$$f(u,\lambda) = \int_{\Omega} F(x,u(x),\nabla u(x);\lambda)\,dx , \tag{5.4.68}$$

$$u(x)\big|_{\partial\Omega} = 0 \quad (0 \leqslant \lambda \leqslant 1) , \tag{5.4.69}$$

which homotopically transform the functional

$$f(\cdot) = f(\cdot;0)$$

into some standard functional

$$f_1(\cdot) = f(\cdot;1) .$$

The family (5.4.68) is constructed in such a way that u_* is a critical point of each functional $f(\cdot;\lambda)$ or, equivalently, u_* is a solution of each boundary-value problem of the one-parameter family

$$\sum_{i=1}^{N}\frac{d}{dx_i}\frac{\partial F(x,u(x),\nabla u(x);\lambda)}{\partial p_i} - \frac{\partial F(x,u(x),\nabla u(x);\lambda)}{\partial u} = 0 , \tag{5.4.70}$$

$$u\big|_{\partial\Omega} = 0 \quad (0 \leqslant \lambda \leqslant 1)\,. \tag{5.4.71}$$

The standard functional f_1 is chosen such that we can decide whether its critical point u_* is a minimizer. Then we verify whether the critical point u_* of the family (5.4.70), (5.4.71) is isolated. If it is isolated (in particular, if it is unique), then u_* is a minimizer of the functional f under investigation if and only if it is a minimizer of f_1. Using this procedure, we can easily establish sufficient conditions for a point to be a minimizer for problems in the classical calculus of variations (Jacobi theorems) and multidimensional calculus of variations, and also in many problems of mathematical physics and in nonlinear programming problems.

5.5 Stability of Solutions of Ordinary Differential Equations

5.5.1 Stability of Gradient Systems

Consider the *gradient system* of ordinary differential equations

$$\frac{dx}{dt} = -\nabla f(x) \quad (x \in \mathbb{R}^N)\,, \tag{5.5.1}$$

where $f : \mathbb{R}^N \to \mathbb{R}$ is a continuously differentiable function. Let

$$\nabla f(0) = 0\,.$$

Then 0 is a state of equilibrium of the system (5.5.1).

The following results hold.

Theorem 5.5.1. *The zero state of equilibrium of the system* (5.5.1) *is Lyapunov stable if and only if* 0 *is a local minimizer for* f.

Theorem 5.5.2. *The zero state of equilibrium of the system* (5.5.1) *is asymptotically stable if and only if* 0 *is an isolated critical point and a local minimizer for* f.

Proof. In order to prove Theorem 5.5.1, it suffices to note that in the case where 0 is a local minimizer of f, this function is a Lyapunov function of the system (5.5.1):

$$\frac{df(x)}{dt} = -|\nabla f(x)|^2 \leqslant 0\,. \tag{5.5.2}$$

Now if 0 is not a local minimizer of f, then f is a Lyapunov–Chetaev function for (5.5.1), and consequently the zero state of equilibrium of (5.5.1) is unstable. The theorem follows. □

Theorem 5.5.2 can be proved similarly.

These two theorems show that the analysis of the stability of equilibrium states of a gradient system can be reduced to the equivalent problem of whether critical points of the corresponding potential are minimizers. Therefore the deformation method can be used to investigate the stability of gradient systems. We shall give some simple examples to illustrate this method.

Consider two gradient systems

$$\frac{dx}{dt} = -\nabla f_0(x) \quad (x \in \mathbb{R}^N)\,, \tag{5.5.3}$$

$$\frac{dx}{dt} = -\nabla f_1(x) \quad (x \in \mathbb{R}^N) \tag{5.5.4}$$

for which 0 is an isolated state of equilibrium. The one-parameter family of gradient systems

$$\frac{dx}{dt} = -\nabla f(x;\lambda) \quad (x \in \mathbb{R}^N,\ 0 \leqslant \lambda \leqslant 1)$$

is said to be a *nondegenerate deformation* of the system (5.5.3) into (5.5.4) if the potential $f : \mathbb{R}^N \times [0,1] \to \mathbb{R}$ and its gradient $\nabla f : \mathbb{R}^N \times [0,1] \to \mathbb{R}^N$ are continuous in the collection of variables, if, for every $\lambda \in [0,1]$, zero is a critical point of the potential $f(\cdot;\lambda)$ isolated uniformly with respect to $\lambda \in [0,1]$ and if

$$f(\cdot;0) = f_0, \quad f(\cdot;1) = f_1\,.$$

Theorems 2.1.1 and 5.5.2 imply

Theorem 5.5.3. *Suppose that there exists a nondegenerate deformation of the gradient system* (5.5.3) *into* (5.5.4). *If* 0 *is an asymptotically stable state of equilibrium of* (5.5.3) *then* 0 *is an asymptotically stable state of equilibrium of* (5.5.4).

Corollary 1. *Suppose that for* $x \in B(\rho) \subset \mathbb{R}^N$, $x \neq 0$, *the gradients* $\nabla f_0(x)$ *and* $\nabla f_1(x)$ *of the potentials* f_0 *and* f_1 *of the systems* (5.5.3) *and* (5.5.4) *are not opposite in direction. If the zero state of equilibrium of* (5.5.3) *is asymptotically stable then so is the zero state of equilibrium of* (5.5.4).

In order to prove the corollary, it suffices to consider the following linear nondegenerate deformation of (5.5.3) into (5.5.4):

$$\frac{dx}{dt} = -((1-\lambda)\nabla f_0(x) + \lambda \nabla f_1(x)) \quad (x \in B(\rho),\ 0 \leqslant \lambda \leqslant 1)\,.$$

5.5.2 Stability of Hamiltonian Systems

Consider a *Hamiltonian system*

$$\begin{cases} \dfrac{dy}{dt} = \dfrac{\partial H(x,y)}{\partial x}\,, \\[2mm] \dfrac{dx}{dt} = -\dfrac{\partial H(x,y)}{\partial y}\,. \end{cases} \tag{5.5.5}$$

Here, x, $y \in \mathbb{R}^N$ and $H : \mathbb{R}^N \times \mathbb{R}^N \to \mathbb{R}$ is a smooth function, called the *Hamiltonian* of the system. Let

$$\frac{\partial H(0,0)}{\partial x} = \frac{\partial H(0,0)}{\partial y} = 0\,.$$

Then the point $(0,0) \in \mathbb{R}^N \times \mathbb{R}^N$ is a state of equilibrium of the system (5.5.5). By Dirichlet's theorem, this state of equilibrium is Lyapunov stable if the point $(0,0)$ is a local minimizer of the Hamiltonian H. It should be emphasized that, in contrast with the case of gradient systems, the presence of a minimizer of H at 0 is only a sufficient condition for stability of the zero state of equilibrium of (5.5.5): it is easy to construct an example of a Hamiltonian system which is stable at 0 but for whose Hamiltonian 0 is not a minimizer.

As in the case of gradient systems, we can use the homotopy method to investigate the stability of equilibrium states of Hamiltonian systems. Here is the corresponding result.

We say that the one-parameter family of Hamiltonian systems

$$\begin{cases} \dfrac{dy}{dt} = \dfrac{\partial H(x,y;\lambda)}{\partial x}\,, \\[2mm] \dfrac{dx}{dt} = -\dfrac{\partial H(x,y;\lambda)}{dy} \end{cases} \tag{5.5.6}$$

$$\Big(x, y \in \mathbb{R}^N, \quad 0 \leqslant \lambda \leqslant 1, \quad H : \mathbb{R}^N \times \mathbb{R}^N \times [0,1] \to \mathbb{R}\,, \\ \frac{\partial H(0,0;\lambda)}{\partial x} = \frac{\partial H(0,0;\lambda)}{\partial y} = 0 \quad (0 \leqslant \lambda \leqslant 1)\Big)$$

is a *nondegenerate deformation* of the Hamiltonian system

$$\begin{cases} \dfrac{dy}{dt} = \dfrac{\partial H_0(x,y)}{\partial x}\,, \\[2mm] \dfrac{dx}{dt} = -\dfrac{\partial H_0(x,y)}{\partial y} \end{cases} \tag{5.5.7}$$

into the Hamiltonian system

$$\begin{cases} \dfrac{dy}{dt} = \dfrac{\partial H_1(x, y)}{\partial x}, \\ \dfrac{dx}{dt} = -\dfrac{\partial H_1(x, y)}{\partial y} \end{cases} \tag{5.5.8}$$

if the Hamiltonian H and its partial derivatives $\partial H/\partial x$ and $\partial H/\partial y$ are continuous on $\mathbb{R}^N \times \mathbb{R}^N \times [0, 1]$, if the point $(0, 0)$ is a state of equilibrium of system (5.5.6) isolated uniformly with respect to $\lambda \in [0, 1]$ for every value of the parameter $\lambda \in [0, 1]$, and if

$$H(\cdot, \cdot; 0) = H_0(\cdot; \cdot), \quad H(\cdot, \cdot; 1) = H_1(\cdot; \cdot) .$$

Theorem 5.5.4. *Suppose that there exists a nondegenerate deformation of the Hamiltonian system* (5.5.7) *into the Hamiltonian system* (5.5.8). *If the point* $(0, 0)$ *is a minimizer of the Hamiltonian* H_0 *then* $(0, 0)$ *is a Lyapunov stable state of equilibrium of the system* (5.5.8).

The proof of this result follows from Dirichlet's theorem and Theorem 2.1.1.

5.5.3 Stability of Gradient Systems in the Large

Recall that the zero state of equilibrium of the dynamical system

$$\frac{dx}{dt} = g(x) \quad (g(0) = 0, \ x \in \mathbb{R}^N) \tag{5.5.9}$$

is *stable in the large* if it is Lyapunov stable and, for any initial condition $x \in \mathbb{R}^N$, the solution $p(\cdot, x)$ of the system (5.5.9) satisfies $\lim_{t\to\infty} p(t, x) = 0$. It follows from the definition of stability in the large that if the state of equilibrium of a dynamical system is stable in the large, then, in particular, it is asymptotically stable.

For the zero state of equilibrium of system (5.5.9) to be stable in the large, it is sufficient that there exists a growing Lyapunov function V for this system, i.e., a C^1-function for which

$$V(x) > V(0) \quad (x \neq 0) ,$$
$$(\nabla V(x), g(x)) < 0 \quad (x \in \mathbb{R}^N, \ x \neq 0) \tag{5.5.10}$$

and

$$\lim_{|x|\to\infty} V(x) = \infty .$$

When we investigate the stability in the large of gradient systems of the form

$$\frac{dx}{dt} = -\nabla f(x) \quad (\nabla f(0) = 0, \ x \in \mathbb{R}^N) , \tag{5.5.11}$$

it is desirable to have information about stability in terms of the potential f.

Theorem 5.5.5. *Suppose that* 0 *is the unique critical point of the potential* f *of the system* (5.5.11) *which is a local minimizer. Suppose that*

$$|\nabla f(x)| \geqslant \alpha(|x|) \quad (x \in \mathbb{R}^N)\,, \tag{5.5.12}$$

where $\alpha : (0, \infty) \to \mathbb{R}^N$ *is a continuous positive function for which*

$$\int_0^\infty \alpha(s)ds = \infty\,. \tag{5.5.13}$$

Then the zero state of equilibrium of system (5.5.11) *is stable in the large.*

Proof. It follows from Theorem 2.1.4 that

$$\lim_{|x|\to\infty} f(x) = \infty\,. \tag{5.5.14}$$

It remains to note that the potential f is a Lyapunov function of the system (5.5.11), and the theorem is proved. □

Note that the condition (5.5.14) on the growth of the potential f is essential for the stability in the large of the zero state of equilibrium of (5.5.11). It cannot be weakened, for example, to the requirement that 0 be a global minimizer of f. It is easy to give an example of a gradient system unstable in the large in $\mathbb{R}^2$ whose potential has unique critical point 0 and such that 0 is a global minimizer.

Combining Theorems 5.5.3 and 5.5.5, we can easily give a deformational criterion for stability of gradient systems in the large.

5.5.4 Conley Index and the Stability of Dynamical Systems

So far we have used the deformation theorems to study the stability of specific types of differential equations (gradient, potential, Hamiltonian). The theory of Conley index allows us to investigate the stability of general systems of differential equations. Here is, for instance, an application to the investigation of the stability of systems of ordinary differential equations.

Consider two systems

$$\frac{dx}{dt} = f_0(x) \quad (x \in \mathbb{R}^N) \tag{5.5.15}$$

and

$$\frac{dx}{dt} = f_1(x) \quad (x \in \mathbb{R}^N)\,. \tag{5.5.16}$$

We assume here that f_0 and f_1 are locally Lipschitzian and that the systems (5.5.15) and (5.5.16) have a zero state of equilibrium, i.e.,

$$f_0(0) = f_1(0) = 0\,.$$

A one-parameter family of systems

$$\frac{dx}{dt} = f(x;\lambda) \quad (x \in \mathbb{R}^N,\ 0 \leqslant \lambda \leqslant 1) \tag{5.5.17}$$

whose right-hand side is continuous on $\mathbb{R}^N \times [0,1]$ and locally Lipschitzian with respect to the variable x, is a *nondegenerate deformation* of (5.5.15) into (5.5.16) if

$$f(0;\lambda) = 0\ , \tag{5.5.18}$$

$$f(x;0) = f_0(x), \quad f(x;1) = f_1(x) \tag{5.5.19}$$

and there exists a ball $B(\rho)$ in $\mathbb{R}^N$ with $\rho > 0$ containing no trajectories for the family (5.5.17) except the zero state of equilibrium. In particular, this ball contains no states of equilibrium of (5.5.17) except 0.

Theorem 5.5.6. *Suppose that there exists a nondegenerate deformation of the system* (5.5.15) *into the system* (5.5.16). *If the zero state of equilibrium of* (5.5.15) *is asymptotically stable then so is the zero state of equilibrium of* (5.5.16).

For the proof we require several preparatory results.
Consider the auxiliary system

$$\frac{dx}{dt} = -x \quad (x \in \mathbb{R}^N)\ . \tag{5.5.20}$$

The zero state of equilibrium of the system (5.5.20) is asymptotically stable and is an isolated invariant set, and so the Conley index h_0 of this state of equilibrium is defined. Along with the system (5.5.20), we consider the system

$$\frac{dx}{dt} = f(x) \quad (f(0) = 0,\ x \in \mathbb{R}^N) \tag{5.5.21}$$

with a locally Lipschitzian function f on the right-hand side. Suppose that the zero state of equilibrium of (5.5.21) is asymptotically stable. We shall show that the Conley index of this state of equilibrium is defined. For this purpose, it suffices to show that there exists a neighborhood of this state of equilibrium containing no trajectories of (5.5.21) except the trajectory $\{0\}$.

Consider a neighborhood U of the zero state of equilibrium such that for every point $x \in U$ the solution $p(t,x)$ with $p(0,x) = x$ of the system (5.5.21) satisfies $\lim_{t\to\infty} p(t,x) = 0$. We shall show that no trajectory of (5.5.21) except $\{0\}$ lies entirely in U. If this is false, then we have

$$p(t,x_0) \in U \quad (t \in (-\infty;\infty))$$

for some $x_0 \in U$. Consider the set of α-limit points of the trajectory $p(\cdot,x_0)$. This set does not contain 0 since otherwise (5.5.21) would have a separatrix emanating from 0 and tending to 0, contradicting the assumption that 0 is asymptotically stable. Since the set of α-limit points is a union of trajectories, U would contain a whole trajectory of (5.5.21) for which 0 is not a limit point. This contradicts the choice of the neighborhood U.

Thus the Conley index of the zero state of equilibrium of the system (5.5.21) is defined; we denote it by $h(f)$.

Lemma 5.5.1. *For any choice of the function f in the system* (5.5.21) *we have*

$$h(f) = h_0 . \tag{5.5.22}$$

Proof. We shall use some constructions that we described when we introduced the Conley index of an isolated invariant set of a dynamical system.

The zero state of equilibrium of (5.5.21) will play the part of this set. As an isolating neighborhood of this state of equilibrium we choose a ball $B(\rho)$ such that

$$\lim_{t\to\infty} p(t,x) = 0 \quad (x \in B(\rho)) . \tag{5.5.23}$$

By Theorem 4.1.1, there exists an index pair $\langle N_1, N_2\rangle$ such that

$$\overline{N_1 \setminus N_2} \subset B\left(\tfrac{1}{2}\rho\right) . \tag{5.5.24}$$

The set N_1 determines the family of sets

$$N_1^t = \{p(t,x) : x \in N_1,\ p(\tau,x) \subset B(\rho) \text{ for } \tau \in [0,t]\} ;$$

(see Sect. 4.1.3). Lemma 4.1.10 shows that all sets N_1^t/N_2 $(t \geqslant 0)$ are homotopy equivalent. We claim that

$$N_1^t \subset B\left(\tfrac{1}{4}\rho\right) \tag{5.5.25}$$

for all sufficiently large t. To prove this, it suffices to show that for any $r \leqslant \rho$ there exists some $T(r) > 0$ such that

$$p(t,x) \in B(r) \quad (t \geqslant T(r),\ x \in B(\rho)) . \tag{5.5.26}$$

Indeed, since the zero state of equilibrium of (5.5.21) is stable we have

$$p(t,x) \in B(r) \quad (x \in B(r_0),\ t \geqslant 0)$$

for some $r_0 < r$. Let $x \in B(\rho)$. By (5.5.23), there exists some $t(x) > 0$ for which

$$p(t(x),x) \in B\left(\tfrac{1}{2}r_0\right) .$$

Since the solutions of (5.5.21) depend continuously on the initial data, it follows that

$$p(t(x),y) \in B(r_0) \quad (y \in U(x))$$

for some neighborhood $U(x)$ of x. The family $\{U(x) : x \in B(\rho)\}$ of open sets forms a cover of the compact set $B(\rho)$; we choose a finite subcover $\{U(x_1), \ldots, U(x_n)\}$ and write

$$T(r) = \max\{t(x_1), \ldots, t(x_n)\} .$$

The inclusions (5.5.26) hold for this choice of $T(r)$.

Thus (5.5.25) holds for all sufficiently large t. It follows from (5.5.24) and (5.5.25) that for large t we have

$$N_1^t \cap N_2 = \varnothing ;$$

but then, for such t, we have

$$N_1^t/N_2 \sim N_1^t/\varnothing .$$

Since the sets N_1^t $(t \geqslant 0)$ can be contracted within themselves to a point, we have

$$h(f) \sim N_1^t/\varnothing \sim \{0; 1\}/\{0\} . \qquad (5.5.27)$$

On the other hand, simple calculations show that

$$h_0 \sim \{0; 1\}/\{0\} . \qquad (5.5.28)$$

From (5.5.27) and (5.5.28) we conclude that (5.5.22) holds, and the lemma follows. □

Let us consider (5.5.21) again, without assuming *a priori* that the zero state of equilibrium is asymptotically stable; instead we assume just that this state of equilibrium is an isolated invariant set of (5.5.21), i.e., that there is a ball $B(\rho)$ containing no nonzero trajectories. Then the Conley index $h(f)$ for the zero state of equilibrium of the system is defined.

Lemma 5.5.2. *If*

$$h(f) = h_0 , \qquad (5.5.29)$$

then the zero state of equilibrium of the system (5.5.21) *is asymptotically stable.*

Proof. We fix an arbitrary index pair $\langle N_1, N_2\rangle$ for the zero state of equilibrium. We shall show that

$$N_2 = \varnothing . \qquad (5.5.30)$$

If this is false, then there exist continuous mappings

$$\varphi : N_1 \to \{0; 1\}, \quad \psi : \{0; 1\} \to N_1 \qquad (5.5.31)$$

such that

$$\varphi(x) = 0 \quad (x \in N_2) , \qquad (5.5.32)$$

$$\psi(0) \in N_2 ,$$

$$\varphi \circ \psi \sim \mathrm{id}_{\{0;1\}} , \tag{5.5.33}$$

$$\psi \circ \varphi \sim \mathrm{id}_{N_1} . \tag{5.5.34}$$

From the continuity of φ and from (5.5.31), (5.5.32) it follows that $\varphi = 0$; but this contradicts (5.5.33) and (5.5.34). Thus $N_2 = \varnothing$.

It follows from the definition of an index pair that

$$p(t,x) \in B(\rho) \quad (x \in N_1,\ t \geqslant 0) . \tag{5.5.35}$$

Since N_1 contains a neighborhood of zero, and since we are free to decrease ρ, condition (5.5.35) ensures that the zero state of equilibrium of the system (5.5.21) is Lyapunov stable. It remains to show that all solutions of (5.5.21) with initial conditions which are sufficiently small in norm tend to zero. If this is not true, then

$$p(t,x) \in B(\rho) \quad (0 \leqslant t < \infty)$$

for some $x \neq 0$, and we have

$$0 \in \overline{\{p(t,x) : t \geqslant 0\}} .$$

Therefore the set of ω-limit points of this trajectory contains a whole nonzero trajectory of (5.5.21). The lemma now follows. □

Now we may prove Theorem 5.5.6. By Lemma 5.5.1 we have

$$h(f_0) = h_0 .$$

The homotopy invariance of Conley index implies that

$$h(f_1) = h_0 ,$$

and the theorem follows from an appeal to Lemma 5.5.2. □

5.6 Optimal Control Problems

5.6.1 Statement of the Problem

Consider the following *optimal control problem* for motion with a free right-hand end and fixed time:

$$\int_0^T F(s,x(s),u(s))ds \to \min , \tag{5.6.1}$$

$$\frac{dx}{dt} = g(t,x,u), \quad x(0) = 0 , \tag{5.6.2}$$

$$\int_0^T u^2(s)ds \leqslant 1 . \tag{5.6.3}$$

We assume that the function $F(t, x, u)$ and the vector function $g(t, x, u)$, and their first derivatives with respect to x_i, u_j ($i = 1, \ldots, N$; $j = 1, \ldots, M$) are continuous on $[0, T] \times \mathbb{R}^N \times \mathbb{R}^M$.

Constraints of the type (5.6.3) frequently arise in problems concerning the correction of the trajectory of a moving body using energy constraints (in particular, control by a small propulsion, see, e.g., [127]). The form of the constraints (5.6.3) naturally suggests that we take $L_2[0, T]$ as the control space.

Suppose that the Cauchy problem (5.6.2) has a unique solution $x = x(t)$ for every control $u = u(t) \in L_2[0, T]$. We denote by φ the operator which associates the control u with the solution x. Then problem (5.6.1)–(5.6.3) is equivalent to the minimization of the functional

$$f(u) = \int_0^T F(s, \varphi(u), u(t))ds \tag{5.6.4}$$

on the unit ball B of $L_2[0, T]$, i.e., to the problem

$$f(u) \to \min, \quad u \in B\,. \tag{5.6.5}$$

Natural constraints on the rate of increase of F and g imply that the functional f is $L_2[0, T]$-regular. We shall use the deformation method to investigate the critical points of the functional (5.6.5) (or, equivalently, problem (5.6.1)–(5.6.3)) when such conditions hold.

5.6.2 Auxiliary Results

Lemma 5.6.1. *Suppose that $c_1(t, x)$ and $c_2(t, x)$ are continuous functions such that*

$$\left|\frac{\partial F(t, x, u)}{\partial x}\right| + \left|\frac{\partial g(t, x, u)}{\partial x}\right| \leqslant c_1(t, x)(1 + |u|^2)\,, \tag{5.6.6}$$

$$\left|\frac{\partial F(t, x, u)}{\partial u}\right| + \left|\frac{\partial g(t, x, u)}{\partial u}\right| \leqslant c_2(t, x)(1 + |u|)\,, \tag{5.6.7}$$

for all $0 \leqslant t \leqslant T$, $x \in \mathbb{R}^N$, $u \in \mathbb{R}^M$. Then the functional (5.6.4) *is continuously differentiable on $L_2[0, T]$ and*

$$\begin{gathered}\nabla f(u) = \frac{\partial F}{\partial u}(t, x, u) \\ + \left(\frac{\partial g}{\partial u}(t, x, u)\right)^* (X^{-1}(t))^* \int_t^T (X(s))^* \frac{\partial F}{\partial u}(s, x, u)\, ds\,.\end{gathered} \tag{5.6.8}$$

Here $*$ is the transposition operation, $x = \varphi(u)$, and $X(t)$ is the fundamental matrix of the linear system

$$\frac{dh}{dt} = \frac{\partial g}{\partial x}(t, x(t), u(t))h \,. \tag{5.6.9}$$

Proof. Write $\mathring{W}_1^1[0,T]$ for the Banach space of absolutely continuous mappings $x : [0,T] \to \mathbb{R}^N$ such that $x(0) = 0$. We define a norm in $\mathring{W}_1^1[0,T]$ by

$$\|x\|_{\mathring{W}_1^1[0,T]} = \int_0^T |x'(s)| \, ds \,.$$

Consider the operator

$$A(x,u) = x(t) - \int_0^t g(s, x(s), u(s)) \, ds$$

on $\mathring{W}_1^1[0,T] \times L_2[0,T]$. By (5.6.6) we have

$$|g(t,x,u)| \leqslant c_3(t,x)(1 + |u|^2) \,,$$

for some continuous function $c_3(t,x)$. Therefore A maps $\mathring{W}_1^1[0,T] \times L_2[0,T]$ into $\mathring{W}_1^1[0,T]$. We claim that A is continuously Fréchet differentiable.

Let $(x,u) \in \mathring{W}_1^1[0,T] \times L_2[0,T]$. We define the linear operator

$$\mathcal{D}(x,u) : \mathring{W}_1^1[0,T] \times L_2[0,T] \to \mathring{W}_1^1[0,T]$$

by

$$\mathcal{D}(x,u)(h,q) = h(t) - \int_0^t \left(\frac{\partial g}{\partial x}(s,x,u)h(s) + \frac{\partial g}{\partial u}(s,x,u)q(s) \right) ds \,.$$

Then

$$\|A(x+h, u+q) - A(x,u) - \mathcal{D}(x,u)(h,q)\|_{\mathring{W}_1^1[0,T]}$$

$$\leqslant \|h\|_{\mathring{W}_1^1[0,T]} \int_0^1 \left\| \frac{\partial g}{\partial x}(t, x + \theta h, u + \theta q) - \frac{\partial g}{\partial x}(t,x,u) \right\|_{\widetilde{L}_1[0,T]} d\theta$$

$$+ \|q\|_{L_2[0,T]} \int_0^1 \left\| \frac{\partial g}{\partial u}(t, x + \theta h, u + \theta q) - \frac{\partial g}{\partial u}(t,x,u) \right\|_{\widetilde{L}_2[0,T]} d\theta \,.$$

In the right-hand side of this inequality, $\widetilde{L}_1[0,T]$ denotes the Banach space of $N \times N$ matrix functions

$$a(t) = [a_{ij}(t)] \quad (0 \leqslant t \leqslant T)$$

with the norm

$$\|a\|_{\widetilde{L}_1[0,T]} = \left(\sum_{i,j=1}^{N} \left(\int_0^T |a_{ij}(s)|\, ds \right)^2 \right)^{1/2},$$

and $\widetilde{L}_2[0,T]$ denotes the Hilbert space of $N \times N$ matrix functions

$$b(t) = [b_{ij}(t)] \quad (0 \leqslant t \leqslant T)$$

with the norm

$$\|b\|_{\widetilde{L}_2[0,T]} = \left(\sum_{i=1}^{N} \sum_{j=1}^{M} \int_0^T b_{ij}^2(s)\, ds \right)^{1/2}.$$

The inequalities (5.6.6) and (5.6.7) imply that the superposition operators

$$R_0(x,u) = \frac{\partial g}{\partial x}(t, x(t), u(t))$$

and

$$R_1(x,u) = \frac{\partial g}{\partial u}(t, x(t), u(t)),$$

from $\overset{\circ}{W}{}_1^1[0,T] \times L_2[0,T]$ to $\widetilde{L}_1[0,T]$ and $\widetilde{L}_2[0,T]$ respectively, are continuous; see Sect. 1.5.4. Therefore

$$\lim_{\|h\|_{\overset{\circ}{W}{}_1^1[0,T]} + \|q\|_{L_2[0,T]} \to 0} \int_0^1 \left\| \frac{\partial g}{\partial x}(t, x+\theta h, u+\theta q) - \frac{\partial g}{\partial x}(t,x,u) \right\|_{\widetilde{L}_1[0,T]} d\theta = 0$$

and

$$\lim_{\|h\|_{\overset{\circ}{W}{}_1^1[0,T]} + \|q\|_{L_2[0,T]} \to 0} \int_0^1 \left\| \frac{\partial g}{\partial u}(t+\theta h, u+\theta q) - \frac{\partial g}{\partial u}(t,x,u) \right\|_{\widetilde{L}_2[0,T]} d\theta = 0 .$$

Consequently,

$$\lim_{\|h\|_{\overset{\circ}{W}{}_1^1[0,T]} + \|q\|_{L_2[0,T] \to 0}} \frac{\|A(x+h, u+q) - A(x,u) - \mathcal{D}(x,u)(h,q)\|_{\overset{\circ}{W}{}_1^1[0,T]}}{\|h\|_{\overset{\circ}{W}{}_1^1[0,T]} + \|q\|_{L_2[0,T]}} = 0 .$$

Thus

$$A'(x,u) = \mathcal{D}(x,u) .$$

The continuity of the derivative follows from (5.6.6) and (5.6.7).

Now we consider the operator $A'(x,u)$. Clearly

$$A'(x,u)h = h(t) - \int_0^t \frac{\partial g}{\partial x}(s,x,u)h(s)\,ds\,. \tag{5.6.10}$$

It follows from this that $A'(x,u)$ is a bounded operator from $\overset{\circ}{W}{}_1^1[0,T]$ to itself, and so by Banach's theorem (Theorem 1.4.2), it has a continuous inverse. It follows from the implicit function theorem (Theorem 1.6.2) that the operator

$$\varphi : L_2[0,T] \to \overset{\circ}{W}{}_1^1[0,T]$$

is continuously differentiable on $L_1[0,T]$ and, at each point $u \in L_2[0,T]$, its derivative is given by

$$\varphi'(u)q = X(t)\int_0^t X^{-1}(s)\frac{\partial g}{\partial u}(s,x,u)q(s)\,ds\,, \tag{5.6.11}$$

where $X(t)$ is the fundamental matrix of the linear system (5.6.9).

Consider the functional

$$\mathcal{L}(x,u) = \int_0^T F(s,x(s),u(s))\,ds\,.$$

The inequalities (5.6.6), (5.6.7) imply that the linear functional $l(x,u)$ defined by

$$l(x,u)(h,q) = \int_0^T \left(\frac{\partial F}{\partial x}(s,x,u)h(s) + \frac{\partial F}{\partial u}(s,x,u)q(s)\right) ds$$

is bounded (for fixed $x \in \overset{\circ}{W}{}_1^1[0,T]$, $u \in L_2[0,T]$) and that

$$\lim_{\|h\|_{\overset{\circ}{W}{}_1^1[0,T]} + \|q\|_{L_2[0,T]} \to 0} \frac{|\mathcal{L}(x+h,u+q) - \mathcal{L}(x,u) - l(x,u)(h,q)|}{\|h\|_{\overset{\circ}{W}{}_1^1[0,T]} + \|q\|_{L_2[0,T]}} = 0\,.$$

Therefore $\mathcal{L}$ is continuously differentiable on $\overset{\circ}{W}{}_1^1[0,T] \times L_2[0,T]$ and

$$\nabla\mathcal{L}(x,u) = l(x,u)\,.$$

Now the functional f defined in (5.6.4) can be written in the form

$$f(u) = \mathcal{L}(\varphi(u),u)\,.$$

Therefore f is continuously differentiable on $L_2[0,T]$ and we have

$$(\nabla f(u),q) = \int_0^T \left(\left(\frac{\partial F}{\partial x}(s,\varphi(u),u), \varphi'(u)q \right) + \left(\frac{\partial F}{\partial u}(s,\varphi(u),u), q(s) \right) \right) ds \,. \tag{5.6.12}$$

The equation (5.6.8) now follows from (5.6.11) and (5.6.12) and the lemma is proved. □

Let $x_*(t)$ be a solution of the Cauchy problem (5.6.2) corresponding to the control $u_*(t)$. We set

$$c_*(t) = \left(X_*^{-1}(t)\right)^* \int_t^T (X_*(s))^* \frac{\partial F}{\partial x}(s, x_*(s), u_*(s))\, ds \,,$$

where $X_*(t)$ is the fundamental matrix of the system (5.6.9) for

$$x = x_*(t), \quad u = u_*(t) \,.$$

Lemma 5.6.2. *Suppose that the conditions of Lemma* 5.6.1 *hold and that for some positive L and α we have*

$$\left(\left(\frac{\partial g}{\partial u}(t,x,u_1) - \frac{\partial g}{\partial u}(t,x,u_2) \right)^* c_*(t), u_1 - u_2 \right) + \left(\frac{\partial F}{\partial u}(t,x,u_1) - \frac{\partial F}{\partial u}(t,x,u_2), u_1 - u_2 \right) \geqslant \alpha \left| u_1 - u_2 \right|^2 , \tag{5.6.13}$$

$$\left| \frac{\partial g}{\partial u}(t,x,u_1) - \frac{\partial g}{\partial u}(t,x,u_2) \right| \leqslant L \left| u_1 - u_2 \right| \tag{5.6.14}$$

for all $t \in [0,T]$, $x \in \mathbb{R}^N$, $u \in \mathbb{R}^M$. *Then the functional* (5.6.4) *is* $L_2[0,T]$*-regular on some neighborhood in* $L_2[0,T]$ *of the control* u_*.

Proof. The operator

$$C(u) = \left(X^{-1}(t)\right)^* \int_t^T (X(s))^* \frac{\partial F}{\partial x}(s,\varphi(u),u)\, ds$$

is a continuous mapping from $L_2[0,T]$ to $\mathring{W}_1^1[0,T]$ and $C(u_*) = c_*(t)$. Therefore we have

$$\sup_{u \in B(\varepsilon_0, u_*)} \max_{0 \leqslant t \leqslant T} |C(u)(t) - c_*(t)| < \frac{\alpha}{2L} \tag{5.6.15}$$

for some $\varepsilon_0 > 0$. We shall show that f is $L_2[0,T]$-regular on the ball $B(\varepsilon_0, u_*)$ in $L_2[0,T]$.

Let (u_n) be a sequence in $B(\varepsilon_0, u_*)$ that converges weakly to u_0 and let

$$\overline{\lim_{n\to\infty}}\,(\nabla f(u_n), u_n - u_0) \leqslant 0\,. \tag{5.6.16}$$

We set $x_n = \varphi(u_n)$. Then

$$(\nabla f(u_n), u_n - u_0) = H_n + I_n + J_n + K_n\,,$$

where

$$H_n = \int_0^T \left(\frac{\partial F}{\partial u}(s, x_n(s), u_n(s)) - \frac{\partial F}{\partial u}(s, x_n(s), u_0(s)), u_n(s) - u_0(s)\right) ds$$
$$+ \int_0^T \left(\left(\frac{\partial g}{\partial u}(s, x_n(s), u_n(s)) - \frac{\partial g}{\partial u}(s, x_n(s), u_0(s))\right)^* c_*(s), u_n(s) - u_0(s)\right) ds\,,$$

$$I_n = \int_0^T \left(\left(\frac{\partial g}{\partial u}(s, x_n(s), u_n(s)) - \frac{\partial g}{\partial u}(s, x_n(s), u_0(s))\right)^* (C(u_n) - c_*(s))\,, \right.$$
$$\left. u_n(s) - u_0(s)\right) ds\,,$$

$$J_n = \int_0^T \left(\frac{\partial F}{\partial u}(s, x_n(s), u_0(s)), u_n(s) - u_0(s)\right) ds\,,$$

$$K_n = \int_0^T \left(\left(\frac{\partial g}{\partial u}(s, x_n(s), u_0(s))\right)^* C(u_n), u_n(s) - u_0(s)\right) ds\,.$$

We shall bound the terms H_n, I_n, J_n, K_n. By (5.6.13), we have

$$H_n \geqslant \alpha\, \|u_n - u_0\|^2_{L_2[0,T]}\,. \tag{5.6.17}$$

The inequalities (5.6.14) and (5.6.15) imply that

$$I_n \leqslant L \max_{0\leqslant t\leqslant T} |C(u_n)(t) - c_*(t)|\, \|u_n - u_0\|^2_{L_2[0,T]} \leqslant \tfrac{1}{2}\alpha\, \|u_n - u_0\|^2_{L_2[0,T]}\,. \tag{5.6.18}$$

Finally, since the sequences

$$\left(\frac{\partial F}{\partial u}(t, x_n(t), u_0(t))\right) \quad \text{and} \quad \left(\left(\frac{\partial g}{\partial u}(t, x_n(t), u_0(t))\right)^* C(u_n)\right)$$

are compact in $L_2[0,T]$ and the sequence (u_n) converges weakly to u_0, it follows that

$$\lim_{n\to\infty} J_n = 0\,, \tag{5.6.19}$$

$$\lim_{n\to\infty} K_n = 0\,. \tag{5.6.20}$$

From (5.6.17)–(5.6.20) we see that

$$\varlimsup_{n\to\infty} (\nabla f(u_n), u_n - u_0) \geqslant \tfrac{1}{2}\alpha \varlimsup_{n\to\infty} \|u_n - u_0\|^2_{L_2[0,T]} \ .$$

This and (5.6.16) imply that

$$\lim_{n\to\infty} \|u_n - u_0\|_{L_2[0,T]} = 0 \ .$$

Hence the functional (5.6.4) is $L_2[0,T]$-regular on the ball $B(\varepsilon_0, u_*)$ in $L_2[0,T]$, and the result follows. □

Now suppose that the control u enters linearly into the dynamical equation (5.6.2), i.e., the optimal control problem has the form

$$\int_0^T F(s, x(s), u(s))ds \to \min \ , \tag{5.6.21}$$

$$\frac{dx}{dt} = g(t,x) + A(t)u, \quad x(0) = 0 \ , \tag{5.6.22}$$

$$\int_0^T u^2(s)ds \leqslant 1 \ , \tag{5.6.23}$$

where

$$A(t) = \begin{bmatrix} a_{11}(t) & \ldots & a_{1M}(t) \\ \ldots & \ldots & \ldots \\ a_{N1}(t) & \ldots & a_{NM}(t) \end{bmatrix} \tag{5.6.24}$$

is a matrix with continuous coefficients $a_{ij}(t)$ $(i = 1, \ldots, N; j = 1, \ldots, M)$. In this case, condition (5.6.13) assumes the form

$$\left(\frac{\partial F}{\partial u}(t, x, u_1) - \frac{\partial F}{\partial u}(t, x, u_2), u_1 - u_2 \right) \geqslant \alpha |u_1 - u_2|^2 \quad (\alpha > 0) \tag{5.6.25}$$

and condition (5.6.14) is automatically satisfied. Conditions (5.6.6) and (5.6.7) assume the forms

$$\left| \frac{\partial F(t,x,u)}{\partial x} \right| \leqslant c_3(t,x)(1 + |u|^2) \ , \tag{5.6.26}$$

$$\left| \frac{\partial F(t,x,u)}{\partial u} \right| \leqslant c_4(t,x)(1 + |u|) \tag{5.6.27}$$

respectively, where $c_3(t,x)$ and $c_4(t,x)$ are continuous functions. The gradient ∇f of the functional (5.6.4) has the form

$$\nabla f(u) = \frac{\partial F(t,x,u)}{\partial u} + (A(t))^*(X^{-1}(t))^* \int_t^T (X(s))^* \frac{\partial F}{\partial u}(s,x,u)\, ds \ , \tag{5.6.28}$$

where

$$x = \varphi(u) , \tag{5.6.29}$$

and $X(t)$ is the fundamental matrix of the linear system

$$\frac{dh}{dt} = \frac{\partial g}{\partial x}(t, x(t))h . \tag{5.6.30}$$

The proof of the following lemma is similar to that of Lemma 5.6.2.

Lemma 5.6.3. *If the conditions* (5.6.25)–(5.6.27) *are satisfied then the functional* (5.6.4) *corresponding to the problem* (5.6.21)–(5.6.23) *is* $L_2[0,T]$-*regular.*

5.6.3 Deformation Theorems

Consider a one-parameter family of optimal control problems

$$\int_0^T F(s, x(s), u(s); \lambda)ds \to \min , \tag{5.6.31}$$

$$\frac{dx}{dt} = g(t, x; \lambda) + A(t; \lambda)u, \quad x(0) = 0 , \tag{5.6.32}$$

$$\int_0^T u^2(s)ds \leqslant 1, \quad 0 \leqslant \lambda \leqslant 1 . \tag{5.6.33}$$

Here we assume that the scalar function $F(t, x, u; \lambda)$ and the vector function $g(t, x; \lambda)$ and their first derivatives with respect to the variables x_i, u_j $(i = 1, \dots, N; j = 1, \dots, M)$ are continuous on $[0,T] \times \mathbb{R}^N \times \mathbb{R}^N$ and that there are continuous functions $c_1(t, x; \lambda)$, $c_2(t, x; \lambda)$ on $[0,T] \times \mathbb{R}^N \times [0,1]$ and $\alpha > 0$ such that

$$\left| \frac{\partial F(t, x, u; \lambda)}{\partial x} \right| \leqslant c_1(t, x; \lambda)(1 + |u|^2) , \tag{5.6.34}$$

$$\left| \frac{\partial F(t, x, u; \lambda)}{\partial u} \right| \leqslant c_2(t, x; \lambda)(1 + |u|) , \tag{5.6.35}$$

$$\left(\frac{\partial F(t, x, u_1; \lambda)}{\partial u} - \frac{\partial F(t, x, u_2; \lambda)}{\partial u}, u_1 - u_2 \right) \geqslant \alpha |u_1 - u_2|^2 ; \tag{5.6.36}$$

the entries of the $N \times M$ matrix

$$A(t; \lambda) = \begin{bmatrix} a_{11}(t; \lambda) & \dots & a_{1M}(t; \lambda) \\ \dots & \dots & \dots \\ a_{N1}(t; \lambda) & \dots & a_{NM}(t; \lambda) \end{bmatrix}$$

are assumed to be continuous on $[0,T] \times [0,1]$.

The control $u_* = u_*(t)$ is *locally optimal* for the problem (5.6.1)–(5.6.3) if u_* is a local minimizer for the problem (5.6.5).

Theorem 5.6.1. *Suppose that for each* $\lambda \in [0,1]$ *the functional* $f(\cdot;\lambda)$ *corresponding to problem* (5.6.31)–(5.6.33) *has a critical point* $u(\lambda) = u(t;\lambda)$ *in* $\operatorname{int} B$ *which continuously depends on* λ *and is uniformly isolated in* $L_2[0,T]$. *Suppose also that for* $\lambda = 0$ *the control*

$$u_0(t) = u(t;0)$$

is locally optimal in the problem

$$\int_0^T F(s, x(s), u(s); 0)ds \to \min , \tag{5.6.37}$$

$$\frac{dx}{dt} = g(t, x; 0) + A(t; 0)u, \quad x(0) = 0 , \tag{5.6.38}$$

$$\int_0^T u^2(s)ds \leqslant 1 . \tag{5.6.39}$$

Then the control

$$u_1(t) = u(t;1)$$

is locally optimal in the problem

$$\int_0^T F(s, x(s), u(s); 1)ds \to \min, \tag{5.6.40}$$

$$\frac{dx}{dt} = g(t, x; 1) + A(t; 1)u, \quad x(0) = 0, \tag{5.6.41}$$

$$\int_0^T u^2(s)ds \leqslant 1 . \tag{5.6.42}$$

The proof of this theorem follows immediately from that of Theorem 3.1.1.

We have $u_* \in \operatorname{int} B$ for optimal control in the problem (5.6.1)–(5.6.3) only in exceptional cases. Generally u_* lies on the boundary of the domain of constraints, i.e., $\|u_*\|_{L_2[0,T]} = 1$. In this case, the necessary condition for optimality takes the form

$$\nabla f(u_*) + \mu u_* = 0 , \tag{5.6.43}$$

where $\mu > 0$. When applied to the problem (5.6.21)–(5.6.23), this condition can be written in the expanded form

$$\frac{\partial F(t, x_*(t), u_*(t))}{\partial u}$$
$$+ (A(t))^*(X^{-1}(t))^* \int_t^T (X(s))^* \frac{\partial F(s, x_*(s), u_*(s))}{\partial u} ds + \mu u_*(t) = 0 ,$$

where $x_*(t)$ is a solution of the Cauchy problem

$$\frac{dx}{dt} = g(t,x) + A(t)u_*(t), \quad x(0) = 0 \,,$$

and $X(t)$ is the fundamental matrix of the linear system

$$\frac{dh}{dt} = \frac{\partial g}{\partial x}(t, x_*(t))h \,.$$

Theorem 3.5.1 implies

Theorem 5.6.2. *Suppose that for each value of the parameter* $\lambda \in [0,1]$ *the system*

$$\frac{\partial F(t,x(t),u(t);\lambda)}{\partial u} + (A(t;\lambda)^*(X^{-1}(t;\lambda))^* \int_t^T (X^{-1}(s;\lambda))^* \frac{\partial F(s,x(s),u(s);\lambda)}{\partial u} ds + \mu u(t) = 0 \,,$$

$$\frac{dX(t;\lambda)}{dt} = \frac{\partial g(t,x(t);\lambda)}{\partial x} X(t;\lambda) \,,$$

$$X(0;\lambda) = I \,,$$

$$\frac{dx(t)}{dt} = g(t,x(t);\lambda) + A(t;\lambda)u(t) \,,$$

$$x(0) = 0 \,,$$

$$\int_0^T u^2(s)ds = 1$$

has a solution

$$\{x(\lambda);u(\lambda)\} = \{x(t;\lambda), u(t;\lambda)\} \quad (0 \leqslant \lambda \leqslant 1)$$

which is uniformly isolated in $L_2[0,T]$ *with respect to* $\lambda \in [0,1]$. *If*

$$u_0(t) = u(t;0)$$

is locally optimal in problem (5.6.32)–(5.6.39), *then the control*

$$u_1(t) = u(t;1)$$

is locally optimal in problem (5.6.40)–(5.6.42).

5.7 Bifurcation of Critical Points in Variational Problems

5.7.1 Statement of the Problem and a Necessary Condition for Bifurcation

Mathematical models of many nonlinear problems lead to equations with nonunique solutions. These include the equilibrium problem for a rotating liquid, the problem of loss of stability of elastic systems, problems of origination of forms of oscillatory states, problems from wave theory, convection, etc. In the most frequently encountered situations, the mathematical description of these phenomena reduces to the study of nonzero solutions of an operator equation of the form

$$A(x;\lambda) = 0\,, \tag{5.7.1}$$

where $A : E_0 \times \mathbb{R} \to E_1$ is an operator, $x \in E_0$ is an element of a Banach space, and λ is a scalar parameter. We assume that

$$A(0;\lambda) = 0\,, \tag{5.7.2}$$

i.e., 0 is a solution of Eq. (5.7.1), and we are interested in whether there are also nonzero solutions.

The notion of a bifurcation point is one of the main ideas related to this problem. A number λ_* is a *bifurcation point* of the zero solution of Eq. (5.7.1) (or simply a bifurcation point for Eq. (5.7.1)) if for all $\varepsilon > 0$ there is some $\lambda \in (\lambda_* - \varepsilon, \lambda_* + \varepsilon)$ such that Eq. (5.7.1) has a nonzero solution in the ball $B(\varepsilon)$ in E_0.

We are interested in the existence of bifurcation points of variational problems, i.e., problems of the form

$$\nabla f(x;\lambda) = 0\,, \tag{5.7.3}$$

where $f(\cdot;\lambda) : E \to \mathbb{R}$ is a differentiable functional which has a zero critical point for every λ and ∇ is the gradient operator with respect to the variable x.

If f is twice Fréchet differentiable, then λ_* can be a bifurcation point for Eq. (5.7.3) only if the operator

$$\nabla^2 f(0;\lambda_*) : E \to E^*$$

does not have a continuous inverse. Indeed, the implicit function theorem (see Sect. 1.6.2) implies

Theorem 5.7.1. *Suppose that for every pair (x,λ) in some neighborhood U of the pair $(0,\lambda_*)$ the operator $\nabla f(x;\lambda)$ has a Fréchet derivative $\nabla^2 f(x;\lambda) : E \to E^*$ which is continuous in the operator norm. Let λ_* be a bifurcation point for Eq. (5.7.3). Then the operator $\nabla^2 f(0;\lambda_*) : E \to E^*$ does not have a continuous inverse.*

Proof. If the conclusion is false then, by Theorem 1.6.2, Eq. (5.7.3) has only the solution $x(\lambda) = 0$ in balls $B(r)$ in E of small radius r for all λ sufficiently close to λ_*, as required. □

In particular, if $E = H$ is a Hilbert space, then the inclusion

$$0 \in \sigma(\nabla^2 f(0; \lambda_*)) \tag{5.7.4}$$

is a necessary condition for λ_* to be a bifurcation point of the zero solution of Eq. (5.7.3).

5.7.2 Deformation Principle for Minimizers in the Study of Bifurcation Points

Let H be a real Hilbert space. Consider a one-parameter family of H-regular (see Sect. 3.1.1) functionals $f(\cdot;\lambda) : H \to \mathbb{R}$ $(0 \leqslant \lambda \leqslant 1)$. Suppose that the functional f and its gradient ∇f are continuous on $H \times \mathbb{R}$ and that $\nabla f(x;\lambda)$ is continuous in λ uniformly with respect to x from some ball $B(\rho)$ in H. Theorem 3.1.1 implies

Theorem 5.7.2. *If* 0 *is a critical point of* $f(\cdot;\lambda)$ *for each* $\lambda \in [0,1]$ *and is a local minimizer for* $f_0 = f(\cdot;0)$ *but not a minimizer for* $f_1 = f(\cdot;1)$, *then the zero solution of Eq.* (5.7.3) *has at least one bifurcation point.*

A similar result holds for bifurcation points of critical points of E-regular functionals.

Let E be a separable reflexive real Banach space and $f(\cdot;\lambda) : E \to \mathbb{R}$ $(0 \leqslant \lambda \leqslant 1)$ a one-parameter family of E-regular (see Sect. 3.2.1) functionals. We assume that the functional f and its gradient are continuous on $E \times [0,1]$ and that the gradient $\nabla f(\cdot;\lambda) : E \to E^*$ is continuous in λ uniformly with respect to x from some ball $B(\rho)$ in E. Theorem 3.2.1 implies

Theorem 5.7.3. *If* 0 *is a critical point of* $f(\cdot;\lambda)$ *for each* $\lambda \in [0,1]$ *and is a local minimizer for* $f_0 = f(\cdot;0)$ *but not a minimizer of* $f_1 = f(\cdot;1)$, *then the zero solution of Eq.* (5.7.3) *has at least one bifurcation point in* $[0,1]$.

We shall illustrate Theorems 5.7.2 and 5.7.3 by discussing bifurcations of critical points in multidimensional variational problems.

Let Ω be a bounded domain in $\mathbb{R}^N$ with smooth boundary. On the Sobolev space $\mathring{W}_2^m(\Omega)$ we consider a functional

$$f(u;\lambda) = \int_{\Omega} F(x, u(x), \ldots, D^m u(x); \lambda)\, dx \tag{5.7.5}$$

depending on the parameter $\lambda \in [0,1]$. Here, as usual,

$$D^k u(x) = \left\{ \mathcal{D}^\alpha u(x) = \frac{\partial^k u(x)}{\partial x_1^{\alpha_1} \ldots \partial x_N^{\alpha_N}} : \alpha = (\alpha_1, \ldots, \alpha_N),\ |\alpha| = k \right\}.$$

We shall assume that the integrand $F(x,\xi;\lambda)$ and its first and second derivatives with respect to all components of the vector ξ are continuous on $\overline{\Omega}\times\mathbb{R}^M\times[0,1]$.

The critical points of $f(\cdot;\lambda)$ coincide with the generalized solutions from $\mathring{W}_2^m(\Omega)$ of the Dirichlet problem for the Euler equation of the functional (5.7.5):

$$\sum_{\alpha\leqslant m}(-1)^{|\alpha|}\mathcal{D}^\alpha\frac{\partial F(x,u(x),\ldots,D^m u(x);\lambda)}{\partial\xi_\alpha}=0\,, \tag{5.7.6}$$

$$\mathcal{D}^\alpha u(x)\big|_{\partial\Omega}=0\quad(|\alpha|\leqslant m-1)\,. \tag{5.7.7}$$

We assume that for every $\lambda\in[0,1]$ the boundary-value problem (5.7.6), (5.7.7) has the zero solution, i.e., the function $u(x)=0$ is a critical point of each functional $f(\cdot;\lambda)$ $(0\leqslant\lambda\leqslant 1)$.

Theorem 5.7.4. *Suppose that we have*

$$\left|\frac{\partial^2F(x,\xi;\lambda)}{\partial\xi_\alpha\partial\xi_\beta}\right|\leqslant C\left(1+\sum_{m-N/2\leqslant|\gamma|\leqslant m}|\xi_\gamma|^{p_\gamma}\right)^{p_{\alpha\beta}},$$

where

p_γ *is an arbitrary positive number if* $\gamma=m-N/2$,
$p_\gamma=2N/(N-2(m-|\gamma|))$ *if* $m-N/2<|\gamma|<m$,
$p_{\alpha\beta}=1-p_\alpha^{-1}-p_\beta^{-1}$ *if* $|\alpha|=|\beta|=m$,
$p_{\alpha\beta}=1-p_\alpha^{-1}$ *if* $m-N/2\leqslant|\alpha|\leqslant m$,
$p_{\alpha\beta}=1$ *if* $|\alpha|,|\beta|<m-N/2$,
$0<p_{\alpha\beta}<1-p_\alpha^{-1}-p_\beta^{-1}$ *if* $|\alpha|,|\beta|>m-N/2$ *and* $|\alpha|+|\beta|<2m$.

Suppose in addition that

$$\sum_{|\alpha|=|\beta|=m}\frac{\partial^2F(x,\xi;\lambda)}{\partial\xi_\alpha\partial\xi_\beta}\eta_\alpha\eta_\beta\geqslant c\sum_{|\alpha|=m}\eta_\alpha^2$$

$$(c>0,\ x\in\Omega,\ \xi\in\mathbb{R}^M,\ 0\leqslant\lambda\leqslant 1,\ \eta=\{\eta_\alpha:|\alpha|=m\})\,.$$

Suppose finally that the zero critical point is a local minimizer on $\mathring{W}_2^m(\Omega)$ *for the functional* $f_0=f(\cdot;0)$ *and is not a local minimizer of* $f_1=f(\cdot;1)$.

Then the zero critical point of the functional (5.7.5) *has at least one bifurcation point in* $[0,1]$.

This result follows immediately from Theorem 5.4.2.

Next we consider the functional (5.7.5) on the Banach space $\mathring{W}_p^m(\Omega)$. We assume that the Dirichlet problem (5.7.6), (5.7.7) for $\mathring{W}_p^m(\Omega)$ has 0 as a solution for every $\lambda\in[0,1]$. Theorem 5.4.3 implies

Theorem 5.7.5. *Suppose that the inequalities*

$$\left|\frac{\partial^2 F(x,\xi;\lambda)}{\partial\xi_\alpha\partial\xi_\beta}\right| \leqslant g_1\left(\sum_{|\gamma|<m-N/p}|\xi_\gamma|\right)\left(1+\sum_{m-N/p\leqslant\gamma\leqslant m}|\xi_\gamma|^{p_\gamma}\right)^{p_{\alpha\beta}}$$

hold, where

g_1 *is a continuous positive nondecreasing function,*
p_γ *is an arbitrary positive number if* $|\gamma| = m - N/p$,

$$p_\gamma = \frac{Np}{N-(m-|\gamma|)p} \quad \text{if } m - N/p < |\gamma| \leqslant m,$$

$p_{\alpha\beta} = p_{\beta\alpha}$ *and are defined by the conditions*

$$p_{\alpha\beta} = 1 - \frac{1}{p_\alpha} - \frac{1}{p_\beta} \quad \text{if } |\alpha| = |\beta| = m\ ,$$

$$p_{\alpha\beta} = 1 - \frac{1}{p_\alpha} \quad \text{if } m - \frac{N}{p} \leqslant \alpha \leqslant m \text{ and } |\beta| < m - \frac{N}{p}\ ,$$

$$p_{\alpha\beta} = 1 \quad \text{if } |\alpha|, |\beta| < m - \frac{N}{p}\ ,$$

$$0 < p_{\alpha\beta} < 1 - \frac{1}{p_\alpha} - \frac{1}{p_\beta} \quad \text{if } |\alpha|, |\beta| \geqslant m - N/p \text{ and } |\alpha| + \beta < 2m\ .$$

Suppose in addition that

$$\sum_{|\alpha|=|\beta|=m}\frac{\partial^2 F(x,\xi;\lambda)}{\partial\xi_\alpha\partial\xi_\beta}\eta_\alpha\eta_\beta \geqslant g_2\left(\sum_{|\gamma|<m-N/p}|\xi_\gamma|\right)\sum_{|\gamma|=m}|\xi_\gamma|^{p-2}\sum_{|\alpha|=m}\eta_\alpha^2\ ,$$

where $x \in \overline{\Omega}$, $\xi \in \mathbb{R}^M$, *and the function* g_2 *is continuous, positive, and nonincreasing. Suppose finally that* 0 *is a local minimizer on* $\mathring{W}_p^m(\Omega)$ *of the functional* $f_0 = f(\cdot;0)$ *but not a minimizer of* $f_1 = f(\cdot;1)$.

Then the zero critical point of the functional (5.7.5) *has at least one bifurcation point in* $[0,1]$.

In Theorems 5.7.4 and 5.7.5 there are strong constraints on the growth of the integrands of the functionals under study. These constraints can be removed if the functionals depend only on the first derivatives.

Let Ω be a bounded connected domain in $\mathbb{R}^N$ with smooth boundary. Consider the family of integral functionals

$$f(u;\lambda) = \int_\Omega F(x,u(x),\nabla u(x);\lambda)dx \quad (0 \leqslant \lambda \leqslant 1) \tag{5.7.8}$$

on the space $\mathring{C}^1(\overline{\Omega})$. The integrands

$$f(x,\xi;\lambda) \quad (x \in \overline{\Omega},\, \xi = (u, p_1, \dots, p_N) \in \mathbb{R}^{N+1},\, 0 \leqslant \lambda \leqslant 1)$$

of the functionals are assumed to be three-times continuously differentiable with respect to the collection of variables. The critical points of (5.7.8) are classical solutions of the Dirichlet problem for the Euler equation of this functional:

$$\sum_{i=1}^{N} \frac{d}{dx_i} \frac{\partial F(x,u(x),\nabla u(x);\lambda)}{\partial p_i} - \frac{\partial F(x,u(x),\nabla u(x);\lambda)}{\partial u} = 0\,, \tag{5.7.9}$$

$$u(x)\,|_{\partial\Omega} = 0\,. \tag{5.7.10}$$

We assume that 0 is a critical point of (5.7.8) for every $\lambda \in [0,1]$, i.e., 0 is a solution of the Dirichlet problem (5.7.9), (5.7.10) for every $\lambda \in [0,1]$. Moreover, we assume that the Legendre condition

$$\sum_{i,j=1}^{N} \frac{\partial^2 F(x,u,p;\lambda)}{\partial p_i \partial p_j} \eta_i \eta_j \geqslant c \sum_{i=1}^{N} \eta_i^2 \tag{5.7.11}$$

holds for all $x \in \overline{\Omega}$, $\eta = (\eta_1, \dots, \eta_N) \in \mathbb{R}^N$, $u \in \mathbb{R}$, $p \in \mathbb{R}^N$, $c > 0$.

Theorem 5.4.4 implies

Theorem 5.7.6. *Suppose that the zero critical point is a weak minimizer for the functional $f_0 = f(\cdot;0)$ but is not a weak minimizer for $f_1 = f(\cdot;1)$. Then the zero critical point of the functional (5.7.8) has at least one bifurcation point in $[0,1]$.*

5.7.3 Conley Index and Bifurcation Points

We now consider bifurcation points of the zero critical points of the one-parameter family of functionals (5.7.8).

We assume that Ω is a bounded domain in $\mathbb{R}^N$ with smooth boundary, and that the integrand $F : \Omega \times \mathbb{R}^{N+1} \times [0,1] \to \mathbb{R}$ is a C^4-function that satisfies the strengthened Legendre condition (5.7.11).

Let $W_p^2(\Omega)$ be the Sobolev space of functions $u : \Omega \to \mathbb{R}$ which have generalized first and second derivatives that are summable with degree p, and let $\mathring{W}_2^1(\Omega)$ be the Sobolev space of functions $u : \Omega \to \mathbb{R}$ which vanish on $\partial\Omega$ and have square-summable generalized first derivatives. Let

$$E = W_p^2(\Omega) \cap \mathring{W}_2^1(\Omega)\,, \tag{5.7.12}$$

$$H = \mathring{W}_2^1(\Omega)\,. \tag{5.7.13}$$

We assume that $p > \max(2, N)$ and we take the norm in E induced by the norm of $W_p^2(\Omega)$.

Lemma 5.7.1. *The functional $f(\cdot;\lambda)$ is (E,H)-regular for every $\lambda \in [0,1]$.*

Proof. Since $p > \max(2,N)$, it follows (see, e.g., [153]) that E can be continuously embedded into $C^1(\overline{\Omega})$. Therefore the functional $f(\cdot,\lambda)$ is twice continuously differentiable on E. In this case,

$$\langle \nabla f(u), h\rangle = \sum_{|\alpha|\leqslant 1} \int_\Omega \frac{\partial F}{\partial \xi_\alpha}(x,u(x),\nabla u(x);\lambda)\mathcal{D}^\alpha h(x)\,dx\ , \tag{5.7.14}$$

$$\langle \nabla^2 f(u)h, g\rangle = \sum_{|\alpha|,|\beta|\leqslant 1} \int_\Omega \frac{\partial^2 F}{\partial \xi_\alpha \partial \xi_\beta}(x,u(x),\nabla u(x);\lambda)\mathcal{D}^\alpha h(x)\mathcal{D}^\beta g(x)\,dx\ . \tag{5.7.15}$$

It follows from (5.7.14) and (5.7.15) that the inequalities (4.2.1), (4.2.2) and the equation (4.2.3) hold on each ball $B(r)$ in E, and the lemma is proved. □

Let $\Delta : E \to W_p^{-1}(\Omega)$ be the Laplace operator: it is an isomorphism from E to $W_p^{-1}(\Omega)$, with inverse $\Delta^{-1} : W_p^{-1}(\Omega) \to E$. We have

$$\nabla_H f(u;\lambda) = \Delta^{-1} \sum_{|\alpha|\leqslant 1} (-1)^{1+|\alpha|}\mathcal{D}^\alpha \frac{\partial F}{\partial \xi_\alpha}(x,u(x),\nabla u(x);\lambda)\ , \tag{5.7.16}$$

$$\nabla_H^2 f(u;\lambda)h = \Delta^{-1} \sum_{|\alpha|,|\beta|\leqslant 1} (-1)^{1+|\alpha|}\mathcal{D}^\alpha \left(\frac{\partial^2 F}{\partial \xi_\alpha \partial \xi_\beta}(x,u(x),\nabla u(x);\lambda)\mathcal{D}^\beta h(x)\right)\ . \tag{5.7.17}$$

We set

$$A(u;\lambda)h = \Delta^{-1} \sum_{|\alpha|,|\beta|=1} (-1)^{1+|\alpha|}\mathcal{D}^\alpha \left(\frac{\partial^2 F}{\partial \xi_\alpha \partial \xi_\beta}(x,u(x),\nabla u(x);\lambda)\mathcal{D}^\beta h(x)\right)\ , \tag{5.7.18}$$

$$B(u;\lambda)h = \Delta^{-1} \sum_{|\alpha|+|\beta|<2} (-1)^{1+|\alpha|}\mathcal{D}^\alpha \left(\frac{\partial^2 F}{\partial \xi_\alpha \partial \xi_\beta}(x,u(x),\nabla u(x);\lambda)\mathcal{D}^\beta h(x)\right)\ . \tag{5.7.19}$$

The linear operators

$$A(u;\lambda) : \mathring{W}_2^1(\Omega) \to \mathring{W}_2^1(\Omega) \quad \text{and} \quad B(u;\lambda) : \mathring{W}_2^1(\Omega) \to \mathring{W}_2^1(\Omega)$$

are bounded. Using the embedding theorems (see Sects. 1.1.6 and 1.4.9), we can easily show that $B(u;\lambda)$ is completely continuous on $\mathring{W}_2^1(\Omega)$. We claim that $A(u;\lambda) : \mathring{W}_2^1(\Omega) \to \mathring{W}_2^1(\Omega)$ has a continuous inverse. Indeed, $A(u;\lambda)$ is self-adjoint on $\mathring{W}_2^1(\Omega)$ and, by condition (5.7.11) we have

$$(A(u;\lambda)h,h)_{\mathring{W}_2^1(\Omega)} \geqslant c(h,h)_{\mathring{W}_2^1(\Omega)}\ . \tag{5.7.20}$$

Therefore

$$0 \notin \sigma(A(u;\lambda)) . \tag{5.7.21}$$

We shall need the following lemma.

Lemma 5.7.2 (Fredholm). *Let $C : H \to H$ be a self-adjoint operator that can be represented in the form*

$$C = A + B ,$$

where A is an invertible operator and B is a completely continuous operator. Then 0 is an isolated point of the spectrum $\sigma(C)$ of C and the kernel of C is finite-dimensional.

Proof. Suppose that 0 is not an isolated point of $\sigma(C)$. Then there exists a sequence (λ_n) of nonzero points of points of $\sigma(C)$ converging to 0. We can assume that $A - \lambda_n I$ is invertible for each n.

We fix an integer $n \geqslant 1$. Since $\lambda_n \in \sigma(C)$ there is a sequence $(h_n^k)_{k\in\mathbb{N}}$ of elements of norm 1 for which

$$\|Ch_n^k - \lambda_n h_n^k\| < \frac{1}{k} \quad (k = 1, 2, \ldots) . \tag{5.7.22}$$

We set

$$Ch_n^k - \lambda_n h_n^k = g_n^k . \tag{5.7.23}$$

Then (5.7.23) yields that

$$h_n^k = -(A - \lambda_n I)^{-1} B h_n^k + (A - \lambda_n I)^{-1} g_n^k . \tag{5.7.24}$$

Since the operator $(A - \lambda_n I)^{-1} B$ is completely continuous, the sequence $((A - \lambda_n I)^{-1} B h_n^k)_{k\in\mathbb{N}}$ is compact and, since by (5.7.22) the sequence $(g_n^k)_{k\in\mathbb{N}}$ converges to 0 in norm, the sequence $(\,(A - \lambda_n I)^{-1} g_n^k)_{k\in\mathbb{N}}$ also converges to 0. It now follows from (5.7.24) that the sequence $(h_n^k)_{k\in\mathbb{N}}$ is compact; therefore we can assume that it converges to an element h_n. Taking the limit as $k \to \infty$ in (5.7.23), we obtain

$$Ch_n = \lambda_n h_n .$$

In this way we may construct elements $h_1, h_2, \ldots$ that form an orthonormal sequence. On the other hand, the equations

$$h_n = -(A - \lambda_n I)^{-1} B h_n \quad (n = 1, 2, \ldots) ,$$

which result from passage to the limit as $k \to \infty$ in (5.7.24) imply that the sequence (h_n) is compact. This is a contradiction and we conclude that 0 is an isolated point of $\sigma(C)$.

To complete the proof, it remains to show that the kernel of C is finite-dimensional.

Each isolated point of the spectrum of a self-adjoint operator is an eigenvalue. Suppose that 0 is an eigenvalue of infinite multiplicity. Then there exists an orthonormal (and, hence, noncompact) sequence (y_n) such that

$$Cy_n = 0 \quad (n = 1, 2, \ldots) . \tag{5.7.25}$$

On the other hand, (y_n) is compact since (5.7.25) implies that

$$y_n = -A^{-1}By_n .$$

The lemma follows from this contradiction. □

Remark. We note that the assumption that C is self-adjoint is essential for the validity of the lemma. Let A be the invertible operator A on the l_2-space of sequences

$$x = (\ldots, \xi_{-1}, \xi_0, \xi_1, \ldots)$$

defined by

$$A(\ldots, \xi_{-1}, \xi_0, \xi_1, \ldots) = (\ldots, \eta_{1-}, \eta_0, \eta_1, \ldots) ,$$

where

$$\eta_i = \xi_{i+1} \quad (i = 0, \pm 1, \pm 2, \ldots) .$$

Let B be the completely continuous operator defined by

$$B(\ldots, \xi_{-1}, \xi_0, \xi_1, \ldots) = (\ldots, \zeta_{-1}, \zeta_0, \zeta_1, \ldots) ,$$

where $\zeta_0 = -\xi_1$ and $\zeta_i = 0$ for $i \neq 0$. It is easy to show that the spectrum of the operator $C = A + B$ contains the interval $[-1, 1]$.

Fix $u \in \overset{\circ}{W}{}_2^1(\Omega)$ and $\lambda \in [0, 1]$ and consider the spectrum of the operator

$$\nabla_H^2 f(u; \lambda) : \overset{\circ}{W}{}_2^1(\Omega) \to \overset{\circ}{W}{}_2^1(\Omega) .$$

Lemma 5.7.3. *The negative part $\sigma(\nabla_H^2 f(u; \lambda)) \cap (-\infty, 0]$ of the spectrum of $\nabla_H^2 f(u; \lambda)$ either is empty or consists of a finite number of eigenvalues of finite multiplicity.*

Proof. Let

$$\mu \in \sigma(\nabla_H^2 f(u; \lambda)) \cap (-\infty, 0] .$$

It suffices to show that 0 is an isolated eigenvalue of finite multiplicity of the operator

$$\nabla_H^2 f(u; \lambda) - \mu I = A(u; \lambda) - \mu I + B(u; \lambda) ,$$

where $A(u; \lambda)$ and $B(u; \lambda)$ are defined by (5.7.18) and (5.7.19) respectively, We set

$$\widetilde{A}(u; \lambda) = A(u; \lambda) - \mu I .$$

The operator $\widetilde{A}(u;\lambda)$ is self-adjoint on $\mathring{W}_2^1(\Omega)$ and

$$(\widetilde{A}(u;\lambda)h,h)_{\mathring{W}_2^1(\Omega)} \geqslant (c-\mu)(h,h)_{\mathring{W}_2^1(\Omega)}\,.$$

Therefore $\widetilde{A}(u;\lambda)$ is invertible on $\mathring{W}_2^1(\Omega)$. We conclude by Lemma 5.7.2 that 0 is an isolated eigenvalue of $\nabla_H^2 f(u;\lambda)-\mu I$ of finite multiplicity, and the lemma follows. □

This lemma shows that, for every $\lambda\in[0,1]$, there exists an admissible subspace L_λ of the functional $f(\cdot;\lambda)$ at the point $u=0$. We shall show that the Conley index of the zero critical point of $f(\cdot;\lambda)$ is defined for every $\lambda\in[0,1]$. To do this, we must verify that the requirements (1) and (2) from Theorem 4.2.2 hold.

Let $u\in W_p^2(\Omega)$. Then we have

$$\frac{\partial F}{\partial\xi_\alpha}(x,u(x),\nabla u(x);\lambda)\in L_p(\Omega)$$

for $|\alpha|\leqslant 1$, and consequently

$$\nabla_H f(u;\lambda)\in W_p^2(\Omega)$$

by (5.7.16). Therefore the image of the operator $\nabla_H f(\cdot;\lambda)$ lies in $W_p^2(\Omega)$. It follows from (5.6.16) that $\nabla_H f(\cdot;\lambda): W_p^2(\Omega)\to W_p^2(\Omega)$ is continuously differentiable and that its Fréchet derivative coincides with $\nabla_H^2 f(\cdot;\lambda)$. Therefore condition (1) of Theorem 4.2.2 holds.

Suppose that L is an admissible subspace for the functional $f(\cdot;\lambda)$ at the point u_*, that P is the orthogonal projection from H to L, and that $w\in E\cap L^\perp$. We must show that any solution u of the equation

$$(I-P)\nabla_H^2 f(0;\lambda)u=w \tag{5.7.26}$$

in $\mathring{W}_2^1(\Omega)$ lies in E. By (5.7.17), we can rewrite Eq. (5.7.26) in the form

$$\sum_{|\alpha|,|\beta|\leqslant 1}(-1)^{1+|\alpha|}\mathcal{D}^\alpha(a_{\alpha\beta}(x)\mathcal{D}^\beta u(x))=v(x)+\Delta w(x)\,, \tag{5.7.27}$$

where

$$v(x)=P\nabla_H^2 f(0;\lambda)u(x)\in W_p^2(\Omega)\,,$$

$$a_{\alpha\beta}(x)=\frac{\partial^2 F}{\partial\xi_\alpha\partial\xi_\beta}(x,0,0;\lambda)\in C^1(\overline{\Omega})\,.$$

Since $w\in W_p^2(\Omega)$, it follows that $\Delta w\in L_p(\Omega)$. Therefore, by (5.7.11), equation (5.7.27) is a linear elliptic second-order equation with continuous coefficients and with right-hand side from $L_p(\Omega)$; its solution lies in $W_p^2(\Omega)$ (see, e.g., [153]).

Therefore condition (2) from Theorem 4.2.2 also holds. Consequently, the Conley index $h(0;f(\cdot;\lambda))$ of the zero critical point of $f(\cdot;\lambda)$ is defined for every $\lambda\in[0,1]$. By Theorem 4.2.2 this index is a homotopy invariant. Thus we have proved the following assertion.

Theorem 5.7.7. *Suppose that for every $\lambda \in [0,1]$ the Dirichlet problem*

$$\sum_{i=1}^{N} \frac{d}{dx_i}\frac{\partial F}{\partial p_i}(x,u(x),\nabla u(x);\lambda) - \frac{\partial F}{\partial u}(x,u(x),\nabla u(x);\lambda) = 0\ , \qquad (5.7.28)$$

$$u(x)\,|_{\partial\Omega} = 0 \qquad (5.7.29)$$

for the Euler equation of the integral functional

$$f(u;\lambda) = \int_{\Omega} F(x,u(x),\nabla u(x);\lambda)\,dx \qquad (5.7.30)$$

has the solution $u(x) = 0$, and suppose that this solution is isolated in the set $E = W_p^2(\Omega) \cap \overset{\circ}{W}{}_2^1(\Omega)$ uniformly with respect to $\lambda \in [0,1]$. Suppose also that the integrand $F(x,u,p;\lambda)$ ($x \in \overline{\Omega}$, $u \in \mathbb{R}$, $p \in \mathbb{R}^N$, $\lambda \in [0,1]$) of this functional satisfies the strengthened Legendre condition

$$\sum_{i,j=1}^{N} \frac{\partial^2 F}{\partial p_i \partial p_j}(x,u,p;\lambda)\eta_i\eta_j \geqslant c\sum_{i=1}^{N}\eta_i^2 \quad (c>0)\ . \qquad (5.7.31)$$

Then the Conley index $h(0;f(\cdot;\lambda))$ of the zero critical point of the functional (5.7.30) is defined for every $\lambda \in [0,1]$ and is independent of λ.

This theorem implies

Theorem 5.7.8. *Suppose that the zero critical point of the functionals*

$$f_0(u) = \int_{\Omega} F(x,u(x),\nabla u(x);0)\,dx \qquad (5.7.32)$$

and

$$f_1(u) = \int_{\Omega} F(x,u(x),\nabla u(x);1)\,dx \qquad (5.7.33)$$

is isolated in $E = W_p^2(\Omega) \cap \overset{\circ}{W}{}_2^1(\Omega)$ and the strengthened Legendre condition (5.7.31) holds. If

$$h(0;f_0) \neq h(0;f_1)\ , \qquad (5.7.34)$$

then the zero critical point of the functional (5.7.30) has at least one bifurcation point in $[0,1]$.

Condition (5.7.34) is particularly effective when the zero critical point of the functionals f_0 and f_1 is nondegenerate. In this case, the Conley index of this critical point can be associated with the sum of the multiplicities $\beta_-(f_0)$ and $\beta_-(f_1)$ of the negative eigenvalues of the self-adjoint operators $\nabla_H^2 f_0(0)$ and $\nabla_H^2 f_1(0)$. In this case, the following theorem is valid.

Theorem 5.7.9. *Suppose that*

$$0 \notin \sigma(\nabla_H^2 f_0(0)) \cup \sigma(\nabla_H^2 f_1(0)) \quad \text{and} \quad \beta_-(f_0) \neq \beta_-(f_1)\,.$$

Suppose also that the strengthened Legendre condition (5.7.31) *holds. Then the zero critical point of* (5.7.30) *has at least one bifurcation point in* $[0,1]$.

Remark. In all assertions of this section on bifurcation points of the zero critical point, the global strengthened Legendre condition (5.7.31) can be relaxed. It suffices to require that it holds just at the zero critical point, i.e., that the following inequality holds:

$$\sum_{i,j=1}^{N} \frac{\partial^2}{\partial p_i \partial p_j} F(x,0,0;\lambda)\eta_i\eta_j \geqslant c \sum_{i=1}^{N} \eta_i^2 \quad (c>0)\,.$$

5.7.4 Bifurcation Points of Critical Points in One-Dimensional Variational Problems

Consider a one-parameter family of functionals

$$f(x;\lambda) = \int_0^T F(t,x(t),x'(t);\lambda)dt \quad (x(0)=x(T)=0,\ 0 \leqslant \lambda \leqslant 1) \tag{5.7.35}$$

from the classical calculus of variations. The reader will easily be able to generalize the above results on bifurcations to this case. We shall only comment here on an analog of Theorem 5.7.9 in this simple situation.

Suppose that the Lagrangian $F(t,x,p;\lambda)$ is four-times continuously differentiable in $[0,T] \times \mathbb{R} \times \mathbb{R} \times [0,1]$, that the two-point boundary-value problem

$$\frac{d}{dt}\frac{\partial F}{\partial p}(t,x(t),x(t);\lambda) - \frac{\partial F}{\partial x}(t,x(t),x(t);\lambda) = 0\,, \tag{5.7.36}$$

$$x(0) = x(T) = 0 \tag{5.7.37}$$

has the zero solution for every $\lambda \in [0,1]$ and that the strengthened Legendre condition

$$F''_{pp}(t,0,0;\lambda) \geqslant c > 0 \tag{5.7.38}$$

holds at the zero critical point. The functional $f(\cdot;\lambda)$ is (E,H)-regular for $E = C^1[0,T],\ H = \overset{\circ}{W}{}_2^1[0,T]$. In this case, the operator $\nabla_H^2 f(0;\lambda)$ has the form

$$\begin{aligned}\nabla^2_H f(0;\lambda)h = &\int\limits_0^t (a(s;\lambda)h(s) + b(s;\lambda)h'(s))\,ds \\ &-\int\limits_0^t\int\limits_0^s (c(\tau;\lambda)h(\tau) + a(\tau;\lambda)h'(\tau))\,d\tau ds \\ &-\frac{t}{T}\int\limits_0^T (a(s;\lambda)h(s) + b(s;\lambda)h'(s))ds \\ &+\frac{t}{T}\int\limits_0^T\int\limits_0^s (c(\tau;\lambda)h(\tau) + a(\tau;\lambda)h'(\tau))\,d\tau ds,\end{aligned} \tag{5.7.39}$$

where

$$a(t;\lambda) = \frac{\partial^2 F}{\partial p \partial x}(t,0,0;\lambda)\,, \quad b(t;\lambda) = \frac{\partial^2 F}{\partial p^2}(t,0,0;\lambda)\,,$$

$$c(t;\lambda) = \frac{\partial^2 F}{\partial x^2}(t,0,0;\lambda)\,.$$

The negative part $\sigma(\nabla^2_H f(0;\lambda)) \cap (-\infty, 0]$ of the spectrum of the self-adjoint operator $\nabla^2_H f(0;\lambda)$ on $\overset{\circ}{W}{}^1_2[0,T]$ consists of simple eigenvalues. Their number is equal to the number of nontrivial zeros in $[0,T]$ of the solution $h(t;\lambda)$ to the linear Cauchy problem

$$\frac{d}{dt}(a(t;\lambda)h + b(t;\lambda)h') - (c(t;\lambda)h + a(t;\lambda)h') = 0\,, \tag{5.7.40}$$

$$h(0;\lambda) = 0, \quad h'_t(0;\lambda) = 1\,. \tag{5.7.41}$$

We have

$$0 \in \sigma(\nabla^2_H f(0;\lambda)) \tag{5.7.42}$$

if and only if

$$h(T;\lambda) = 0\,. \tag{5.7.43}$$

Let

$$h(T;0) \neq 0, \quad h(T;1) \neq 0\,, \tag{5.7.44}$$

and let β_0, β_1 be the number of nontrivial zeros of $h(\cdot;0)$ and $h(\cdot;1)$ on $[0,T]$. The following theorem is valid.

Theorem 5.7.10. *If $\beta_0 \neq \beta_1$ then the zero critical point of the functional* (5.7.35) *has at least one bifurcation point in* $[0,1]$.

5.7.5 Analysis of Bifurcation Values of a Parameter

Theorem 5.7.1 given in Sect. 5.7.1 only gives necessary conditions for bifurcation, namely, λ_* can only be a bifurcation point of the zero solution of the equation

$$\nabla f(x;\lambda) = 0 \quad (x \in E,\ \nabla f(0;\lambda) = 0,\ 0 \leqslant \lambda \leqslant 1) \tag{5.7.45}$$

if the operator $\nabla^2 f(0;\lambda_*) : E \to E^*$ is non-invertible. Therefore we must examine all values of λ_* for which this operator is non-invertible. The results of this section indicate one way to carry out the examination. Here are several criteria for bifurcation.

Let $f(\cdot;\lambda) : H \to \mathbb{R}$, $(\nabla f(0;\lambda) = 0,\ 0 \leqslant \lambda \leqslant 1)$ be a one-parameter family of H-regular functionals continuous on $H \times [0,1]$ together with the gradient $\nabla f(\cdot;\lambda)$ and let the gradient $\nabla f(\cdot;\lambda)$ be continuous in λ uniformly with respect to x from some ball $B(\rho)$ in H. Theorem 5.7.2 implies

Theorem 5.7.11. *Suppose that $\lambda_* \in [0,1]$ and that every neighborhood of λ_* contains points λ_1 and λ_2 such that 0 is a local minimizer of the functional $f(\cdot;\lambda_1)$ but not a minimizer of $f(\cdot;\lambda_2)$. Then λ_* is a bifurcation point of the zero critical point of the functional $f(\cdot;\lambda)$.*

A similar result holds for E-regular functionals on separable reflexive Banach spaces.

Suppose that E is a real Banach space, H is a real Hilbert space, and that E is continuously and densely embedded into H. Consider a one-parameter family $f(\cdot;\lambda) : B(r) \to \mathbb{R}$ $(0 \leqslant \lambda \leqslant 1)$ of (E,H)-regular functionals. Let

$$\nabla f(0;\lambda) = 0 \quad (0 \leqslant \lambda \leqslant 1)\ ,$$

and let f, $\nabla_H f$, $\nabla_H^2 f$ be continuous on $B(\rho) \times [0,1]$, and $f(\cdot;\lambda)$, $\nabla_H f(\cdot;\lambda)$, $\nabla^2 f(\cdot;\lambda)$ be continuous in λ uniformly with respect to $x \in B(\rho)$. Theorem 4.2.2. implies

Theorem 5.7.12. *Suppose that the following conditions hold:*

(1) for each $\lambda \in [0,1]$ the operator $\nabla_H f(\cdot;\lambda)$ is a continuously differentiable mapping from $B(r)$ into E,

(2) for each $\lambda \in [0,1]$ there is an admissible finite-dimensional subspace L_λ of E such that for every $w \in E \cap L_\lambda^\perp$ the solution x of the linear equation

$$Q_\lambda \nabla_H^2 f(0;\lambda)x = w\ ,$$

where Q_λ is the orthogonal projection onto $L_\lambda^\perp$ in H, lies in E,

(3) every neighborhood of the point $\lambda_* \in [0,1]$ contains points λ_1 and λ_2 such that 0 is an isolated extremal point of the functionals $f(\cdot;\lambda_1)$ and $f(\cdot;\lambda_2)$ and the Conley indices $h(0;f(\cdot;\lambda_1))$ and $h(0;f(\cdot;\lambda_2))$ of this critical point are not equal.

Then λ_* is a bifurcation point of the zero critical point of the functional $f(\cdot;\lambda)$.

Suppose that 0 is an isolated point of the spectrum $\sigma(\nabla_H^2 f(0;\lambda_*))$ which is an eigenvalue of finite multiplicity k. Then, for values of λ close to λ_*, the intersection of $\sigma(\nabla_H^2 f(0;\lambda))$ with a small neighborhood U of 0 consists of a finite number of eigenvalues of total multiplicity k. Let H_λ be the subspace spanned by the corresponding eigenvectors and $P_\lambda : H \to H_\lambda$ the orthogonal projection onto H_λ. Let $\beta_-(\lambda)$ and $\beta_+(\lambda)$, respectively, be the sums of the multiplicities of the negative and positive eigenvalues of the operator $\nabla_H^2 f(0;\lambda)$ which lie in U. Theorem 5.7.12 implies

Theorem 5.7.13. *Suppose that the following conditions hold:*

(1) *for each $\lambda \in [0,1]$ the operator $\nabla_H f(\cdot;\lambda)$ is a continuously differentiable mapping from $B(r)$ to E,*

(2) *for each $y \in E$, the linear equation*

$$P_\lambda \nabla_H^2 f(0;\lambda)x + (I - P_\lambda)x = y$$

has a unique solution in H and it lies in E,

(3) *every neighborhood of λ_* contains points λ_1 and λ_2 such that*

$$\beta_-(\lambda_1) + \beta_+(\lambda_1) = \beta_-(\lambda_2) + \beta_+(\lambda_2) = k \,, \tag{5.7.46}$$

$$\beta_-(\lambda_1) \neq \beta_-(\lambda_2) \,. \tag{5.7.47}$$

Then λ_* is a bifurcation point of the zero critical point of the functional $f(\cdot;\lambda)$.

We shall apply Theorems 5.7.11 and 5.7.13 to study bifurcation points of the zero critical point of the functional

$$f(u;\lambda) = \int_\Omega F(x, u(x), \nabla u(x); \lambda)\, dx \tag{5.7.48}$$

considered on the space

$$W_{0,p}^2(\Omega) = W_p^2(\Omega) \cap \mathring{W}_2^1(\Omega) \,.$$

Let Ω be a bounded domain in $\mathbb{R}^N$, let $p > \max(2, N)$, and suppose that the integrand $F(x,u,p;\lambda)$ is sufficiently smooth and that

$$\sum_{i,j=1}^N \frac{\partial^2 F(x,u,p;\lambda)}{\partial p_i \partial p_j} \xi_i \xi_j \geqslant c \sum_{i=1}^N \xi_i^2 \quad (c > 0) \,.$$

Then the functional $f(\cdot;\lambda)$ is $(W_{0,p}^2(\Omega), \mathring{W}_2^1(\Omega))$-regular.

We assume that for each $\lambda \in [0,1]$ the zero function is a critical point of (5.7.48) and that for $\lambda = \lambda_* \in [0,1]$ the linear Dirichlet problem

$$\sum_{i=1}^{N} \frac{d}{dx_i}\left(\sum_{j=1}^{N} \frac{\partial^2 F(x,0,0;\lambda_*)}{\partial p_i \partial p_j}\frac{\partial u(x)}{\partial x_j}\right) + \sum_{i=1}^{N} \frac{d}{dx_i}\left(\frac{\partial^2 F(x,0,0;\lambda_*)}{\partial u \partial p_i} u(x)\right) - \sum_{i=1}^{N} \frac{\partial^2 F(x,0,0;\lambda_*)}{\partial u \partial p_i}\frac{\partial u(x)}{\partial x_i} - \frac{\partial^2 F(x,0,0;\lambda_*)}{\partial u^2} u(x) = 0\,, \tag{5.7.49}$$

$$u(x)\big|_{\partial\Omega} = 0 \tag{5.7.50}$$

has a nontrivial solution. Theorems 5.7.10 and 5.7.12 imply the following assertion.

Theorem 5.7.14. *Suppose that in every neighborhood of the number λ_* there exist λ_1 and λ_2 such that the critical point $u(x) = 0$ is a minimizer for* (5.7.48) *for $\lambda = \lambda_1$ and is not a minimizer for $\lambda = \lambda_2$. Then λ_* is a bifurcation point of the zero critical point of* (5.7.48).

Suppose that the problem (5.7.49), (5.7.50) has exactly k linearly independent solutions. Let $\beta_-(\lambda)$ and $\beta_+(\lambda)$ be the sums of the multiplicities of the negative and positive eigenvalues of the elliptic differential operator defined by the left-hand side of Eq. (5.7.49) under the boundary condition (5.7.50).

Theorem 5.7.15. *Suppose that in every neighborhood of the number λ_* there exist λ_1 and λ_2 such that* (5.7.46) *and* (5.7.47) *hold. Then λ_* is a bifurcation point of the zero critical point of* (5.7.48).

5.7.6 Bifurcations of Solutions of the Ginzburg–Landau Equations

The Ginzburg–Landau phenomenological theory of superconductivity is concerned with the behavior of a superconductor in an external magnetic field. The state of the superconductor is described by the solutions of the *Ginzburg–Landau equations.* After an appropriate choice of measurement units, these equations and the boundary conditions satisfied by their solutions take the form

$$(-i\nabla - A)^2\psi + \mu\,|\psi|^2\,\psi - \lambda\psi = 0\,, \tag{5.7.51}$$

$$-\operatorname{rot}\operatorname{rot} A = A\,|\psi|^2 + i(\psi^*\nabla\psi - \psi\nabla\psi^*)\,, \tag{5.7.52}$$

$$(n, (-i\nabla\psi - A\psi))\big|_{\partial\Omega} = 0\,, \tag{5.7.53}$$

$$\operatorname{rot} A \times n\big|_{\partial\Omega} = 0\,. \tag{5.7.54}$$

Here, Ω is a bounded convex domain in $\mathbb{R}^3$ with boundary $\partial\Omega$, n is a normal vector to $\partial\Omega$, ψ is a complex function called the *order parameter* ($|\psi|^2$ is proportional to the density of superconducting electrons),

$$A = (A_1, A_2, A_3) \tag{5.7.55}$$

is the vector potential of the magnetic induction vector, ∇ is the gradient operator in $\mathbb{R}^3$,

$$\nabla = \left(\frac{\partial}{\partial x_1}, \frac{\partial}{\partial x_2}, \frac{\partial}{\partial x_3}\right),$$

rot is rotation,

$$\mathrm{rot} = \left(\frac{\partial}{\partial x_2} - \frac{\partial}{\partial x_3}, \frac{\partial}{\partial x_3} - \frac{\partial}{\partial x_1}, \frac{\partial}{\partial x_1} - \frac{\partial}{\partial x_2}\right),$$

$*$ is complex conjugation, $(\cdot,\cdot)$ is the inner product in $\mathbb{R}^3$, and λ and μ are real parameters. The positive parameter μ depends only on the density of the substance and the parameter λ is proportional to the difference in temperatures $T_c - T$, where T_c is the *phase transition point.*

The boundary-value problem (5.7.51)–(5.7.54) clearly has the zero solution. We are interested in nonzero solutions.

The Ginzburg–Landau equations are the Euler equations for the functional of the free energy of the superconductor which is defined on the pairs $u = (\psi, A)$ by

$$f(u;\lambda) = \frac{1}{2}\int_\Omega \left(|\mathrm{rot}\, A|^2 + |\nabla\psi - iA\psi|^2 + \tfrac{1}{2}\mu\, |\psi|^4 - \lambda\, |\psi|^2\right) dx. \tag{5.7.56}$$

Write $\mathcal{H}_1$ for the Hilbert space of complex functions whose real and imaginary parts are in $W_2^1(\Omega)$. We define the inner product in $\mathcal{H}_1$ by

$$(\psi, \varphi)_{\mathcal{H}_1} = \Re \int_\Omega (\psi(x)\varphi^*(x) + (\nabla\psi(x), \nabla\varphi^*(x)))\, dx\, .$$

We emphasize that we consider $\mathcal{H}_1$ as a real vector space, because we require the differentiability of the functional f.

Let $\mathcal{H}_2$ be the Hilbert space of vector functions $A{=}(A_1, A_2, A_3)$ whose components belong to $W_2^1(\Omega)$. The inner product of two vector functions $A = (A_1, A_2, A_3)$ and $B = (B_1, B_2, B_3)$ in $\mathcal{H}_2$ is defined by

$$(A, B)_{\mathcal{H}_2} = \int_\Omega \left(\sum_{i=1}^{3} A_i B_i + \sum_{i,k=1}^{3} \frac{\partial A_k}{\partial x_i}\frac{\partial B_k}{\partial x_i}\right) dx\, .$$

We set

$$\mathcal{H} = \mathcal{H}_1 \times \mathcal{H}_2$$

and define the inner product of elements $u = (\psi, A)$, $v = (\varphi, B)$ of $\mathcal{H}$ by

$$(u, v)_{\mathcal{H}} = (\psi, \varphi)_{\mathcal{H}_1} + (A, B)_{\mathcal{H}_2} .$$

Direct calculations show that the functional f is continuously Fréchet differentiable on $\mathcal{H}$ and that

$$\begin{aligned}(\nabla f(u), v)_{\mathcal{H}} \\ &= \int_{\Omega} \Big((\operatorname{rot} A, \operatorname{rot} B) + ((A\,|\psi|^2 + i\psi^*\nabla\psi - i\psi\nabla\psi^*), B)\Big)\,dx \\ &\quad + \Re \int_{\Omega} \Big(((\nabla\psi - iA\psi), (\nabla\varphi^* + iA\varphi^*)) + (\mu\,|\psi|^2\,\psi - \lambda\psi)\varphi^*\Big)\,dx\ ;\end{aligned} \tag{5.7.57}$$

here $u = (\psi, A)$, $v = (\varphi, B)$, and $(\cdot, \cdot)$ is the inner product in $\mathbb{R}^3$.

Let $P_1 : \mathcal{H} \to \mathcal{H}_1$ and $P_2 : \mathcal{H} \to \mathcal{H}_2$ be the orthogonal projections of $\mathcal{H}$ onto $\mathcal{H}_1$ and $\mathcal{H}_2$.

Clearly the point $u = (\psi, A)$ of $\mathcal{H}$ is a critical point of f if and only we have

$$P_1 \nabla f(u) = 0\ , \tag{5.7.58}$$

$$P_2 \nabla f(u) = 0\ . \tag{5.7.59}$$

We shall call the operator equations (5.7.58), (5.7.59) the *Ginzburg–Landau equations* and their solutions the *generalized solutions* of the boundary-value problem (5.7.51)–(5.7.54). If the generalized solution $u = (\psi, A)$ is sufficiently smooth, then the pair (ψ, A) is the classical solution of the boundary-value problem (5.7.51)–(5.7.54) for the Ginzburg–Landau equations.

Let E be the subspace of $\mathcal{H}_2$ consisting of vector functions A representable in the form

$$A = \nabla p \quad (p \in W_2^2(\Omega)) \tag{5.7.60}$$

and by F the subspace of $\mathcal{H}_2$ consisting of the vector functions A such that

$$\operatorname{div} A = 0, \quad (A, n)\big|_{\partial\Omega} = 0\ . \tag{5.7.61}$$

Then

$$\mathcal{H}_2 = E \oplus F\ , \tag{5.7.62}$$

the subspaces E and F are orthogonal (in the sense of $L_2(\Omega)$) and the inequality

$$\|A\|_{\mathcal{H}_2} \leqslant C\,\|\operatorname{rot} A\|_{L_2(\Omega)}\ , \tag{5.7.63}$$

where $C > 0$ is a constant, holds for all $A \in F$.

Clearly the direct product H of the spaces $\mathcal{H}_1$ and F is a closed subspace of $\mathcal{H}$. By (5.7.63), the norm $\|\cdot\|_{\mathcal{H}}$ on H is equivalent to the norm

$$\|u\|_H = \|(\psi, A)\|_H = \big(\,\|\psi\|^2_{\mathcal{H}_1} + \|\operatorname{rot} A\|^2_{L_2(\Omega)}\,\big)^{1/2}\ .$$

On the space H, the functional (5.7.56) can be represented in the form

$$f(u;\lambda) = \tfrac{1}{2} \|u\|_H^2 + g(u;\lambda) \,. \tag{5.7.64}$$

It follows from (5.7.57) that the functional $g(\cdot;\lambda)$ is Fréchet differentiable on H and its gradient $\nabla g(\cdot;\lambda) : H \to H$ is completely continuous. Therefore the gradient $\nabla f(\cdot;\lambda)$ of the functional $f(\cdot;\lambda)$ on H is the sum of the identity operator and a completely continuous operator. Consequently, $f(\cdot;\lambda)$ is H-regular.

Lemma 5.7.4. *If $u = (\psi, A)$ is a critical point of the functional $f(\cdot;\lambda)$ regarded as a function on H, then u is also a critical point of this functional on $\mathcal{H}$.*

Proof. Direct verification shows that the functional $f(\cdot;\lambda)$ possesses the property of *gauge invariance*, i.e., satisfies

$$f((\psi, A);\lambda) = f((\psi \exp ip, A + \nabla p);\lambda)$$

for any function $p \in W_2^2(\Omega)$. This implies the statement of the lemma. □

Thus the problem of finding the critical points of the functional $f(\cdot;\lambda)$ on $\mathcal{H}$ has been reduced to the problem of finding the critical points on the smaller space H.

Consider the boundary-value problem

$$-\Delta v = \nu v, \quad n \cdot \nabla v|_{\partial\Omega} = 0 \,, \tag{5.7.65}$$

where Δ is the Laplace operator. Simple calculations show that the eigenvalues μ_k of the operator $\nabla^2 f(0;\lambda)$ are related to the eigenvalues ν_k of problem (5.7.65) by the equation

$$\mu_k = \frac{\nu_k - \lambda}{1 + \nu_k} \quad (k = 0, 1, \ldots) \,. \tag{5.7.66}$$

Therefore, if the parameter λ lies in the interval $(0, \nu_1)$, then the operator $\nabla^2 f(0;\lambda)$ has one simple negative eigenvalue and the rest of the spectrum is positive, and so the point 0 is not a minimizer of the functional $f(\cdot;\lambda)$. Moreover since 0 is clearly is a minimizer of $f(\cdot;0)$ for $\lambda = 0$, Theorem 5.7.11 implies that the point $\lambda_* = 0$ is a bifurcation point of the zero critical point of the functional (5.7.56). Thus we have proved the following theorem.

Theorem 5.7.16. *For every sufficiently small $\lambda > 0$ the system of Ginzburg–Landau equations has a nontrivial solution*

$$u(t) = \{\psi(\lambda), A(\lambda)\}$$

and

$$\lim_{\lambda \to 0} \|u(\lambda)\|_{\mathcal{H}_1} = 0 \,.$$

Remark. Using the notion of topological index, we can show that the Ginzburg–Landau equations have at least two nontrivial gauge inequivalent solutions for small $\lambda > 0$.

Additional Remarks and Bibliographic Comments

Chapter 1

Basic properties of metric and normed spaces can be found in almost any textbook on functional analysis (see, e.g., [139, 144, 165]). For information about specific function spaces and, in particular, embedding theorems, see [181]. Differential calculus in infinite-dimensional spaces is developed, for instance, in [165]. The main facts about ordinary differential equations in finite-dimensional and infinite-dimensional spaces (existence and uniqueness of the solution of the Cauchy problem, nonlocal extendability, Lyapunov stability, asymptotic stability, stability in the large, etc.) can be found in the monographs [7, 87, 88, 131]. Information about convex and Lipschitzian functionals is given in [4, 65, 91, 134]. The properties of the specific linear and nonlinear operators encountered in this monograph are presented in [145].

The second-order necessary and sufficient conditions in extremal problems can be found in [2, 4, 16]. The monographs and papers [65, 87, 88, 91, 221] are devoted to the generalization of the classical theory of extremal problems to nonsmooth problems (convex and Lipschitzian). Various generalizations of the classical Weierstrass theorems can be found in [223, 224]. The fundamental results concerning monotonic operators and potential operators are given in [7, 91, 224].

Chapter 2

The deformation principle for a minimizer for smooth functions of finitely many variables (Theorem 2.1.1) is presented in [19], and Theorem 2.1.4 on global minimizers is proved in [18]. We outline briefly a generalization of Theorem 2.1.4 given in [8]. It is useful in problems concerning dissipativity conditions for infinite-dimensional dynamical systems.

Let H be a real Hilbert space and $f : H \to \mathbb{R}$ a continuously Fréchet differentiable functional whose gradient ∇f is locally Lipschitzian. Suppose that

$$\inf_{\|x\| \geqslant r} \|\nabla f(x)\| \geqslant \alpha(r) \quad (r \geqslant r_0)$$

for some positive monotone increasing function $\alpha(r)$ ($r_0 \leqslant r < \infty$). The following result holds.

Theorem. *If*

$$\inf_{x \in S(r_1)} f(x) > \sup_{x \in B(r_0)} |f(x)|$$

for some $r_1 > r_0$ *then* f *is growing, that is,*

$$\lim_{\|x\| \to \infty} f(x) = \infty .$$

The deformation principle for a minimizer for Lipschitzian functions is given in [35]. Regular and critical points for arbitrary continuous functions are introduced in [20] in connection with topological characteristics of critical points of nonsmooth functions.

The following definitions seem natural in relation to the constructions in Sect. 2.2.5.

Let $f : \mathbb{R}^N \to \mathbb{R}$ be a continuous function. We call a point $x_0 \in \mathbb{R}^N$ a *regular point* of f if there exist a neighborhood $U(x_0)$ of x_0, a sequence of continuously differentiable functions $f_n : U(x_0) \to \mathbb{R}^N$, and some $a > 0$ such that

$$\lim_{n \to \infty} \sup_{x \in U(x_0)} |f(x) - f_n(x)| = 0 ,$$

and

$$|\nabla f_n(x)| > a \quad (x \in U(x_0),\ n = 1, 2, \ldots) .$$

We use the term *critical points* for non-regular points.

We call a compact set K in $\mathbb{R}^N$ a *regular set* of f if there exist an open set $U(K)$ containing K, a sequence of continuously differentiable functions $f_n : U(K) \to \mathbb{R}^N$, and some $a > 0$ for which

$$\lim_{n \to \infty} \sup_{x \in U(K)} |f(x) - f_n(x)| = 0 ,$$

and

$$|\nabla f_n(x)| > a \quad (x \in U(K),\ n = 1, 2, \ldots) .$$

Conjecture. Suppose that every point of a compact set K in $\mathbb{R}^N$ is a regular point of the continuous function $f : \mathbb{R}^N \to \mathbb{R}$. Then K is a regular set of f.

If this conjecture were true, then it would provide a basis for the construction of a substantive theory of critical points of arbitrary continuous functions of finitely many variables.

The choice of curves lemma (Lemma 2.3.1), on which the converse of the deformation principle for real polynomials depends, can be found, for instance, in [171], and the proof of the Lojasiewicz inequality, the key result for proving Theorem 2.3.2, in [166]. The monograph [170] is devoted to the h-cobordism theorem (Theorem 2.3.4). Theorem 2.3.5 is proved in [1].

The degree theory for mappings in finite-dimensional spaces can be found in [151, 172]. Our proof of Hopf's classification theorem (Theorem 2.4.10) differs only in minor details from Hopf's original proof. A different proof, based on the notion of a rigid bordism introduced by Pontryagin, is presented in [172]. Parusinski's remarkable theorem (Theorem 2.4.11) from [186] generalizes Hopf's theorem to gradient vector fields. It turns out that nondegenerate gradient fields defined on a sphere are gradient homotopic if and only if their rotations are equal.

Chapter 3

The $(S)_+$-property for nonlinear operators was introduced by Browder [51, 52] and, independently, by Skrypnik [214], in connection both with the solvability of boundary-value problems for nonlinear elliptic equations and with convergence problems for Galerkin's method. Operators which are close to those which possess the $(S)_+$-property (strongly closed operators) were considered by Pokhozhaev [190] and Brezis [49].

The deformation principle for minimizers of smooth functionals on Hilbert spaces (Theorem 3.1.1) is given in [21, 22] and the generalization to the case of functionals defined on reflexive Banach spaces (Theorem 3.2.1) in [9]. Theorem 3.3.1 can be found in [18]. Applications of the deformation principle to problems in the calculus of variations are described in [22]. The invariance of a weak minimizer under nondegenerate deformations of variational problems is proved in [30].

The deformation principle and its generalizations to Lipschitzian functionals on infinite-dimensional spaces as well as to problems with constraints are developed in [208–213]. The main results from these papers are presented in Sects. 3.4, 3.5.

The paper [135] gives a generalization of the deformation principle to functionals defined on metric spaces. Here is its main result.

Let M be a metric space with metric d and let $f : M \to \mathbb{R}$ be a continuous functional. Then $x_* \in M$ is a *regular* point of f if there exist a neighborhood U of x_* and a continuous mapping $\varphi : U \times [0,1] \to M$ for which the following conditions hold for all $(x,t) \in U \times [0,1]$:

(1) $\varphi(x,0) \equiv x$;
(2) for $t > 0$ we have

$$f(x) - f(\varphi(x,t)) > 0 .$$

Points which are not regular are called *critical points* of f.

Suppose that $\delta > 0$. The functional f is said to be *δ-regular* at the regular point x_* if

$$f(x) - f(\varphi(x,t)) \geqslant \delta d(x, \varphi(x,t)) \quad ((x,t) \in U \times [0,1]) .$$

Write $\delta(f, x_*)$ for the least upper bound of the numbers δ for which this inequality holds.

Let G be an open set in M and $x_* \in G$. A continuous nondecreasing function $\mu : \mathbb{R}_+ \to \mathbb{R}_+$ which is positive for positive values of its argument is a *modulus of regularity* of f on G with respect to x_* if

$$\delta(f, x) \geqslant \mu(d(x, x_*))$$

for all $x \in G$.

Let $f_0, f_1 : \mathring{B}(x_*) \to \mathbb{R}$ be two continuous functionals defined on an open unit ball $\mathring{B}(x_*)$ in a complete locally connected metric space and suppose that x_* is the only critical point of these functionals.

A one-parameter family of continuous functionals $f(\cdot;\lambda) : \mathring{B}(x_*) \to \mathbb{R}$ $(0 \leqslant \lambda \leqslant 1)$ is called a *nondegenerate deformation* of $f_0 = f(\cdot;0)$ into $f_1 = f(\cdot;1)$ if

(1) for every $\lambda \in [0,1]$ the point x_* is the unique critical point of the functional $f(\cdot;\lambda)$ on $\mathring{B}(x_*)$,

(2) there exists a modulus of continuity $\mu : \mathbb{R}_+ \to \mathbb{R}_+$ on $\mathring{B}(x_*) \setminus \{x_*\}$ with respect to x_* which is common to all functionals $f(\cdot;\lambda)$ $(0 \leqslant \lambda \leqslant 1)$,

(3) $f(x;\lambda)$ is continuous in λ uniformly with respect to $x \in \mathring{B}(x_*)$.

Theorem. *Suppose that there exists a nondegenerate deformation of the functional $f_0 : \mathring{B}(x_*) \to \mathbb{R}$ into the functional $f_1 : \mathring{B}(x_*) \to \mathbb{R}$. If x_* is a local minimizer of f_0, then x_* is a local minimizer of f_1.*

The author of [41] introduced the notion of a normal deformation. Here is one of the modifications of the deformation principle.

Let E be a Banach space that is continuous and densely embedded into a Hilbert space H. Consider a family of functionals $f(\cdot;\lambda)$ $(0 \leqslant \lambda \leqslant 1)$ defined on E and possessing the following properties:

(1) for all $x \in E$ and $\lambda \in [0,1]$ the functional $f(\cdot;\lambda)$ is twice Fréchet differentiable at x and

$$\nabla f(x;\lambda) \in H \quad (x \in E, 0 \leqslant \lambda \leqslant 1) ,$$

$$\nabla^2 f(x;\lambda) h \in H \quad (x, h \in E, 0 \leqslant \lambda \leqslant 1) ;$$

(2) in the expansion

$$f(x+h,\lambda) = f(x,\lambda) + (\nabla f(x,\lambda), h)_H + \tfrac{1}{2}(\nabla^2 f(x,\lambda)h, h) + \omega(x,h,\lambda) ,$$

the remainder term $\omega(x,h,\lambda)$ satisfies

$$\lim_{\|h\|_E \to 0} \sup_{\|x\|_E \leqslant \rho, \|h\|_E \leqslant 1, 0 \leqslant \lambda \leqslant 1} \|h\|_H^{-2} \omega(x,h,\lambda) = 0$$

for each $\rho > 0$;

(3) for every λ the operator $\nabla f(\cdot;\lambda)$ is a mapping from E to E and is continuous in the collection of variables;

(4) the operator function $\nabla^2 f(\cdot;\lambda)$ maps E into $\mathcal{L}(E)$ and is continuous in the collection of variables $x \in E$, $\lambda \in [0,1]$ with respect to the norm of the operators acting in E;

(5) for all $x \in E$ and $\lambda \in E$ the operator $\nabla^2 f(x;\lambda)$ has an extension to a continuous self-adjoint operator

$$\overline{\nabla}^2 f(x;\lambda) : H \to H ,$$

and the operator function $\overline{\nabla}^2 f(\cdot;\lambda) : H \to \mathcal{L}(H)$ is continuous in the collection of variables with respect to the norm of the operators on H.

We call such a family an (E,H)-*normal deformation* of the functional $f_0 = f(\cdot;0)$ into $f_1 = f(\cdot;1)$ if the following conditions hold.

(I) The spectrum of each operator

$$\nabla^2 f(x;\lambda) : E \to E$$

coincides with the spectrum $\sigma_H(\overline{\nabla}^2 f(x;\lambda))$ of the operator

$$\overline{\nabla}^2 f(x;\lambda) : H \to H .$$

(II) For each $\lambda \in [0,1]$ the functional $f(\cdot;\lambda)$ has a unique critical point $x(\lambda)$ in E which depends continuously on $\lambda \in [0,1]$.

(III) For each $\lambda \in [0,1]$ either

$$0 \notin \sigma_H(\overline{\nabla}^2 f(x(\lambda);\lambda))$$

or 0 is an isolated point of $\sigma_H(\overline{\nabla}^2 f(x(\lambda);\lambda))$.

(IV) If

$$0 \in \sigma_H(\overline{\nabla}^2 f(x(\lambda);\lambda),$$

then 0 is an eigenvalue of finite multiplicity k of the operators

$$\nabla^2 f(x(\lambda);\lambda) : E \to E \quad \text{and} \quad \overline{\nabla}^2 f(x(\lambda);\lambda) : H \to H .$$

Theorem. *Suppose that there exists an (E,H)-normal deformation of the functional f_0 into the functional f_1. If the point $x_0 = x(0)$ is a local minimizer of f_0 in E then $x_1 = x(1)$ is a local minimizer of f_1 in E.*

Chapter 4

Our description of the theory of Conley index generally follows Conley's original monograph [69]. The concept of an (E,H)-regular functional was introduced in [43] and the constructions used to generalize Conley theory to infinite-dimensional problems were also taken from [43]. The approach to Conley index in [188] is close to that in Sect. 4.2.2.

Interesting applications of Conley index to the study of solitary waves on the surface of an ideal liquid are given in [188].

Chapter 5

Some applications of the deformation principle for a minimizer to problems of classical analysis are given in [22, 30, 40]. The deformation principle for nonlinear programming problems is established in [28, 29]. These papers describe applications to the determination of sufficient conditions for a minimizer in nonlinear programming problems, to the investigation of degenerate extremals, and to the proof of nonsmooth inequalities and inequalities with constraints.

The deformation principle for finite-dimensional multicriteria problems is given in [38], and the generalization to infinite-dimensional multicriteria problems in [213].

The deformation principle for one-dimensional variational calculus problems is established in [22] and the multidimensional version is given in [30].

The following problem is of interest in connection with Theorem 5.5.3.

Consider a one-parameter family of *potential systems*

$$\frac{d^2x}{dt^2} = -\nabla f(x;\lambda) \quad (x \in \mathbb{R}^N,\ 0 \leqslant \lambda \leqslant 1)\,.$$

Assume that the potential f is sufficiently smooth and that the origin is the only state of equilibrium in the unit ball B in $\mathbb{R}^N$ for all $\lambda \in [0, 1]$.

Conjecture. Suppose that 0 is a Lyapunov stable state of equilibrium of the system

$$\frac{d^2x}{dt^2} = -\nabla f(x;0) \quad (x \in \mathbb{R}^N)\,.$$

Then 0 is a Lyapunov stable state of equilibiurm of the system

$$\frac{d^2x}{dt^2} = -\nabla f(x;1) \quad (x \in \mathbb{R}^N)\,.$$

This conjecture would hold if we could prove that the Lyapunov stability of the isolated zero state of equilibrium of the potential system

$$\frac{d^2x}{dt^2} = -\nabla f(x) \quad (x \in \mathbb{R}^N)$$

implies that 0 is a local minimizer of f. This is true for $N = 1$ and $N = 2$ (see [185]). For $N \geqslant 3$, it is still an open question.

Theorem 5.5.6 was announced in [149]. The constructions used in the proof of Theorems 5.6.1 and 5.6.2 are taken from [43].

The publications [10, 31, 42, 61, 63, 112, 140, 142, 150, 183, 200–202, 214, 222] are devoted to questions about bifurcations of extremals of variational problems.

Theorem 5.7.1 gives necessary conditions for bifurcation for the parameter λ. The necessary conditions become sufficient if λ enters into the family of functionals in a special way. We shall describe the corresponding result, which is due to Skrypnik [214].

Let f and g be nonlinear differentiable functionals defined in a neighborhood U of 0 in a real separable Hilbert space H. Suppose that 0 is a critical point of f and g, i.e., $\nabla f(0) = \nabla g(0) = 0$.

Consider the problem of bifurcation points of the equation

$$\nabla f(x) - \lambda \nabla g(x) = 0 . \tag{1}$$

We assume that the following conditions hold.

1. The functional $g : U \to \mathbb{R}$ is weakly continuous and uniformly Fréchet differentiable on U.

2. The gradient $\nabla g : U \to H$ of g is Fréchet differentiable at 0 and its Fréchet derivative $\nabla^2 g(0) : H \to H$ is a bounded self-adjoint operator.

3. The functional $f : U \to \mathbb{R}$ is H-regular and

$$(\nabla f(x), x) \geqslant \nu \|x\|^2$$

for some constant $\nu > 0$.

4. At each point $x \in U$ the gradient $\nabla f : U \to H$ of f is Gâteaux differentiable and its Gâteaux derivative $\nabla^2 f(x)$ has the following property: if

$$\lim_{n\to\infty} \|x_n\| = 0 ,$$

then

$$\lim_{n\to\infty} \left\|\nabla^2 f(x_n)h - \nabla^2 f(0)h\right\| = 0$$

for all $h \in H$.

Theorem. *For the point λ_* to be a bifurcation point of Eq.* (1), *it is necessary and sufficient that the equation*

$$\nabla^2 f(0)h - \lambda_* \nabla^2 g(0)h = 0$$

have nonzero solutions.

A study of bifurcation values of parameters in problems with completely continuous operators can be found in Krasnosel'skii's monograph [145].

The bifurcation method is applied to problems on bifurcation points in variational problems in [57]. Applications of Conley theory to bifurcations of critical points of certain functionals arising in hydrodynamics are given in [188]. Bifurcations of solutions of the Ginzburg–Landau equations are studied in [25, 31, 143, 183].

In this monograph we have not touched on the applications of the deformation method to the approximate construction of solutions of nonlinear problems. There are many publications devoted to these applications and the reader can find a reasonably complete bibliography in [5, 32, 33, 79–83, 121, 155, 161, 184, 191, 194, 195, 207, 228].

References

1. Abduvaitov, H. (1987): On the problem of the stability of gradient systems. Avtomat. i Telemekh., 6, 3–6.
2. Akhiezer, N.I. (1962): The calculus of variations. Blaisdell Publishing Co., New York–London.
3. Aleksandrov, P.S. (1998): Combinatorial Topology. (Reprint of the 1956, 1957 and 1960 translations.) Dover Publications, Inc., New York.
4. Alekseev, I.M., Tikhomirov, V.M., Fomin, S.V. (1987): Optimal Control. Contemporary Soviet Mathematics. Consultants Bureau, New York.
5. Allgower, E.L., Georg, K. (1980): Simplicial and continuation methods for approximating fixed points and solutions to systems of equations. SIAM Review, **22**, 28–85.
6. André, N., Shafrir, I. (1995): Minimization of a Ginzburg–Landau functional with weight. C. R. Acad. Sci. Paris Sér. I Math., **321**, 999–1004.
7. Aubin, J.-P., Ekeland, I. (1986): Applied Nonlinear Analysis. J. Wiley & Sons, New York.
8. Barbakadze, T.N. (1989): Invariance of the minimum under deformations of smooth functionals. Soobshch. Akad. Nauk Gruz. SSR, **136**, 281–284.
9. Barbakadze, T.N., Bobylev, N.A., Kondakov, G.V. (1990): Deformation principle of minimum for functionals on Banach spaces. Sb. Trudov VNII Sist. Issled., 13, 6–15.
10. Bartsch, T., Clapp, M. (1990): Bifurcation theory for symmetric potential operators and the equivariant cup-length. Math. Z., **204**, 341–356.
11. Beeson, M.J., Tromba, A.J. (1984): The cusp catastrophe of Toma in the bifurcation of minimal surfaces. Manuscripta Math., **46**, 273–308.
12. Benci, V. (1987): A new approach to Morse–Conley theory. Recent Adv. Hamiltonian Systems. Proc. Int. Conf. L'Aguila, 10–13 June 1986, Singapore, 1–52.
13. Benci, V., Giannoni, F. (1992): Morse theory for functional of class C^1. C. R. Acad. Sci. Paris Sér. I Math., **315**, 883–888.
14. Benci, V., Pacella, F. (1985): Morse theory for symmetric functionals on the sphere and an application to a bifurcation problem. Nonlinear Anal., **9**, 763–773.
15. Bernstein, S.N. (1960): Equations of Variational Calculus. In: Bernstein, S.N., Collected Works (in Russian), vol. III, Fizmatgiz, Moscow.

16. Bliss, G. (1946): Lectures on the Calculus of Variations. University of Chicago Press, Chicago.
17. Bobylev, N.A. (1977): An approximation method for introducting rotation for certain classes of continuous vector fields. In: Qualitative and Approximate Methods for the Investigation of Vector Equations, 2 (in Russian). Yaroslavl Univ. Press, Yaroslavl, 25–31.
18. Bobylev, N.A. (1979): Lyapunov functions and global extremum problems. Avtom. i Telemekh., 11, 5–9.
19. Bobylev, N.A. (1980): On a technique for investigating the stability of gradient systems. Avtomat. i Telemekh., 8, 33–35.
20. Bobylev, N.A. (1980): Regular and critical points of continuous functions. In: Qualitative and Approximate Methods of Investigation of Operator Equations (in Russian). Yaroslavl Univ. Press, Yaroslavl, 30–42.
21. Bobylev, N.A. (1981): On the deformation method for investigating performance functionals of systems with infinitely many degrees of freedom. Avtomat. i Telemekh., 7, 11–18.
22. Bobylev, N.A. (1983): On deformation of functionals having a unique critical point. Mat. Zametki, **34**, 3, 387–398.
23. Bobylev, N.A. (1985): Solvability of boundary value problems and criteria of minimum of integral functionals. Ukrain. Mat. Zh., **37**, 4, 418–424.
24. Bobylev, N.A. (1985): Topological characteristics of optimal controls. Avtomat. i Telemekh., 7, 64–72.
25. Bobylev, N.A. (1986): On the topological degree of extremals in multidimensional variational problems. Funktsional Anal. Prilozhen, **20**, 2, 8–13.
26. Bobylev, N.A. (1986): Topological degree of extremals in optimization problems. Chapter 2 in: Bobylev, N. A. et al. (eds.) Mathematical Systems Theory (in Russian). Nauka, Moscow, 25–44.
27. Bobylev, N.A. (1987): On a characteristic of extremals in variational problems. Dokl. Akad. Nauk SSSR, **292**, 3, 521–524.
28. Bobylev, N.A. (1989): The deformation method for investigating nonlinear programming problems. I. Avtomat. i Telemekh., 7, 82–90.
29. Bobylev, N.A. (1989): The deformation method for investigating nonlinear programming problems. II. Avtomat. i Telemekh., 8, 24–33.
30. Bobylev, N.A. (1993): Homotopy methods in variational problems. J. Soviet Math., **67**, 1, 2657–2712.
31. Bobylev, N. A., Emel'yanov, S. V., Korovin, S. K. (1999): Geometrical Methods in Variational Problems. Kluwer Acad. Publ., Dordrecht.
32. Bobylev, N. A., Ismailov, I. G., Propoi, A. I. (1990): Deformation-iterational Procedures for the Approximate Solution of Extremal Problems. Preprint. Inst. of Control Problems of the USSR Acad. Sci., Moscow.
33. Bobylev, N. A., Ismailov, I. G., Propoi, A. I. (1990): Extension method in the problem of approximate construction of solutions to infinite-dimensional optimization problems. Soviet Phys. Dokl., **35**, 15, 406–408.
34. Bobylev, N.A., Klimov, V.S. (1992): Methods of Nonlinear Analysis in Nonsmooth Optimization Problems (in Russian), Nauka, Moscow.

35. Bobylev, N.A., Kondakov, G.V. (1991): Deformation method for investigating nonsmooth optimization problems. Automat. Remote Control, **52**, 5, part 1, 628–638.

36. Bobylev, N.A., Korovin, S.K. (1996): Topological characteristics of extremals of variational problems. J. Math. Sci., **78**, 4, 363–432.

37. Bobylev, N.A., Korovin, S.K. (1997): Topological Methods in Variational Problems (in Russian), Dialog–MGU Press, Moscow.

38. Bobylev, N.A., Korovin, S.K., Skalyga, V.I. (1996): A homotopic method for studying multicriteria problems. Automat. Remote Control, **57**, 10, part 2, 1513–1521 (1997).

39. Bobylev, N.A., Krasnosel'skii, M.A. (1974): Deformations of systems into asymptotically stable systems. Avtomat. i Telemekh., 7, 5–8.

40. Bobylev, N.A., Krasnosel'skii, M.A. (1981): Analysis for Extremum. Degenerate Cases. Preprint (in Russian). Inst. of Control Problems of USSR Acad. Sci., Moscow.

41. Bobylev, N.A., Krasnosel'skii, M.A. (1989): Normal deformations of variational problems. Soviet Math. Dokl., **40**, 1, 141–144.

42. Bobylev, N.A., Krasnosel'skii, M.A. (1990): Bifurcation of extremals of variational problems. Soviet Math. Dokl., **42**, 2, 420–423.

43. Bobylev, N.A., Krasnosel'skii, M.A. (1994): A scheme for investigating extremals in multidimensional variational problems. Funct. Anal. Appl, **28**, 4, 227–237.

44. Bobylev, N.A., Krasnosel'skii, M.A. (1997): Principles of affinity in nonlinear problems. J. Math. Sci., **83**, 4, 485–521.

45. Boltyanskii, V.G. (1971): Mathematical Methods of Optimal Control. Holt, Rinehart & Winston, Inc., New York.

46. Bolza, O. (1920): Calculus of Variations. Dover.

47. Borisovich, Yu.G., Bliznyakov, N.M., Izrailevich, Ya.A., Fomenko, T.N. (1995): Introduction to Topology (in Russian). Nauka, Moscow.

48. Bosarge, W. (1968): Infinite dimensional iterative methods and applications. IBM Houston Sci. Center Rept., **320**, 2347.

49. Brezis, H. (1968): Équations et inéquations non linéaires dans les espaces vectoriels en dualité. Ann. Inst. Fourier (Grenoble), **18**, 1, 115–175.

50. Browder, F.E. (1963): Nonlinear elliptic boundary value problems. Bull. Amer. Math. Soc., **69**, 6, 862–874.

51. Browder, F.E. (1965): Variational methods for nonlinear elliptic eigenvalue problems. Bull. Amer. Math. Soc., **71**, 1, 176–183.

52. Browder, F.E. (1968): Nonlinear eigenvalue problems and Galerkin approximations. Bull. Amer. Math. Soc., **74**, 4, 651–656.

53. Browder, F.E. (1970): Nonlinear elliptic boundary value problems and the generalized topological degree. Bull. Amer. Math. Soc., **76**, 5, 999–1005.

54. Browder, F.E. (1970): Pseudo-monotone operators and the direct method of the calculus of variations. Arch. Ration. Mech. Anal., **38**, 4, 268–277.

55. Browder, F.E. (1984): The theory of degree of mapping for nonlinear mappings of monotone type. Nonlinear Partial Differential Equations and their Applications. Collège de France Seminar, 6, 165–177, Pitman, Boston.
56. Browder, F.E., Petryshyn, W.V. (1968): The topological degree and Galerkin approximations for noncompact operators in Banach spaces. Bull. Amer. Math. Soc., **74**, 4, 641–646.
57. Butkovskii, A.G. (1975): Control Methods for Distributed Parameter Systems (in Russian). Nauka, Moscow.
58. Carbou, G. (1995): Unicité et minimalité des solutions d'une équation de Ginzburg–Landau. Ann. Inst. H. Poincaré. Anal. Non Linéaire, **12**, 3, 305–318
59. Chang, G.Q.. (1981): On a bifurcation theorem due to Rabinowitz. J. Systems Sci. Math. Sci., **4**, 191–195.
60. Chow, S.N., Hale, J.K. (1982): Methods of Bifurcation Theory. Springer-Verlag, New York–Berlin.
61. Chow, S.N., Lauterbach, R.A. (1988): A bifurcation theorem for critical points of variational problems. Nonlinear Anal., **12**, 1, 51–61.
62. Chow, S.N., Mallet-Paret, J., Yorke, J.A. (1978): Finding zeroes of maps: homotopy methods that are constructive with probability one. Math. Comp., **32**, 143, 881–899.
63. Clark, D.C. (1975): Eigenvalue bifurcation of odd gradient operator. Rocky Mountain J. Math., **5**, 317–336.
64. Clarke, F.H. (1975): Generalized gradients and applications. Trans. Amer. Math. Soc., **205**, 247–262.
65. Clarke, F.H. (1983): Optimization and Nonsmooth Analysis. John Wiley & Sons, New York.
66. Coffman, C.V. (1988): Lusternik–Schnirelman theory: complementary principles and the Morse index. Nonlinear Anal., **12**, 5, 507–529.
67. Comte, M., Mironescu, P. (1995): Sur quelques propriétés des minimiseurs de l'energie de Ginzburg–Landau. C. R. Acad. Sci. Paris Sér. I Math., **320**, 1323–1326.
68. Conley, C.C. (1971): On a generalization of the Morse index. In: Ordinary Differential Equations, 27–33. Academic Press, New York.
69. Conley, C.C. (1978): Isolated invariant sets and the Morse index. CBMS Reg. Conf. Ser. Math., Amer. Math. Soc., Providence, USA.
70. Conley, C.C., Easton, R.W. (1971): Isolated invariant sets and isolating blocks. Trans. Amer. Math. Soc., **158**, 1, 35–61.
71. Conley, C.C., Gardner, R. (1989): An application of the generalized Morse index to travelling wave solutions of a competitive reaction diffusion model. Indiana Univ. Math. J., **33**, 319–343.
72. Conley, C.C., Zehnder, E. (1983): The Birkhoff–Lewis fixed point theorem and a conjecture of V.I. Arnold. Invent. Math., **73**, 1, 33–49.
73. Conley, C.C., Zehnder, E. (1984): Morse-type index for flows and periodic solutions for Hamiltonian equations. Comm. Pure Appl. Math., **37**, 2, 207–253.

74. Conley, C.C., Zehnder, E. (1984): Subharmonic solutions and Morse theory. Physica A, **124**, 649–658.

75. Conley, C.C., Zehnder, E. (1986): A global fixed point theorem for symplectic maps and subharmonic solutions of Hamiltonian equations on tori. Proc. Symp. Pure Math., **45**, 283–299.

76. Dancer, E.N. (1984): Degenerate critical points, homotopy indices and Morse inequalities. I. J. Reine Angew. Math., **350**, 1–22.

77. Dancer, E.N. (1987): Degenerate critical points, homotopy indices and Morse inequalities. II. J. Reine Angew. Math., **382**, 145–164.

78. Dancer, E.N. (1989): Degenerate critical points, homotopy indices and Morse inequalities. III. Bull. Austral. Math. Soc., **40**, 1, 97–108.

79. Davidenko, D.F. (1953): On a method of numerical solution for systems of nonlinear equations. Dokl. Akad. Nauk SSSR, **88**, 5, 601–602.

80. Davidenko, D.F. (1953): On the approximate solution of systems of nonlinear equations. Ukrain. Mat. Zh., **5**, 196–206.

81. Davidenko, D.F. (1955): On application of the method of variation of parameters to the theory of nonlinear functional equations. Ukrain. Mat. Zh., **7**, 18–28.

82. Davidenko, D.F. (1965): The application of the method of variation of a parameter to the construction of iteration formulae of heightened precision for the determination of numerical solutions of nonlinear integral equations. Soviet Math. Dokl., **6**, 702–706.

83. Davidenko, D.F. (1965): On the application of the method of variation of a parameter to the construction of iterative formulae of higher accuracy for matrix inversion. Soviet Math. Dokl., **6**, 738–742.

84. Deimling, K. (1985): Nonlinear Functional Analysis. Springer, Berlin.

85. Del Pino, M., Felmer, P. (1995): Locally energy-minimizing solution of the Ginzburg–Landau equation. C. R. Acad. Sci. Paris Sér. I Math., **321**, 9, 1207–1211.

86. Demidovich, B.P. (1967): Lectures on the Mathematical Theory of Stability (in Russian). Nauka, Moscow.

87. Demyanov, V.F. (1982): Nonsmooth Analysis (in Russian), LGU Press, Leningrad.

88. Demyanov, V.F., Vasilyev, L.V. (1985): Nondifferentiable Optimization. Optimization Software, Inc., New York.

89. Du, Y.H. (1991): A deformation lemma and some critical point theorems. Bull. Austral. Math. Soc., **43**, 1, 161–168.

90. Dunford, N.J., Schwartz, J.T. (1958): Linear Operators. I., Interscience Publ., New York.

91. Ekeland, I., Temam, R. (1976): Convex Analysis and Variational Problems. Elsevier North-Holland, Amsterdam.

92. Emelyanov, S.V. (1967): Automatic Control Systems with Variable Structure (in Russian). Nauka, Moscow.

93. Emelyanov, S.V. (1971): Automatische Regelsysteme mit veränderlicher Struktur. (German translation of [92]). Akademie-Verlag, Berlin.
94. Emelyanov, S.V. (1978): Variable Structure Automatic Control Systems (in Russian). Nauka, Moscow.
95. Emelyanov, S.V. (1984): Automatic Control Binary Systems. IRIMS Press, Moscow.
96. Emelyanov, S.V., Burovoi, I.A., Levada, F.Yu. (1998): Control of Indefinite Nonlinear Dynamic Systems. Induced Internal Feedback. Springer, Berlin.
97. Emelyanov, S.V., Korovin, S.K. (2000): Control of complex and uncertain systems. New types of feedback. Springer-Verlag London, Ltd., London.
98. Emelyanov, S.V., Korovin, S.K., Bobylev, N.A. (1997): On a problem from the theory of robustness. Dokl. Ross. Akad. Nauk, **357**, 6, 623–627.
99. Emelyanov, S.V., Korovin, S.K., Bobylev, N.A. (1997): On the robustness of stable dynamical systems. Dokl. Ross. Akad. Nauk, **357**, 4, 421–424.
100. Emelyanov, S.V., Korovin, S.K., Bobylev, N.A. (2000): Attractors of discrete controlled systems. Dokl. Ross. Akad. Nauk, **370**, 5, 521–525.
101. Emelyanov, S.V., Korovin, S.K., Levantovskii, L.V. (1986): Second-order sliding modes in the control of uncertain systems. Soviet J. Comput. Systems Sci., **24**, 4, 63–68.
102. Emelyanov, S.V., Korovin, S.K., Mamedov, I.G. (1994): Sampled-data Variable Structure Systems. CRC Press, Boca Raton, USA.
103. Emelyanov, S.V., Korovin, S.K., Nersisyan, A.L. (1990): On asymptotic properties of observers of the state of uncertain systems with an explicit stationary linear part. Dokl. Akad. Nauk SSSR, **311**, 4, 807–811.
104. Emelyanov, S.V., Korovin, S.K., Nersisyan, A.L. (1990): Stabilization of undetermined neutral objects by a variable structure regulator. Soviet Phys. Dokl., **35**, 6, 496–498.
105. Emelyanov, S.V., Korovin, S.K., Nersisian, A.L., Nisenzon, Y.E. (1992): Output feedback stabilization of uncertain plants: a variable structure systems approach. Internat. J. Control, **55**, 1, 61–81.
106. Emelyanov, S.V., Korovin, S.K., Nikitin, S.V. (1987): Controllability of nonlinear systems. Two-dimensional systems. Itogi Nauki i Tekhn. Ser. Tekh. Kibern., **21**, 3–67.
107. Emelyanov, S.V., Korovin, S.K., Nikitin, S.V. (1988): Investigation of controllability by the methods of the theory of approximation groups and systems on a foliation. Soviet Phys. Dokl., **33**, 6, 400–401.
108. Emelyanov, S.V., Korovin, S.K., Nikitin, S.V. (1988): Nonlinear systems. Controllability, stabilizability, invariance. Itogi Nauki i Tekhniki. Ser. Tekh. Kibern., **23**, 3–107.
109. Emelyanov, S.V., Korovin, S.K., Nikitin, S.V. (1992): Application of foliation theory and approximative groups to controllability analysis of nonlinear systems. Comput. Math. Model., **3**, 4, 387–392.
110. Emelyanov, S.V., Korovin, S.K., Nikitin, S.V. (1992): Nonlinear systems: construction of observers and output differentiators. Comput. Math. Model., **3**, 4, 353–361.

111. Emelyanov, S.V., Korovin, S.K., Nikitin, S.V., Sizikov, V.I. (1992): Stabilization of two-dimensional systems. Comput. Math. Model., **3**, 4, 362–365.

112. Fadell, E., Rabinowitz, P.H. (1977): Bifurcation for odd potential operators and an alternative topological index. J. Functional Analysis, **26**, 48–67.

113. Fiedler, B., Mischaikow, K. (1992): Dynamics of bifurcations for variational problems with $O(3)$ equivariance: a Conley index approach. Arch. Rational Mech. Anal., **119**, 145–196.

114. Floer, A. (1987): A refinement of the Conley index and an application to the stability of hyperbolic invariant sets. Ergodic Theory Dynam. Systems, **7**, 93–103.

115. Floer, A., Zehnder, E. (1988): The equivariant Conley index and bifurcation of periodic solutions of Hamiltonian systems. Ergodic. Theory Dynam. Systems, **8***, 87–98.

116. Fomenko, A.T., Fuks, D.B. (1989): Course in Homotopic Topology (in Russian). Nauka, Moscow.

117. Franzosa, R. (1988): The continuation theory for Morse decompositions and connection matrices. Trans. Amer. Math. Soc., **310**, 781–803.

118. Franzosa, R. (1989): The connection matrix theory for Morse decompositions. Trans. Amer. Math. Soc., **311**, 561–592.

119. Franzosa, R., Mischaikow, K. (1998): Algebraic transition matrices in the Conley index theory. Trans. Amer. Math. Soc., **350**, 889–912.

120. Franzosa, R., Mischaikow, K. (1988): The connection matrix theory for semiflows on (not necessarily locally compact) metric spaces. J. Differential Equations, **71**, 270–287.

121. Freudenstein, F., Roth, B. (1963): Numerical solution of systems of nonlinear equations. J. Assoc. Comput. Math., **10**, 550–556.

122. Gardner, R. (1984): Existence of travelling wave solutions of predator–prey systems via the connection index. SIAM J. Appl. Math., **44**, 1, 56–76.

123. Gavurin, M.K. (1958): Nonlinear functional equations and continuous analogs of iteration processes. Izv. Viss. Uchebn. Zaved. Mat., **5**, 18–31.

124. Ginzburg, V., Landau, L. (1950): Phenomenological theory of superconductivity. Zh. Eksper. Teor. Fiz., **20**, 1064.

125. Gokhberg, I.Ts., Krein, M.G. (1965): Introduction to the Theory of Linear Self-Adjoint Operators (in Russian). Nauka, Moscow.

126. Gordeichuk, A.V., Moroz, V.B. (1998): An example of the gradient homotopy of gradient vector fields. Tr. Inst. Mat. (Minsk), **1**, 30–33.

127. Grodzovskii, G.L., lvanov, Yu.N., Tokarev, V.V. (1975): Mechanics of Space Flight. Optimization Problems (in Russian). Nauka, Moscow.

128. Gromoll, D., Meyer, W. (1969): On differentiable functions with isolated critical points. Topology, **8**, 361–369.

129. Hadamard, J. (1910): Calcul des variations. Hermann, Paris.

130. Hardy, G.H., Littlewood, J.E., Polya, G. (1951): Inequalities. Cambridge University Press, Cambridge.

131. Hartman, P. (1964): Ordinary Differential Equations. J. Wiley & Sons, New York.

132. Hestenes, M.R. (1966): Calculus of Variations and Optimal Control Theory. John Wiley & Sons, New York.

133. Hirsch, M., Smale, S. (1979): On algorithms for solving $f(x) = 0$. Comm. Pure Appl. Math., **32**, 281–313.

134. Ioffe, A.D., Tihomirov, V.M. (1979): Theory of Extremal Problems. North-Holland, Amsterdam.

135. Ioffe, A., Schwartzman, E. (1996): Metric critical point theory. 1. Morse regularity and homotopic stability of minimum. J. Math. Pures Appl. (9), **75**, 2, 125–153.

136. John, F. (1948): Extremum problems with inequalities as subsidiary conditions. In: Studies and Essays, Courant Anniversary Volume, 187–204. Interscience, New York.

137. Kaczinsky, T., Mrozek, M. (1995): Conley index for discrete multi-valued dynamical systems. Topology Appl., **65**, 83–96.

138. Kaczinsky, T., Mrozek, M. (1997): Stable index pairs for discrete dynamical systems. Canad. Math. Bull., **40**, 448–445.

139. Kantorovich, L.V., Akilov, G.P. (1982): Functional Analysis, 2nd edn. Pergamon Press, New York.

140. Keller, J.B., Antman, S. (eds.) (1969): Bifurcation Theory and Nonlinear Eigenvalue Problems. Benjamin/Cummings, New York.

141. Kellogg, R.B., Li, T.Y., Yorke, J. (1976): A constructive proof of the Brouwer fixed-point theorem and computational results. SIAM J. Numer. Anal., **13**, 4, 473–483.

142. Kielhöfer, H. (1988): A bifurcation theorem for potential operators. J. Funct. Anal., **77**, 1, 1–8.

143. Klimov, V.S. (1982): Nontrivial solutions of the Ginzburg–Landau equations. Teoret. Mat. Fiz., **50**, 3, 383–389.

144. Kolmogorov, A.N., Fomin, S.V. (1976): Elements of the Theory of Functions and Functional Analysis (in Russian). Nauka, Moscow.

145. Krasnosel'skii, M.A. (1956): Topological Methods in the Theory of Nonlinear Integral Equations (in Russian). Gosudarstv. Izdat, Moscow.

146. Krasnosel'skii, M.A. (1966): The Operator of Translation along the Trajectories of Differential Equations. American Mathematical Society, Providence, USA.

147. Krasnosel'skii, M.A., Bobylev, N.A., Mukhamadiev, E.M. (1978): A scheme for the study of degenerate extremals of functionals in the classical calculus of variations. Dokl. Akad. Nauk SSSR, **240**, 3, 530–533.

148. Krasnosel'skii, M.A., Bobylev, N.A., Mukhamadiev, E.M. (1978): On the topological index of extremals of functionals in the calculus of variations. Dokl. Akad. Nauk Tadzhik. SSR, **21**, 8, 8–12.

149. Krasnosel'skii, M.A., Mukhamadiev, E.M., Pokrovskii, A.V. (1980): Bifurcation values of parameters in variational problems. Dokl. Akad. Nauk SSSR, **255**, 2, 282–286.

150. Krasnosel'skii, M.A., Povolotskii, A.I. (1953): On variational methods in the problem of bifurcation points. Dokl. Akad. Nauk SSSR, **91**, 1, 19–22.

151. Krasnosel'skii, M.A., Zabreiko, P.P. (1984): Geometric Methods of Nonlinear Analysis. Springer-Verlag, Berlin.

152. Krasovskii, N.N. (1968): Theory of Control of Motion. Linear Systems (in Russian). Nauka, Moscow.

153. Ladyzhenskaya, O.A., Ural'tseva, N.N. (1973): Linear and Quasilinear Equations of Elliptic Type (in Russian). Nauka, Moscow.

154. Lahaye, E. (1934): Une méthode de résolution d'une catégorie d'équations transcendantes. C. R. Acad. Sci. Paris, **198**, 1840–1842.

155. Lahaye, E. (1948): Sur la résolution des systèmes d'équations transcendantes. Acad. Roy. Belgique. Bull. Cl. Sci. (5) **34**, 805–822.

156. Landau, L.D. (1969): Collected Works, vol. 2 (in Russian). Nauka, Moscow.

157. Lang, S. (1962): Introduction to Differentiable Manifolds. Interscience, New York.

158. Lavrentyev, M.A., Lyusternik, L.A. (1935): Fundamentals of Variational Calculus (in Russian). ONTI, Moscow.

159. Leray, J., Schauder, J. (1934): Topologie et équations fonctionelles. Ann. Sci. École Norm. Sup. (3), **51**, 45–78.

160. Lettenmeyer, F. (1942): Über die von einem Punkt ausgehenden Integralkurven einer Differentialgleichung zweiter Ordnung. Deutsche Math., **7**, 56–74.

161. Li, T.Y. (1983): On Chow, Mallet-Paret and Yorke homotopy for solving systems of polynomials. Bull. Inst. Math. Acad. Sinica, **11**, 433–437.

162. Lieb, E.H., Loss, M. (1994): Symmetry of the Ginzburg–Landau minimizer in a disc. Math. Res. Lett., **1**, 6, 701–715.

163. Lifshitz, E.M., Pitaevskii, L.P. (1980): Statistical Physics. Pergamon Press, New York.

164. Lin, F.H. (1996): Some dynamical properties of Ginzburg–Landau vortices. Comm. Pure Appl. Math., **49**, 4, 323–359.

165. Lusternik, L.A., Sobolev, V.J. (1971): Elements of Functional Analysis. Hindustan Publishing Corp., Delhi.

166. Malgrange, B. (1966): Ideals of Differentiable Functions. Oxford University Press, Oxford.

167. Marino, A. (1973): La biforcazione sul caso variazianale. Confer. Sem. Mat. Univ. Bari, 132.

168. Mawhin, J., Ward., J.R., Willem, M. (1986): Variational methods and semilinear elliptic equations. Arch. Rational Mech. Anal., **95**, 3, 269–277.

169. Milnor, J. (1963): Morse Theory. Princeton University Press, Princeton, USA.

170. Milnor, J. (1965): Lectures on the h-Cobordism Theorem. Princeton University Press, Princeton, USA.

171. Milnor, J. (1968): Singular Points of Complex Hypersurfaces. Princeton University Press, Princeton, USA.

172. Milnor, J., Wallace, A. (1968): Topology from the Differentiable Viewpoint. Differential Topology. Princeton University Press, Princeton, USA.

173. Morse, M. (1931): The critical point of a function of n variables. Trans. Amer. Math. Soc., **33**, 72–91.

174. Morse, M. (1938): Functional topology and abstract variational theory. Mem. Sci. Math. 32, Gauthier-Villars, Paris.

175. Morse, M. (1938): The index theorem in the calculus of variations. Duke Math. J., **4**, 1, 231–246.

176. Morse, M. (1996): The Calculus of Variations in the Large. American Mathematical Society, New York.

177. Mrozek, M. (1990): The Conley index of compact ANRs is of finite type. Results. Math., **18**, 306–313.

178. Mrozek. M., Rybakowski, K.P. (1991): A cohomological Conley index for maps on metric spaces. J. Differential Equations, **90**, 1, 143–171.

179. Mukhamadiev, E.M. (1983): Homology groups of critical points of smooth functions. Dokl. Akad. Nauk Tadzhik. SSR, **26**, 9, 553–557.

180. Nagumo, M. (1937): Über die Differentialgleichung $y'' = f(x, y, y')$. Proc. Phys. Math. Soc. Japan, **19**, 861–866.

181. Nikol'skii, S.M. (1977): Approximation of Functions of Many Variables and Embedding Theorems (in Russian). Nauka, Moscow.

182. Nirenberg L. (1974): Topics in Nonlinear Functional Analysis. Courant Institute of Mathematics, New York.

183. Odeh, F. (1969): Bifurcation problem in superconductivity theory. In [140].

184. Ortega, J., Rheinboldt, W. (1970): Iterative Solution of Nonlinear Equations in Several Variables. Academic Press, New York.

185. Palamodov, V.P. (1977): Equilibrium stability in a potential field. Functional Anal. Appl., **11**, 4, 277–289.

186. Parusinski, A. (1990): Gradient homotopies of gradient vector fields. Studia Math., **96**, 1, 73–80.

187. Picard, E. (1893): Sur l'application des méthodes d'approximations succesives à l'étude de certaines équations differentialles ordinaires. J. Math. Pures Appl. (4), **9**, 217–271.

188. Plotnikov, P.I. (1992): Nonuniqueness of solutions of a problem on the solitary waves, and bifurcations of critical points of smooth functionals. Math. USSR-Izv., **38**, 2, 333–357.

189. Pokhozhaev, S.I. (1967): The solvability of nonlinear equations with odd operators. Funkcional. Anal. i Prilozen., **1**, 3, 66–73.

190. Pokhozhaev, S.I. (1979): An approach to nonlinear equations. Soviet Math. Dokl., **20**, 4, 912–916.

191. Polyak, B.T. (1964): Gradient methods for solving equations and inequalities. Zh. Vychisl. Mat. i Mat. Fiz., **4**, 995–1005.

192. Pontryagin, L.S., Boltyanskii, V.G., Gamkrelidze, R.V., Mishchenko, E.F. (1962): Mathematical Theory of Optimal Processes. Interscience, J. Wiley & Sons, New York.

193. Postnikov, M.M. (1971): Introduction to Morse Theory (in Russian). Nauka, Moscow.

194. Roberts, S.M., Shipman, J.S. (1967): Continuation in shooting methods for two-point boundary value problems. J. Math. Anal. Appl., **18**, 45–58.

195. Rosenbloom, P. (1956): The method of steepest descent. In: Proc. Symp. Appl. Math. 6, 127–176. McGraw-Hill Book Company, New York.

196. Rybakowski, K.P. (1985): On a relation between the Browder degree and Conley index for gradient flows. Bull. Math. Belgique Sér B, **37**, 2, 87–96.

197. Rybakowski, K.P. (1987): The Homotopy Index and Partial Differential Equations. Springer-Verlag, Berlin.

198. Rybakowski, K.P., Zehnder, E.A. (1985): A Morse equation in Conley's index theory for semiflows on metric spaces. Ergodic Theory Dynam. Systems, **5**, 1, 123–143.

199. Salamon, D. (1990): Morse theory, the Conley index and Floer homology. Bull. London Math. Soc., **22**, 2, 113–140.

200. Sapronov, Yu.I. (1988): Multimode bifurcations of elastic equilibria. J. Appl. Math. Mech., **52**, 6, 778–785.

201. Sapronov, Yu.I. (1989): Two-mode bifurcation of solutions of the von Kármán equations. Differentsial'nye Uravneniya, **25**, 6, 1078–1081.

202. Sapronov, Yu.I. (1991): Nonlocal finite-dimensional reductions in variational boundary value problems. Math. Notes, **49**, 1–2, 67–73.

203. Sard, A. (1942): The measure of the critical values of differentiable maps. Bull. Amer. Math. Soc., **48**, 883–890.

204. Semenov, M.P. (1965): On a two-point boundary value problem (in Russian). Math. Sb. (N.S.), **66(108)**, 1, 97–108.

205. Shafrir, I. (1991): A deformation lemma. C.R. Acad. Sci. Paris Sér. I Math., **313**, 9, 599–602.

206. Shalasilin, V.I., Kuznetsov, E.B. (1999): The Continuation Method with Respect to a Parameter and the Best Parametrization in Applied Mathematics and Mechanics (in Russian). Editorial URSS, Moscow.

207. Shidlovskaya, N.A. (1958): Application of the differentiation method with respect to a parameter to solving nonlinear equations in Banach spaces. Leningrad. Gos. Univ. Ucen. Zap. Ser. Mat. Nauk, **33**, 3–17.

208. Skalyga, V.I. (1991): The deformation method for studying extremal problems with constraints. In: Dynamics of Inhomogeneous Systems. Collection of papers of VNIISI (in Russian), **14**, 59–63.

209. Skalyga, V.I. (1991): The deformation method for studying the conditional minimum of quality functionals of systems with an infinite number of degrees of freedom. Automat. Remote Control, **50**, 6, part 1, 781–789.

210. Skalyga, V.I. (1993): The deformation method for studying infinite-dimensional optimization problems. Math. Notes, **53**, 240–241.

211. Skalyga, V.I. (1993): The deformation method for studying nonsmooth infinite-dimensional optimization problems. Automat. Remote Control, **54**, 11, part 1, 1630–1632 (1994).

212. Skalyga, V.I. (1995): Deformations of nonsmooth optimization problems that have an isolated extremal. Russian Acad. Sci. Izv. Math., **45**, 1, 187–195.

213. Skalyga, V.I. (1997): On a homotopy method in infinite-dimensional multicriteria problems. Izv. Math., **61**, 4, 813–830.

214. Skrypnik, I.V. (1973): Nonlinear Higher-Order Elliptic Equations (in Russian). Naukova Dumka, Kiev.

215. Skrypnik, I.V. (1976): Solvability and properties of solutions of nonlinear elliptic equations. Itogi Nauki i Tekhniki. Sovr. Prob. Mat., **9**, 131–254. Izd. VINITI, Moscow.

216. Smale, S. (1976): A convergent process of price adjustment and global Newton method. J. Math. Econom., **3**, 1, 1–14.

217. Struwe, M. (1985): Variationsmethoden der Nichtlinearen Funktional Analysis. Universität Bonn.

218. Struwe, M. (1990): Variational Methods, Springer-Verlag, Berlin.

219. Szymczak, A. (1995): The Conley index for decompositions of isolated invariant sets. Fund. Math., **148**, 71–90.

220. Szymczak, A. (1995): The Conley index for discrete dynamical systems. Topology Appl., **66**, 215–240.

221. Tikhomirov, V.M. (1987): Convex analysis. Itogi Nauki i Tekhniki, Sovr. Prob. Mat. Fundam. Napravl., **14**, 5–101. Izd. VINITI, Moscow.

222. Trenogin, V.A., Sidorov, N.A., Loginov, B.V. (1990): The bifurcation equation: potentiality, bifurcations, symmetry. Soviet Math. Dokl., **40**, 3, 517–520.

223. Vainberg, M.M. (1964): Variational Methods for the Study of Nonlinear Operators. Holden-Day, Inc., San Francisco.

224. Vainberg, M.M. (1973): Variational Methods and Method of Monotone Operators in the Theory of Nonlinear Equations. Halsted Press, New York.

225. Vainberg, M.M., Trenogin, V.A. (1974): Theory of Branching of Solutions of Nonlinear Equations. Noordhof International Publishing, Leyden.

226. Vasilyev, F.P. (1981): Methods of Solving Extremal Problems (in Russian). Nauka, Moscow.

227. Voinov, Yu.N. (1974): Boundedness near the boundary of the gradients of generalized solutions of boundary value problems for quasilinear elliptic equations. Vestnik Leningrad. Univ. 7, Mat. Meh. Astronom. Vyp. 2, 5–13.

228. Wacker, H. (ed.) (1978): Continuation Methods. Academic Press, New York.

229. Zhang, De-Tong. (1990): A homotopy method for solving a class of two-point boundary value problems. Numer. Math. J. Chinese Univ. **12**, 4, 304–315.

Notation

$\overline{A}$	closure of the set A
$\operatorname{int} A$, $\mathring{A}$	interior of the set A
∂A	boundary of the set A
$\operatorname{co} A$	convex hull of the set A
$\{x : P\}$	set of elements x with property P
$F \circ G$	composite of the maps F and G
$F(A)$	image of the set A under the mapping F
$F^{-1}(A)$	preimage of the set A under the mapping F
$F\big\|_A$	restriction of the map F to the set A
id_A	identity mapping on the set A
S_1/S_2	quotient space of S_1 with respect to S_2
$\mathbb{R}$	real line
$\mathbb{R}^N$	N-dimensional Euclidean space
$\mathbb{R}^N_+$	nonnegative orthant in $\mathbb{R}^N$
$\lvert x\rvert$	Euclidean norm of $x \in \mathbb{R}^N$
$\lVert x\rVert$	norm of the element x of a Banach space
$d(x, y)$	distance between the points x, y of a metric space
$[a, b]$	closed interval $\{t \in \mathbb{R} : a \leqslant t \leqslant b\}$
(a, b)	open interval $\{t \in \mathbb{R} : a < t < b\}$
$(a, b]$	half-open interval $\{t \in \mathbb{R} : a < t \leqslant b\}$
$[a, b)$	half-closed interval $\{t \in \mathbb{R} : a \leqslant t < b\}$
$B(\rho, x)$	closed ball of radius ρ with center x
$B(\rho)$	closed ball of radius ρ with center 0
B	closed ball of unit radius with center 0
$S(\rho, x)$	sphere of radius ρ with center x
$S(\rho)$	sphere of radius ρ with center 0
S	unit sphere with center 0

B^N	N-dimensional unit ball in $\mathbb{R}^N$ with center 0
S^N	N-dimensional unit sphere in $\mathbb{R}^{N+1}$ with center 0
$T_M(x)$	tangent cone to the set M at x
$N_M(x)$	normal cone to the set M at x
E^*	dual of the space E
$\langle \cdot, \cdot \rangle$	the canonical bilinear form on $E \times E^*$
$\langle y, x \rangle$	value of $y \in E^*$ at $x \in E$
$(\cdot, \cdot)$	inner product in $\mathbb{R}^N$ and in a Hilbert space
$C(\overline{\Omega})$	functional space (1.1.6)
$C^k(\overline{\Omega})$	functional space (1.1.6)
$\mathring{C}^k(\overline{\Omega})$	functional space (1.1.6)
$C^{k,\delta}(\overline{\Omega})$	functional space (1.1.6)
$L_p(\Omega)$	functional space (1.1.6)
$L_\infty(\Omega)$	functional space (1.1.6)
$L_2[0, T]$	functional space (5.6.1)
$W_p^m(\Omega)$	functional space (1.1.6)
$\mathring{W}_p^m(\Omega)$	functional space (1.1.6)
$\widetilde{L}_2[0, T]$	functional space (5.6.2)
$\widetilde{L}_1[0, T]$	functional space (5.6.2)
$\operatorname{Ker} A$	kernel of the linear operator A
$\operatorname{Im} A$	image of the linear operator A
$\|A\|$	operator norm of the linear operator A
$\mathcal{L}(E, F)$	space of bounded linear operators $A : E \to F$
$A'(x)$	derivative of the operator A at x
$\nabla f(x)$	gradient of the functional f at x
$\nabla^2 f(x)$	Hessian of the functional f at x
$f^0(x, v)$	generalized derivative of the functional f at x in the direction of the vector v
$\partial f(x)$	generalized gradient of the functional f at the point x
$\operatorname{mes} A$	Lebesgue measure of the set A
$\operatorname{supp} \varphi$	support of the function φ
div	divergence
rot	rotation
Δ	Laplace operator
$\deg f$	degree of the mapping f
$\gamma(\Phi; \partial\Omega)$	rotation of the vector field Φ on the boundary $\partial\Omega$ of Ω

$\mathrm{ind}(x_0; \Phi)$	topological index of the zero x_0 of the field Φ
$\langle N_1, N_2 \rangle$	index pair
$f \sim g$	f, g are homotopic mappings
$A \sim B$	the sets A, B are homotopy equivalent
$h(S; p)$	Conley index of the invariant set S relative to the flow p
$h(x; f)$	Conley index of the critical point x of the functional f

Name Index

Allgower, E.L. 3
Arzelà, C. 14
Ascoli, J. 14

Banach, S. 10, 16, 19, 20, 31
Bernstein, S.N. 204, 232–233
Bosarge, W. 3
Brezis, H. 275
Brouwer, L.G. 73
Browder, F.E. 275
Brown, A. 87
Bunyakovskii, V.Ya. 234

Cauchy, A.-L. 1, 8, 33, 70, 159, 180–183, 234, 245, 249, 254, 266, 273
Chetaev, N.G. 236
Chow, S.N. 3
Clarke, F.H. 26, 129
Conley, C.C. 6, 135, 142–171, 240–244, 259, 263–264, 267, 277, 279

Davidenko, D.F. 3
Dirichlet, P.G.L. 220–222, 226, 230–231, 238–239, 257–259, 264, 269

Euler, L. 112, 217–220, 222, 226, 230–232, 234, 257, 259, 264, 270

Fréchet, M.R. 22–25, 29, 32, 35, 43, 105, 112, 163, 218, 221, 246, 255, 263, 271–273, 276, 279
Fredholm, E.I. 261
Freudenstein, F. 3
Friedrichs, K.O. 229–231

Gâteaux, R.E. 23–25, 27, 34, 121, 279
Gavurin, M.K. 3
Gel'fand, I.M. 21
Georg, K. 3
Ginzburg, V. 269–272, 279
Goloumb, M. 42
Green, G. 230

Hadamard, J.S. 98
Hahn, H. 16
Hammerstein, A. 29, 42
Hausdorff, F. 13
Hirsch, M. 3
Hölder, L.O. 11
Hopf, H. 86–87, 91–97, 100, 275

Jacobi, C.G.J. 236
Jensen, J.L. 28, 179
John, F. 185, 192

Kellogg, R.B. 3
Krasnosel'skii, M.A. 234, 279
Kronecker, L. 10, 90
Kuhn, H. 185, 192

Lagrange, J.L. 25, 37, 184, 190, 193–195, 201, 216, 232
Landau, L.D. 269–272, 279
Laplace, P.S. 219, 260, 272

Lebesgue, H.L. 11, 40, 74, 83, 87, 183
Legendre, A.M. 222, 229, 232, 234, 259, 264, 265
Leray, J. 3
Lettenmeyer, F. 234
Li, E. 3
Lipschitz, R. 25–26, 37, 56, 105, 117, 121, 159, 163, 170, 183, 185–186, 204, 213, 217–218, 226, 228, 240–241, 273–275
Lojasiewicz, S. 78, 274
Lyapunov, A.M. 33, 47, 50, 118, 236–239, 244, 278

Mallet-Paret, J. 3
Mazur, S. 14
Minkowski, H. 178–179
Morse, M. 81–82, 102

Nagumo, M. 234
Nemytsky, V.V. 29, 42
Newton, I. 2

Palais, R.S. 121–122, 129
Pareto, M. 207, 210
Parusinski, A. 86, 98, 275
Picard, C.E. 234
Poincaré, J.H. 229–231
Pokhozhaev, S.I. 275
Polyak, B.T. 3
Pontryagin, L.S. 275

Riesz, F. 16, 163
Riesz, M. 14
Roberts, S. 3
Rosenbloom, P. 3
Roth, B. 3

Sard, A. 87
Schauder, J. 3
Schwarz, K. 234
Semenov, M.P. 234
Shidlovskaya, N.A. 3
Shipman, J.S. 3
Skrypnik, I.V. 275, 279
Smale, S. 3, 79, 83, 121–122, 129
Sobolev, S.L. 12, 22, 218, 259
Steinhaus, H. 19
Sylvester, J.J. 174–175

Taylor, B. 24
Tietze, H. 30
Tucker, A.W. 185, 192

Urysohn, P.S. 30

Wacker, H. 3
Weierstrass, K.T.W. 39, 106, 273
Wirtinger, W. 234

Young, W.H. 177
Yorke, J. 3

Zhang, De-Tong 3

Subject Index

Arithmetic mean 181
Attraction set 109

Ball
  closed 8
  open 8
Basis, orthogonal 10
Boundary of a manifold 81

Center of a ball 8
Closure of a set 8
Condition
  of complementary slackness 128, 185
  Legendre 222, 232, 264–265
  Lipschitz 25, 56, 159, 228
  of minimizer 35–36
  Palais–Smale 121
  regularity 189
Cone
  normal 38
  tangent 38
Constant of strong convexity 40
Constraint, active 184
Convergence
  in norm 10
  uniform 10
  weak 17
Coordinate system 80
Cover, open 13

Deformation
  continuous 90
  (E, H)-normal 277
  linear 54, 68, 132
  nondegenerate 45, 56, 86, 122, 129, 185, 222, 237–238, 241, 276
    generalized 86
    global 53, 64, 119
Deformation principle for a minimizer 45, 106, 122, 220, 256, 274–276, 278
Deformation retract 82
Degree of mapping 88–89
Derivative
  directional 26
  Fréchet 23
  Gâteaux 25
  generalized 56
  generalized Clarke 26
  Taylor 24
Diffeomorphism 79
Direct sum 9
  of vector fields 91
Domain of definition 15
Embeddings of Sobolev spaces 13, 22

Euler's equation 112, 229–231, 257, 259
Extremal 184–185
Extremum 132

Family
  centered 14
  closed 10
  complete 10

Flow 135
Form, homogeneous 24
Formula
  finite increments 25
  Lagrange 25
Function
  Green's 230
  implicit 32
  Lagrangian 184, 193, 201
  locally Lipschitzian 59
  Lyapunov 50, 236, 239
  Lyapunov–Chetaev 236
  Morse 81
  real analytic 78
  regular 68
Functional
  additive 15
  bounded 15
  convex 28, 42
  distance 37
  E-regular 112
  (E, H)-regular 163
  Fréchet differentiable 24
  Gâteaux differentiable 25
  growing 116
  H-regular 105
  Hammerstein–Goloumb 42
  homogeneous 15
  linear 15
  locally Lipschitzian 26
  lower semicontinuous 39
  Lyapunov 33
  (P, S)-regular 121
  regular 27
  strictly convex 28
  strongly convex 40
  upper semicontinuous 39
  weakly lower semicontinuous 39
  weakly upper semicontinuous 39

Gauge invariance 272
Geometric mean 181
Ginzburg–Landau equations 269
Gradient
  Fréchet 25
  Gâteaux 25
  generalized 26, 56

Hamiltonian 238
Harmonic mean 183
Hessian of a functional 25
Homeomorphism 79
Homotopy 90
Homotopy bridge 90
Hull 14

Index
  Conley 142, 240, 259
    of a critical point 158, 166
  Morse 81
  topological 90
Index pair 139
Inequality
  Bernstein 204
  Cauchy 180–182, 199
  Cauchy–Schwarz–Bunyakovskii 234
  Hölder 11
  Jensen 28, 179
  Lojasiewicz 78
  Minkowski 178
  Poincaré–Friedrichs 229–230
  Young 177
Inner product 10
Interior of a manifold 81

Lagrange multiplier rule 37
Lemma
  on choice of curves 77, 274
  on the factorization of a continuous mapping 147
  Fredholm 261
  Hadamard 98
  Sard 87
Limit
  of a sequence 7
  weak 17

Manifold
oriented 87
smooth 80
with boundary 81
without boundary 80
Mapping
continuous 6
contraction 31
nondegenerate 92
open 20
smooth 79
Mappings, smoothly homotopic 88
Matrix
nonnegative definite 174
positive definite 174
Maximum, essential 11
Method
homotopy 1
of Lyapunov functionals 33
Newton 2
of successive approximations 32
Metric 7
Minimizer
absolute 35
global 35, 49
strict absolute 35
strict global 35
local 34
weak 221
Minor, principal 174
Modulus of regularity 276
Morse number 82

Neighborhood
isolating 136
of a point 8
Node of a spline 123
Norm
of an element 9
of a functional 15
of an operator 19
of a multilinear operator 24
Norms, equivalent 21

Operator
adjoint 21
analytic 24
bounded 19, 22
completely continuous 21–22
continuous 18, 22
continuously differentiable 23
demicontinuous 22
embedding 22
Fréchet differentiable 22
Gâteaux differentiable 23
Hammerstein 29
hemicontinuous 22
inverse 19
Laplace 219, 260
linear 18
monotonic 41
multilinear 23
Nemytskii 29, 42
potential 42
strictly monotonic 41
strongly continuous 22
strongly monotonic 41
superposition 29
Taylor differentiable 24
unbounded 20
uniformly differentiable 23
weakly continuous 22
Order parameter 270
Orientation 87
standard 87
Orthogonal
basis 10
complement 10
projection 10

Parametrization of neighborhood 80
Point
of absolute minimum 35
bifurcation 255
of conditional minimum 35
critical 34, 38, 56, 70, 81, 105, 208, 274–275
nondegenerate 81

fixed 31
of global minimum 35
limit 8
of local minimum 34
locally optimal 219
Pareto optimal 207
of phase transition 270
regular 70, 274–275
singular 89
of strict absolute minimum 35
of strict local minimum 35
Polynomial, real 77
Potential, of an operator 42
Principle
of contraction mappings 31
of nonzero rotation 90
Problem
Cauchy 1, 33
Dirichlet 221–222, 226, 229, 231, 257, 259, 264, 269
mathematical programming 183
nonlinear programming 183–184
optimal control 244
of weak minimum 221
Product of domains 91
Product of quotient spaces 7
Projection, orthogonal 10

Quotient
space 6, 9
topology 6

Radius
of a ball 8
spectral 21
Range of values 15
Relation
equivalence 5
Gel'fand (Gel'fand's formula) 21
Retract, deformation 82
Retraction
deformation 82
strict 82
Rotation of vector field 90

$(S)_+$-property 105
Sequence
Cauchy 8
convergent 7
weakly convergent 17
$*$-weakly convergent 18
Set
admissible 210
bounded 8
closed 5, 8
compact 13
dense 8
invariant 136
isolated 136
open 5
positive invariant 137
precompact 13
real algebraic 77
regular 70, 274
weakly closed 17
weakly compact 17
$*$-weakly closed 18
$*$-weakly compact 18
Sets, diffeomorphic 79
Solution
generalized 219, 271
nonlocally extendable 33
Space
Banach 10
with base point 6
complete 8
complex 9
dual 15–16
Hilbert 10
of linear operators 21
metric 7
normed 9
real 9
reflexive 17
separable 8
Sobolev 12
tangent 80
topological 5, 82
vector 8

$C(\overline{\Omega})$ 11, 14
$C[0,T]$ 217
$C^1(\overline{\Omega})$ 260
$\mathring{C}^1(\overline{\Omega})$ 221, 259
$C^k(\overline{\Omega})$ 12
$\mathring{C}^k(\overline{\Omega})$ 12
$C^2(\overline{\Omega})$ 223
$C^{k,\delta}(\overline{\Omega})$ 12
$C^{2+\alpha}(\overline{\Omega})$ 221, 234
$\mathring{C}^1[0,T]$ 233
$L_p(\Omega)$ 11, 29
$L_\infty(\Omega)$ 11
$L_2[0,T]$ 245
$\widetilde{L}_1[0,T]$ 247
$\widetilde{L}_2[0,T]$ 247
$W_p^m(\Omega)$ 12
$\mathring{W}_p^m(\Omega)$ 12
$W_2^1(\Omega)$ 270
$\mathring{W}_2^m(\Omega)$ 220
$\mathring{W}_2^1(\Omega)$ 226, 229
$W_p^2(\Omega)$ 259
$W_p^{-1}(\Omega)$ 260
$W_2^1(\mathbb{R})$ 112
$\mathring{W}_2^1[0,T]$ 216
$\mathring{W}_1^1[0,T]$ 246
Spaces, homotopically equivalent 6
Spectrum of an operator 20
Spline, piecewise-linear 122
Stability
asymptotic 33, 236
in the large 34
Lyapunov 33, 236
State of equilibrium 33
Subdifferential 28
Subspace 9
admissible 165
Sum, orthogonal 10
Sylvester's criterion 174
System
gradient 159, 167, 236, 239
Hamiltonian 238
potential 278

Theorem
Arzelà–Ascoli 14
Banach 20, 31
Banach–Steinhaus 19
Bernstein 232
Dirichlet 238–239
Hahn–Banach 16
Hausdorff 13
h-cobordism 79, 83, 274
Hopf 92
implicit function 32
John–Kuhn–Tucker 185, 192
Kronecker 90
Lojasiewicz 78
Mazur 14
open mapping 20
partition of unity 31
Parusinski 98
Riesz, F. 16
Riesz, M. 14
separability 16
Smale 79, 83
on a triple 20
Tietze–Urysohn 30
Weierstrass 39, 106
Topology 5
induced 5
Topological pair 6
Trajectory 135
Triad of smooth manifolds 81

Value
critical 87
regular 20, 87
Variation, second 36
Vector field
linear 90
nondegenerate 89
Vector fields, gradient homotopic 97

Young conjugate functions 177

Zero
of a mapping 92
of a vector field 89

Bei Fragen zur Produktsicherheit wenden Sie sich bitte an:
If you have any questions regarding product safety,
please contact:

Walter de Gruyter GmbH
Genthiner Straße 13
10785 Berlin
productsafety@degruyterbrill.com